MOUNTAIN TORRENT
OF THE TIEN SHAN

MONOGRAPHIAE BIOLOGICAE

Editor

J. ILLIES

Schlitz

VOLUME 39

Dr W Junk bv Publishers The Hague–Boston–London 1980

MOUNTAIN TORRENT OF THE TIEN SHAN

A faunistic-ecology essay

K. A. BRODSKY

Translated from Russian by

V. V. GOLOSOV

Dr W Junk bv Publishers The Hague–Boston–London 1980

Distributors:

for the United States and Canada

Kluwer Boston, Inc.
160 Old Derby Street
Hingham, MA 02043
USA

for all other countries

Kluwer Academic Publishers Group
Distribution Center
P.O. Box 322
3300 AH Dordrecht
The Netherlands

Library of Congress Cataloging in Publication Data   CIP

Brodsky, Konstantin Abramovich.
   Mountain torrent of the Tien Shan.

   (Monographiae biologicae; 39)
   Translation of Gornyi potok Tian'-Shania.
   Bibliography
   Includes index.
   1. Stream   ecology—Thian   Shan   Mountains.   2, Animal   ecology—Thian   Shan
Mountains.   I. Title.   II. Series: Monographiae biologicae; v. 39.

QP1.P37   vol. 39   [QH191]   574s
ISBN 90-6193-091-X   (cloth)   [591.52′6323′09516]   80-12664

Contents

Preface

Mountain torrents are objects worthy of great attention from scientists and practical workers. So far they have been considered only as power resources (large streams) or as water sources for irrigation. But the mountain torrents, especially in a region as arid as Middle Asia, represent a much more important object than some standing waters which have been studied more intensively.

Mountain torrents are unique phenomena of nature which require strict conservation and rational management. In this connection it is especially important to implement the International Biological Programme and the Government's decrees concerning nature conservation.

Beyond all question is the aesthetic significance of torrents. Very promising is their use for breeding valuable fishes like trout, and for irrigation of clap and other crop fields on the sub-montane plains. Mountain torrents are true natural laboratories to study animal adaptations to the extremal conditions.

In recent decades, especially in the fifties and the sixties, biologists in Western Europe and North America have again turned to the study of invertebrate fauna of swift waters.[1] Some European investigators have revived the classical works devoted to rivers, brooks and streams (Steinmann, 1907; Thienemann, 1912, 1926; Hubault, 1927). There have been also some general works concerning the classification of swift waters (Illies and Botosaneanu, 1963) and the ecology of running waters in general (Hynes, 1970). Such interest in swift waters is due to the construction of numerous reservoirs, the acclimatization of fishes of the salmon family and others in rivers and brooks, the development of bionics and ecology which inquire the peculiar and highly specialized adaptations of aquatic organisms to life in swift waters, and, lastly, the urgent necessity for nature conservation.

Illies and Botosaneanu (1963), the authors of a work on classification of swift waters, proposed new terms for designation of some specialized branches of the hydrobiological science. Thus, they have used the term *crenobiology* for a division of hydrobiology studying springs; *rhithrobiology* for a division concerned with brooks, streams and mountain rivers or the corresponding sections of large rivers; and *potamology* for a division devoted to the lowland rivers and the lower reaches of mountain rivers. Therefore, the suggested essay on the Tienshanian mountain torrent refers to rhithrobiology.

---

[1] In the Preface and in the chapters which follow we use words like rivers, brooks and streams. These terms have a general meaning without any differentiation between these fast-flowing watercourses, which will be made only in Chapter Five.

The progress of rhithrobiology has been and still is very irregular. Illies and Botosaneanu give in their work a list of rhithrobiological publications in different countries of the world. Although not complete, especially on the U.S.S.R., this list gives an idea of the available knowledge of swift waters from a number of mountain countries. In 1963 it included the following numbers of publication

Germany 21	Holland 1	Yugoslavia........2
Great Britain.....13	Italy 3	India 3
Austria 7	Poland..........3	Japan............3
Belgium 3	Rumania 3	North America. . . .16
Denmark 5	Sweden 4	South America. . . .3
Bulgaria 1	Czechoslovakia....6	Africa 5
France 2	USSR 2	New Zealand 1

Of course, the list is now out-of-date, as it includes the works published prior to 1963; it contains only general works in rhithrobiology, but no papers on specific groups of invertebrates from lotic waters. Only two works[2] by Soviet authors were included, but there are no new publications on swift waters, for example, there are no data on the affluents of the Amur, on the Chirchik and the Zeravshan Rivers, *etc.* Besides, some of the author's publications, in particular, the monograph on the Issyk River, 1935, and others seem to have been unknown. However, even if we take into account the Soviet works in rhithrobiology which were not cited by Illies and Botosaneanu, the total number of publications on mountain torrents will be as small as ten, which is quite insufficient for the vast territory of the Soviet Union with its large mountain regions. The small mountain area in Central Europe has given life to about 50 general works on the ecology of the fauna of mountain rivers and streams, whereas for the vast area of mountains in Central and Middle Asia one may count only 10 or 15 works. This hampers the effective solution of practical and theoretical problems in rhithrobiology. In particular, the classification of running waters can not be made with data obtained only for the running waters of Central Europe and North America. The study of mountain rivers and streams of the Soviet Union is undoubtedly of great practical importance in a wider context of hydrotechnical work, the acclimatization measures, conservation of natural waters and landscapes, *etc.* in this country. The gap in the knowledge of running waters in the mountain regions of the Soviet Union can not be filled through investigations made in adjacent areas. While for western Soviet Union one can use the data from Eastern Europe (outside of the Soviet Union), for Middle Asia this is out of the question, since there have been no comprehensive monographs considering the ecology of the fauna in streams of the Hindu Kush, the Himalaya and the adjacent mountain regions. We can list very few works published in the twenties and the thirties on specific groups of invertebrates of fast-flowing mountain watercourses (Annandale,

---

[2] These are on the rivers of the Teletskoe Lake basin (Lepneva, 1949) and on the Kondara River (Shadin *et al.*, 1951) which was erroneously related by the authors to the Caucasus.

1919; Annandale and Hora, 1922; Hora, 1923, 1930; Tonnoir, 1930, 1931, 1932; Ulmer, 1935).

In recent years there has been a number of papers based on the field data of a few expeditions to the Hindu Kush, the Karakoram, the Himalaya and the Nepal (Traver, 1939; Gibbies, 1949; Kimmins, 1950; Uéno, 1955, 1966; Kapur and Kripalani, 1961; Santokh, 1961; Kawai, 1963, 1966; Mani, 1968; Kaul, 1970; Dubey and Kaul, 1971). However these works still describe single groups of torrential inhabitants (mostly, their systematics) and do not offer an integral view of the fauna of the visited streams, or its ecological characteristics.

Now as never before (if it is not too late[3]), there is the pressing need for the serious study of the fauna and its ecology in the swift waters of mountain regions of the Soviet Union, in particular, of Middle Asia.

It is known, that the mountain massives in the Soviet Union occupy great areas, and Middle Asia (more correctly, the Tien Shan) is the largest among them. The recent glaciation in the glacier regions of the Soviet Union is characterized by the following figures: the total area of glaciation, excluding the Arctic, is $18,510\ \mathrm{km}^2$; this includes the Suntar-Haiata Range $206\ \mathrm{km}^2$, the Alai Mts. (within the Soviet Union) $596\ \mathrm{km}^2$, the Great Caucasus $1,730\ \mathrm{km}^2$ and Middle Asia, with the Dzhungarskii Alatau Range, $15,978\ \mathrm{km}^2$ (Middle Asia, 1968). The large area of the perrenial snow and ice is certainly a source for highly developed drainage system. In Middle Asia there are not less than 10–12 thousands of rivers (Schultz, 1965).

Until recently, the only published monograph devoted to Middle Asia was our work on one stream of the Northern Tien Shan, the Issyk River on the Zailiiskii Alatau Range, published abridged as far back as 1935. It contained the data only for the upper and the middle reaches of the river. The data for the lower reach and for some other adjacent streams were available only as manuscript. In spite of the incompleteness of the description of the Issyk River because of the inadequate knowledge of the invertebrate fauna from Middle Asian mountain torrents (this is still true of some faunal groups), the most important torrential groups and their ecology were characterized together with a general description of the Tienshanian mountain torrent and its microbiotopes.

Without mentioning here the works which develop the systematics of the lithorheophylous fauna of streams or consider the quantitative data on river zoobenthos, we should list those offering a general account of the fauna and its ecology in Middle Asian mountain rivers. A team of workers of the Institute of Zoology, the USSR Academy of Sciences, surveyed some

---

[3] In the Soviet Union and other countries the remaking and the total utilization of mountain streams are going on at a high rate. Because of dam construction, large stretches of streams turn into lakes or slow rivers; many streams are expended totally for irrigation, and this affects, in a growing degree, the middle and even the upper reaches of rivers and streams. Their pollution is growing rapidly because of rafting, construction of enterprises, fast growth of settlements in the mountain zone, *etc*. Unless prompt measures are to be taken to insure the conservation and the rational use of streams as natural complexes, then the streams, being of unique amenity and as breeding sites for most valuable freshwater fishes, will disappear entirely or will become only museum items in the reservations under strict control.

affluents of the Amu-Darya River (Papers on hydrobiology..., 1950), as well as one river of the Western Pamirs (The Kondara River Gorge, 1951). The results of investigations of the Chirchik River (the Chatkalskii Range) and the Zeravshan River (the Turkestanskii and the Zeravshanskii Ranges) were published in a number of papers and in the abstract of the thesis (Sibirtzeva, 1961, 1964; Sibirtzeva *et al.*, 1961). In the 1970's a number of papers by Brodsky and Omorov have reported results for the Akbura River (the Alai Mts.), then a summary of these data was given in the abstract of the thesis by Omorov (1973). Lately, some data on the Turgen and Chilik Rivers (the Zailiiskii and the Kungei Alatau Ranges; Kurmangalieva, 1976b) have become available.

The studies in the invertebrate fauna from mountain streams of the Tien shan made by us in the 1920–1930's and later, after 1967, as well as the results of the above authors, and particularly the careful investigations of the invertebrate fauna and its ecology in the Akbura River made under our guidance, allow to present all the available data in the form of essay. The need for such an essay has been long felt, but now it has become imperative.

As we have pointed out, the systematics of some invertebrate groups from mountain torrents of the Tien Shan is, unfortunately, inadequately known, which prevents a detailed characterization of all groups of the litheoheophylous fauna. Since we are convinced that only accurately identified forms and well-known groups may be the basis for description of the ecological situation in a water-body (Brodsky, 1965), we confine ourselves in this book to the description of most representative torrential groups (the hymarobionts), *viz.* mayflies, caddisflies, stoneflies and some dipteran families. The mayflies and the dipterans were a subject of our taxonomic examination. More recently the systematics of mayflies, stoneflies, caddisflies and some dipteran families has been given in papers and monographs by I. A. Rubzov, S. G. Lepneva, V. Ya. Pankratova, L. A. Zhiltzova, L. K. Sibirtzeva, E. O. Konurbaev, L. A. Kustareva, O. A. Tshernova, N. D. Sinitchenkova and others.

The similarity of ecological conditions in torrents of the mountains bordering on the Tien Shan from the east (Chinese Tien Shan), south-east, south and south-west, and a certain genetic affinity of their faunas (and presence of some common genera and even species) suggest that the value of this book is not limited to the Tien Shan alone.

Acknowledgements

In conclusion, I should like to use this opportunity to express my thanks to all of my colleagues and various institutions for the assistance in selection of the material and the design of the book. First of all, my sincere thanks are due to the editor of this book in Russian, Prof. A. A. Strelkov, now late, who made a number of valuable suggestions on the content. His advice and sincere interest in my work considerably helped me during the writing of the text.

I am very grateful to the authorities of the Biological Station of the

Issyk-kul Lake, Academy of Sciences, Kirghiz SSR, especially to the former manager of the Station, A. O. Konurbaev, to the present manager, L. A. Folian, and to the research workers L. A. Kustareva and V. Bukin for facilities and help in field surveys and for the allocation of original data. During surveys of the Tienshanian torrents I was greatly assisted by a team of members of the State Pedagogical Institute, Osh City, and I am very grateful for this to the former Vice Director of the Institute, E. O. Konurbaev, and to the senior instructor, E. O. Omorov, who ensured the sampling not only in the middle and the lower reaches of streams, but also in the hardly accessible headwater areas.

For the allocation of data on the lithorheophylous fauna from streams on the Central, Northern and Western Tien Shan, I am greatly indebted to Prof. V. F. Gurvitsch and M. F. Vundtzetel of the State University, Tashkent, to L. K. Sibirtzeva of the Academy of Sciences, Uzbek SSR, to E. O. Konurbaev and particularly E. O. Omorov of the State Pedagogical Institute, Osh City.

The drawings of insects were made by A. Lyakhova and the photos of live objects by V. Bukin.

Shipping Instructions for BOOKSRUN

Print this Packing Slip and enclose inside the front cover of the book.

Please ship this item no later than Thu Jul 16, 2026.

Ship to:

ALIBRIS APEX DC 76524203-44
APEX
800 AVONDALE AVE.
GRANDVIEW HEIGHTS, OH 43212-3473
UNITED STATES

PN #	Item ID	Alibris ID	Media Type	Title / Author	Seller List Price	Order Date
76524203-44	Mountainvol39{{{K-B0836717168		BOOK	Mountain Torrent of the Tien Shan: a Faunistic-Ecology Essay Brodsky, K.A.	$31.99	Jul, 6 2026

Good Former xlibrary 1980 hardcover no dust jacket as issued volume 39 withdrawn stamp in book clean pages stamp at top edge of pages has library card pocket and call label at bottom of bind published by Dr W Junk bv Publishers 311 pages{{{ K-8. Ha

A general account of the Tien Shan and its streams

"Middle Asia is a region of immense sand deserts and mountains, rearing beyond the clouds and covered by perpetual snows and ice. These create abundant rivers running down to a plain to irrigate oases there. . . . A system of mountains, inclined piedmont plains, and uninhabited lowlands are the main types of natural landscapes in Middle Asia, striking in their contrasts. . . . Everything is mighty, inimitable and peculiar. The Middle Asian mountains are highest in the Soviet Union (the Communism Peak and the Pobeda Peak rise up to 7495 and 7439 m, respectively). In mid-latitudes, neither in the Soviet Union, nor the whole world are there as powerful icy rivers as the Middle Asian rivers. The largest glaciers are located in the mountains of Kirghizia and Tadjikstan (the Fedchenko and Inylchek glaciers). The mountain ice is a large storage of moisture (2000 km^3)" (Middle Asia, 1968, p. 8).

"The nature of Middle Asia is very diverse. There, we find vast deserts, such as the Transcaspian Region which does not yield in size to the Sahara. . . , and huge mountain ranges up to 20,000 feet or more with noisy streams and wild alpine scenes covered by permanent snows and abundant in enormous glaciers, showing signs of a previous, even more powerful glaciation" (Lipskii, 1902, p. 1). "Investigation of the Middle Asian rivers becomes of particularly great importance owing to the intensive irrigative and hydrotechnical work" (Middle Asia, 1968).

Orography. The Middle Asian mountains are the highest mountain system of the Eurasian mountain belt crossing the whole continent from the Pacific to the Atlantic ocean. These are the western part of the Tien Shan, the Pamirs, and the Kopetdagh. In its physiographical features, the Tien Shan is divided into four regions (provinces). a). The Northern Tien Shan. Its northern limit lies at the Semirechie plains and the southern one extends along the southern foot of the Terskei Alatau. The intermontane depressions are closed by mountains not only from the north and the south, but from the west or the east. b). The Interior Tien Shan. Along its western borders are the Ferghana Range and the Atoinakskii Range and the crest of the Talaskii Alatau; the southern border is formed by the Kokshaaltau Range; to the east lies the crest of the Meridional Range. A part of the Interior Tien Shan is known as the Central Tien Shan. c). The Western Tien Shan.[1] It is bordered by the western fringe of the Interior Tien Shan from the northwest, by the Syrdarya's depression from the west, and by the Ferghana Valley from the south. d). The Southern Tien Shan. It lies south of the Western Tien Shan as far as the Pamirs.

[1] For the floral identification of this region see: Pavlov, 1972.

"The eastern part. . . is formed by the Alaiskii Range". To include this range, at least its southern slopes, in the Tien Shan system is open to question on biogeographical grounds. It appears that this question should be discussed after the comparison of the fauna of the Akbura River in the Alaiskii Range and that of the Issyk River in the Northern Tien Shan. Some authors (Tshupakhin, 1964) name the Southern Tien Shan as the Turkestano-Alaiskaia, or Hissaro-Alaiskaia mountain system and separate it from the Tien Shan, which, in the authors' opinion (Middle Asia, 1968), is of little validity.

Three main tiers of relief have been distinguished on the Tien Shan: (i) the upper tier is the one of upland plains; (ii) the middle tier includes weakly- or moderately-dissected watershed-adjacent areas of the mountains: (iii) the lower tier is of steep dipping mountains. As we shall see below, such a relief stratification is well-defined also in the vertical dissection of the mountain streams.

The Tien Shan retains signs of two glaciations. The late Quaternary one was the valley glaciation, the glaciers being 3–5 times longer than at present. Very extensive dislocation of debris taking frequently the form of a catastrophic phenomenon is characteristic of the Middle Asian mountains. The Tienshanian mountains and the Pamirs have many marks of mighty landslides that occurred in the postglacial time. They dammed many rivers to form numerous lakes.

Climate. As to temperature, the Middle Asian mountainous regions are characterized by cool and cold summers and by cold and very cold winters. Compared with the lowland area, this region has more abundant precipitation, the annual amounts being 400–800 mm in the moderate-mountain area and over 800 mm in the high-mountain area.

Hydrography. The Tien Shan location in the desert zone with a very dry climate, its orographic and hypsometric features are responsible for the highest position of the snowline in Middle Asia, which gradually ascends from north-west to south-east (it approaches a maximum of 5200–5240 m in the highest peak belt, from the Lenin Peak across the Muskol Range towards the Karl Marx and the Engels Peaks). The snowline descends to its lowest level at the northern and north-western outer fringe of the Tien Shan, on the Kirghizskii and Zailiiskii Alatau. The line rises towards the central Tien Shan from 3600 m to 4600 m. Most of the Southern Tien Shan lies in the zone of high snowline from 4600 to 5200 m.

From ten to twelve thousand rivers run in the Middle Asian mountains (Figs. 1–23). Coming out on the plains, the mountain waters are expended for irrigation, filtration, evaporation and run short quickly. Only the Amudarya and the Syrdarya reach the Aral Sea.

Now we shall shortly review the hydrological patterns which depict a general ecological background to life in mountain rivers and streams. The first and still valid classification of the Middle Asian rivers was made by E. M. Oldekop (1917). He divided the rivers into three types with respect to

2

Fig. 1. A glacier at the upper Talgar River (northern slope of the Zailiiskii Alatau) (Photo by the author).

the main source of their alimentation: glacial, mixed and snowy types. Rivers with melt-water of permanent snows and ice as the main source of alimentation were included in the first type. Maximum discharge in these rivers occurs in July and August, the warmest time of the year when the snow and ice melting is most intensive. Rivers referred by the author to the mixed alimentation type are located at lower altitudes and have their

Fig. 2. The icy table on the Talgar Glacier. (Photo by the author.)

Fig. 3. A glacier at the upper Issyk River (northern slope of the Zailiiskii Alatau). Summits on the left rise to 4300 m above sea level (Photo by the author).

watersheds just below the permanent snowline. Rivers of snow alimentation (that is, generated because of melting of the seasonal snows) have maximum discharge in May or early June. On the same grounds, that is according to the alimentation type, another classification was made by L. K. Davydov (1927, 1933, 1953) who retained Oldekop's three types. M. I. Lvovich

Fig. 4. Headwaters of the Cholpon-Ata River tributaries (southern slope of the Kungei-Alatau) (Photo by C. A. Yudin).

4

Fig. 5. The Aksai River valley (Central Tien Shan). Upper course (Photo by E. O. Konurbaev).

Fig. 6. The Aksai River valley (Central Tien Shan). Middle course (Photo by E. O. Konurbaev).

Fig. 7. A mountain stream at the Dolon Pass (Central Tien Shan) (Photo by E. O. Konurbaev).

(1938) proposed a river classification also based on the alimentation principle. However, he differentiated between four, not three types, including in
the fourth rivers with underground alimentation, and he took into account
the water discharge during different seasons. The other three types are the
rivers with ice, snow, and rain water sources.

Fig. 8. The Atbashi River (Central Tien Shan) (Photo by E. O. Konurbaev).

6

Fig. 9. A tributary of the Atbashi River (Central Tien Shan) (Photo by E. O. Konurbaev).

V. L. Shultz (1944, 1965) in some of his papers and the monograph "Rivers of Middle Asia" emphasizes the "most dominant role" of snowmelt water in the total alimentation of the Middle Asian rivers. In contrast, the rain alimentation of the rivers is of little importance (1965, p. 50). According to their alimentation he classifies the rivers into four types: (i) of glacial-snowy alimentation with the maximum discharge in July and August; (ii) of snowy-glacial alimentation with the maximum discharge in May and

Fig. 10. A tributary of the Naryn River (Central Tien Shan) (Photo by E. O. Konurbaev).

Fig. 11. The Zailiiskii Alatau, Alma-Ata gorge (Photo by I. A. Neifeldt).

Fig. 12. The Zailiiskii Alatau, headwaters of the Sernebulak River (moraines) (Photo by I. A. Neifeldt).

June; (iii) of snowy alimentation with the maximum discharge in April and May; and (iv) the rivers of snowy-rainy alimentation with the maximum discharge in March, April and May. One can see that this classification is based on the same principles as E. M. Oldekop's, with the exception for the mixed type. O. P. Shcheglova (1960) made a classification on the same basis. Of interest to biologists are the author's data on the effect of vertical zonality on the type of alimentation. The author suggests four rough limits of the average-weighed elevations of the river types in her classification: the rivers with (i) glacial and (ii) snowy-glacial alimentation have mean watershed elevation above 3000 m; (iii) the rivers with snowmelt alimentation lie at 2000 to 3000 m; and (iv) most rivers with snowy-rainy alimentation lie below 2000 m.

A. M. Muzafarov who examined the flora of the Middle Asian rivers, streams and other waterbodies makes the following observations on river classification: "Considering. . . the classifications and taking into account the significance for aquatic organisms, particularly for algae, of such vital factors as water transparency, the amount and composition of salts dissolved in the water, the amount of suspended loads, water temperature, pH and others" (1965, p. 14, 15). He proposes dividing the Middle Asian rivers, with respect to their alimentation, into the following groups: (i) streams and rivers with glacial and snowy alimentation (due to the permanent snow melting); (ii) those with snowy alimentation (the watershed lies below the permanent snowline); (iii) watercourses with mixed and (iv) lacustrine and underground alimentation. Concerning this classification, which

Fig. 13. A mountain stream on the northern slope of the Zailiiskii Alatau (Photo by I. A. Neifeldt).

seems to meet the needs of botanists ("hydrobotanists", Raspopov, 1963), one should note that although the water genesis is of much value in classification of rivers, streams and other watercourses, the distribution of aquatic invertebrates, mainly, of lithorheophylous fauna, gives good reasons for identification of groups, or types, of watercourses not only by their source of alimentation, but also by the "volume" of water, *i.e.* water discharge, especially during the vegetation period. Therefore, the watercourses with similar alimentation but with different discharges should be placed into different groups. For example, the rivers and streams having glacial alimentation only are difficult to join by the fauna into one type. (For

10

Fig. 14. The Turgen River (Northern Tien Shan). Lower reaches of headwaters, 2300–2500 m above sea level (Photo by E. O. Konurbaev).

details, see the section on classification by fauna where the vertical zonality of watercourses is discussed.)

On the latter subject we shall present the following schemes. Shultz, Timofeeva and Nadezhdin (1936) divide rivers into drainage sections (areas).

1. The humid section is an area of rainfall condensation (mountain area).
2. The sub-arid section is an area of dispersed drainage (predesert, dryish section), from 1000 to 2000 m.
3. The arid section is a desert, dry area.

Such a scheme is too general, of course, and cannot be used for ecologo-faunistic purposes. Muzafarov (1965, p. 19) proposes a somewhat more detailed scheme:

1. The high-mountain section[2] is a belt (yailau[3]) from 2700 to 5000 m and higher (snowfields, glaciers, rocks, syrty, alpine meadows).
2. The mountain section is a belt (tau) from 1200 to 2700 m (the belt of broad-leaved and coniferous forests).
3. The fore-mountain section is a belt (adyr) from 500 to 1200 m.
4. The desert section is a belt (chul) from the delta up to 500 m.
5. The delta.

[2] Description of the sections is given in Chapter 2 devoted to torrential flora.
[3] Here and below, the local names of vertical belts will be used: Yailau is for alpine meadows, Tau is for broad-leaved and coniferous forests, Sazy is for marshes, Chul is for desert lowlands, Adyry is for bared piedmonts, Syrty is for high plains.

11

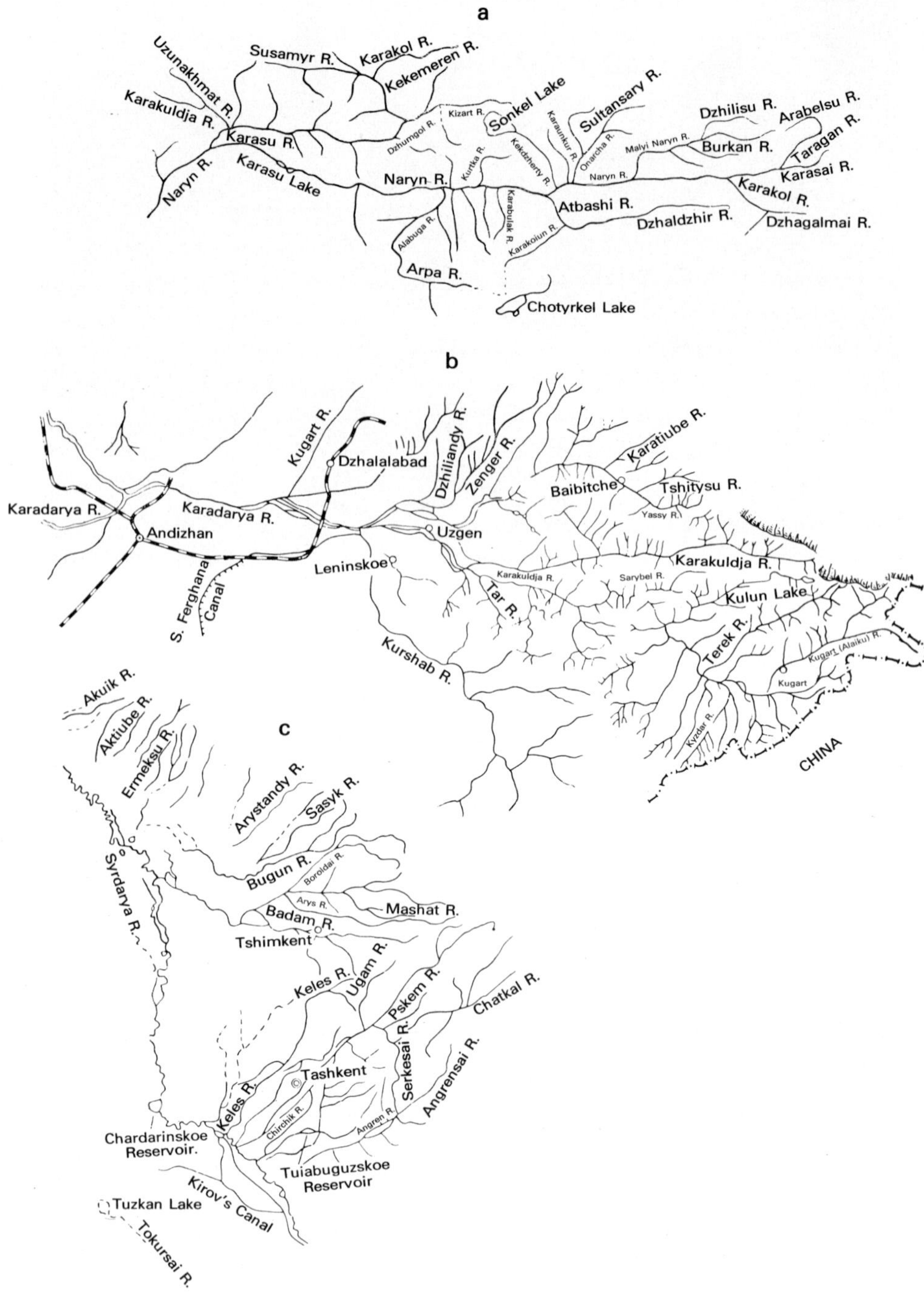

Fig. 15. Schematic maps of the drainage network for some rivers of Middle Asia (Muzafarov, 1965, 1968).
a. the Naryn River; b. the Karadarya River; c. the middle Syrdarya River.

12

Fig. 16. The Cholpon-Ata River (southern slope of the Kungei Alatau). Beginning of the lower river (Photo by C. A. Yudin).

To understand the ecological situation of torrential fauna it is essential to characterize the elements of the annual pattern of runoff (Kuzin, 1960; cited briefly according to: Middle Asia, 1968, Table 1). Whereas a typical feature of the runoff regime in lowland rivers is a flash flood, the rivers of moderate mountains and high mountains are characterized by a slower, or spring, flood. In those belts most of the annual runoff occurs during the spring flood reaching 70–75% of the annual runoff in the moderate-mountain belt of the Tien Shan and Alai and 80–85% in the high-mountain belt, which is due to high specific water-content of rivers during the spring flood and its long duration. In the high-mountain belt, where snow and ice persist during the whole year, the flood duration is 145–150 days and in the moderate-mountain belt it is 95–125 days. During this time, the seasonal snows melt away completely. Low-water stages (when the snowy and glacial alimentation is near minimum or absent) in the vertical belts differ both in the time of onset, end and duration. In low mountain rivers, where seasonal snows melt away in spring, the low waters set in as early as July-August. In the moderate-mountain belt, the low waters in rivers with snowy and snowy-glacial alimentation set in during September, after the melting of seasonal snows. In the high-mountain rivers with glacial-snowy alimentation the stable low waters begin in November–December, when stable negative air temperatures set in. The main types and subtypes of runoff regime in rivers of the mountain region are listed in Table 2.

The mountain streams in Middle Asia show a distinct enough dependence of the normal annual water temperature on the elevation. In the high-mountain belts, where rivers emerge from glaciers, the water temperature is

13

Fig. 17. A torrent in a narrow gorge (southern slope of the Kungei Alatau) (Photo by C. A. Yudin).

about 0°C. Down the stream, the waters become warmer and, near the river outlet from the mountains, the temperature increases up to 10–12°C.

Waters of most rivers running in the high-mountain belts are of bicarbonate type with salt content, generally, under 500 mg/l. This is caused by: appreciable humidification of the mountains by atmospheric precipitation washing out the salts from the soil; by the absence, in the upper mountain belt, of any considerable areas with salt deposits; and by low mineralization of underground waters feeding the rivers. The lowest salt content is typical for rivers falling into the Issyk-kul Lake and for upstream waters of

14

Fig. 18. The Chon-Aksu River (southern slope of the Kungei Alatau). The river bed from large boulders, 2700 m above sea level (Photo by the author).

Fig. 19. The Chon-Aksu River. A section with rip current (Photo by the author).

Fig. 20. The Chon-Aksu River. A section with more "easy-tempered" current (Photo by the author).

Fig. 21. The Cholpon-Ata River (southern slope of the Kungei Alatau). Lower limit of coniferous forest (of *Picea schrenkiana*) (Photo by the author).

16

Fig. 22. The Cholpon-Ata River. Downstream from the coniferous forest zone (Photo by the author).

Fig. 23. The lower Cholpon-Ata River (Photo by the author).

Table 1. Indices of annual regime of runoff (After Kuzin, 1960, abridged).

Hydrological zone	Hydrological region	Mean elevation (m)	date of onset	Spring flood date of the flood crest	duration (number of days)
Mountain–arctic	Western Pamirs	4500	15.V	17.VII	145
	Alai Mts., highest	3500	8.V	15.VII	150
	Tienshan, highest	3500	30.IV	5.VII	150
Mountain–woodland	Tienshan, mountain–woodland	2500	27.III	11.V	125
	Alai Mts., mountain–woodland	2500	14.III	14.IV	95

rivers which drain highlands of the Tien Shan and the Alai. Table 3 lists data on the dissolved salt content by the altitudinal belts in the Syrdarya and the Amudarya basins.

Vegetation. In the Tienshanian mountains (except for the southern range) the vegetation is subdivided as follows: sagebrush and saltwort deserts; the deserts change into sagebrush-bunchgrass steppe of low mountains followed by bunchgrass mountain (fescue-feather) steppe, which in turn change into moderate-mountain meadow steppe and meadows; just above there is a belt of hardwood forest followed by dark coniferous mountain forest; the highest mountain areas are occupied by subalpine and alpine meadows and wastes. This order of vertical stratification of vegetation corresponds, in general, to the geographical latitudinal zones. The southern Tien Shan ranges have their own system of belts; the more favourable to vegetation the orographic and climatic conditions the better the definition of the belts. Thus, in a typical case, the following belts are clearly distinguished on the western slopes of the Ferghana Range: ephemeral sagebrush fore-mountain deserts, tall-grass subtropical steepe (semisavanna), walnut and hardwood mountain forests and shrubs, juniper stands, bunchgrass (fescue) mountain steppe, subalpine and alpine meadows. Dark coniferous forests with the Tienshanian spruce are very characteristic of the Middle Asian mountains. Spruce forests consist of the two main altitudinally alternating groups: the moderate-mountain group and the high-mountain, or subalpine, group. The first prevails at elevations from 1400 to 2500–2600 m, the second at elevations above 2500 m. The juniper open stands are widespread in mountains and fore-mountains of the Kopetdagh on the Tien Shan as well in the Badakshan. They are typical for moderate belts of mountains 1200–1400 to 2000 m high. Above them (from 2000 to 2500 m) are juniper stands consisting of *Juniperus semiglobosa* followed (2500–3200 m) by *Juniperus turkestanica* only.

It is of interest to compare the vertical zonality of the Tienshan vegetation with that in similar regions of the Himalaya. The snowline is much higher in the latter: in the Nepal at 4480 m, in the Ladak at 5790 m, in the Tibet at 6000 m and in the Karakoram at 5500–5640 m (Mani, 1968).

18

Table 2. Main types and subtypes of runoff regime in mountain areas (from "Middle Asia", 1968, abridged).

Type	Subtype	Mean-weighted watershed elevation, in meters	Alimentation (% of mean annual runoff)				Runoff distribution between phase-homogeneous periods (%)		
			glaciers	snow	rain	underground waters	spring III–IV	summer VII–IX	autumn-winter X–II
High-mountain	Glacial	—	<80	—	—	—	—	80–100	—
	Glacial–snowy	>3000	20–35	35	—	—	<30	>55	<18
Moderate mountain	Snowy	2000–3000	10	45	—	—	<45	>40	>20
Low-mountain	Snowy–rainy	950–2000	—	30–55	<15	—	>64	>15	>20
	Rainy–snowy	—	0	30–50	—	50	>55	>15	<30

Table 3. Salt contents (mg/l) in altitudinal belts of drainage areas
of the Syrdarya and the Amudarya Rivers (Muzafarov, 1958).

| | Basin | |
Altitudinal belt	Syrdarya River	Amudarya River
High-mountain, near snowfields and glaciers	40–60	40–60
Mountain	308	150–300
Fore-mountain	387–400	300–500
Desert	380–576	380–766

South of the main Himalaya range, hills and low mountains are covered
by the broad-leaved rain forest (up to 900–1000 m above sea level) and just
above that zone by the evergreen forest. Beginning with 2440–3050 m the
broad-leaved sclerophylous forest grows (oaks and rhododendrons). Still
higher is the zone of birch and juniper ("archa"). In the Western Himalaya
at altitudes 3000–3660 m and in the eastern area at 4260–4575 m there lies
a zone of *Abies spectabilis* with rhododendron brakes followed by abundant
pygmean rhododendron. The zone between 3960 and 6000 m above sea
level is named, generally, as alpine zone. The alpine zone in the Himalaya is
the best defined and highest in the world. The birch and juniper belt (*Betula-
Juniperus*) corresponds to the transitional subalpine zone. The upper forest
limit in the Eastern Himalaya extends near 4000 m and descends to 3000 m
towards the north-eastern Himalaya. The zone above the forest line is rich
in the following plants: *Rhododendron setosum, Rh. anthropogon, Juniperus
squamata, Gentiana, Primula, Arenaria, Leontopodium, Androsace,
Ephedra*, grasses (s.s.) and sedge (Mani, 1968).

Flora of mountain streams in the Tien Shan

Although this volume is devoted to the predominant groups of invertebrates living in the mountain torrents (hymarobionts: mayflies, stoneflies, caddisflies and some Diptera families), it is nevertheless necessary to give some attention to the aquatic vegetation. Firstly, a direct relation exists between the above groups of invertebrates and the microscopic algae as a source of their nutrition. Secondly, of much importance is the role which the algae play as a component of the invertebrate environment.

As for the taxonomic knowledge of the aquatic vegetation, it is definitely in a more favourable position than the aquatic invertebrates.

The results of comprehensive investigations on the vegetation of running waters and mires of Middle Asia are presented in three monographs. In two of them Muzafarov (1958, 1965) gives data on algae. The book by Taubaev (1970) is devoted to macrophytes, flora and vegetation in waterbodies of Middle Asia. One can also get useful information from the papers by Kiselev and Vozjennikova (1950) on the algal flora of the Amudarya watershed, but these data, together with some others, are summarized in the monographs mentioned.

After the monographs had been published, a paper become available on algae in the streams of the Alai Valley and in the Kurshab River (the Alaiskii Range; Karimova, 1972). Although the lists of algal species and of higher plants in Middle Asian waterbodies are relatively complete, the plant ecology is, unfortunately, described by botanists in a very general way, the effect of microbiotopes on the distributive pattern of the torrential plants is not treated, and the role of the "biocenotic factor" in the life of aquatic plants is not considered at all.

Before we present any data on aquatic flora, we consider it useful to give a summary of characteristics for some rivers in the Tien Shan (Table 4) based mainly on the studies by Muzafarov (1968), Sibirtzeva (1961, the Chirchik River), and Sibirtzeva and co-workers (1961, the Zeravshan River).

At first, however, it is necessary to make some preliminary comments. The data in Table 4 are given in an abridged form and in a different sequence than in Muzafarov's monograph where the data are arranged according to the river systems. However, we find it useful to illustrate the major effect of water transparency on the abundance and even on the occurrence of algae. One can see from Muzafarov's conclusions that he also suggests this factor to be determining and even surpassing the altitudinal effect. As to the distribution of species, the absolute height plays in the mountain regions a decisive role. Some information on vertical distribution of algae will be given below in this chapter. In Table 4 the altitudes are not shown, as they fall within the relatively small range of 1700–2000 to

Table 4. Algae and riverside vegetation of the Tienshanien streams (a short account). (Sources: Muzafarov, 1958, 1965; Sibirtzeva, 1961, 1964 (the Chirchik River); Sibirtzeva, Kiseleva and Abdullaev, 1961 (the Zeravshan River)

Swift streams	Short description	Vegetation
	Streams abounding in algae in summertime	
Brooks of subalpine belt (Ferghanskii and Alaiskii Ranges)	Width about 1 m, depth from 3–4 to 20 cm; t about 8°C (August). Transparent water	Some brooks are rich in algae. Mosses on stones. *Vaucheria, Ulothrix zonata, Spirogyra* sp., *Microcystis testacea,* less frequently *Phormidium incrustatum, Hydrurus foetidus.* Diatoms
Kendersu brook (Ferghanskii range)	Small, but typical mountain stream with clear, transparent and cold water. Width 1–2 or 3–4 m, depth 20 cm, t 8°C (July). Stony bed	Massive colonies of *Hydrurus foetidus* on stones, less frequently film formations and thin mucous thicket of diatoms. Representative forms are periphyton, epiphytes, and rheophylous and coldwater mountain forms.
A brook at the Petrov Glacier (Eastern Tien Shan)	Width 1–2 m, depth from 5–6 to 15–20 cm; V up to 1 m/sec; Q 0.5; t 8.5°C (6 August, 1600) Transparent water.	Thin mucous film of blue-green algae (*Hydrurus foetidus*) and diatoms on stones
Kashkasu brook (Yassy River system, Ferghanskii Range)	Typical high-mountain brook. Swift current above cataracts. Much foam. Width 3–4 m, depth from 3–4 to 10–30 cm; t 5°C (September). Transparent water	*Hydrurus foetidus, Ulothrix zonata, U. aequalis, Spirogyra communis, Chamaesiphon curvatus, Calothrix parietina f. brevis* growing on stones. Diatoms of the genera *Diatoma, Ceratoneis, Synedra, Achnanthes, Navicula* and others
Kaindybulak River (an affluent of Alaiku River, Ferghanskii Range)	Length 35 km; affluents – two brooks, both 3–4 m in width, depth 10–30 cm; t 5–10°C (August); large stones on bed. Transparency down to the bottom	On stones numerous *Hydrurus foetidus, Ulothrix tenuissima, Homocothrix schizothichoides, Calothrix parietina,* f. *brevis, Phormidium favosum, Ph. unicinatum* and mucous films of diatoms. Thick coniferous forest (spruce, juniper) on slopes: birch, rowan-tree, willow, shrub on the flood plain
Yassy River in deciduous forest belt (Ferghanskii Range)	Width 15–20 m; depth 30–70 cm; V 1.5 and more; t 5–12°C (September). Transparent water	Circular and semicircular colonies of *Didimosphenia geminata* on stones. Diatoms as mucous films. On shores: black and silver poplars, birch, maple, appletree, alycha, rowan-tree, walnut, sweet brier, barberry, honey suckle, osier-bed

Djaukuchak River (an affluent of the Kumtor River, Terskei Alatau Range)	Width 6–7 m; depth up to 1 m; V 0.25–0.50; t 14–16°C (August); dissolved salts 388 mg/l. Transparent water	Abundance of algae: 13 green, 12 blue-green, 33 species of diatoms
Kuksai River (Shahimardansai River system, Alaiskii Range). Near Chamza-Abad village	Width 3–4 m; depth up to 1 m; V 2.06 (12 June, 1200); t 9–10°C (May), 9.2–11.0°C (July), 10°C (August), 9.9°C (November): O$_2$ 122% (12 July), 93% (30 July); pH 7.5; pebbles, large stones in bed. Transparent water of light blue colour	Periphyton on stones. Green algae: *Cladophora glomerata, Ulothrix aequalis, Ul. variabilis, Ul. zonata, Spirogyra communis, Sp.* sp., *Zygnema* sp., *Phormidium automnale, Ph. favosum.* Diatoms: *Navicula, Cymbella, Synedra* and others. In late autumn: blue-green *Stratonostoc verrucosum*; numerous diatoms (*Synedra, Eucocconeis, Cymbella, Gomphonema, Rhopalodium, Nitzshia, Surirella*); red algae – *Bangia atropurpurea, Batrachospermum moniliforme, Chantransia* sp. More abundant in algae than the Aksai River or the Shahimardansai River. On shores: poplar, walnut, maple, mulberry and others. Wildgrowing plants: hawthorn, sweetbrier, barberry, birch and others.
Affluent of the Isfairamsai River, the Tegirmachsai River (Alaiskii Range)	Width 2–3 m, depth from 20–30 to 60 cm; V 2–2.5 (9 August 1938); t 12.5°C (August, 1100), pH 7.5. Transparent down to bottom	Peryphytom on stones. Green: *Ulothrix zonata, Ul. variabilis, Oedogonium* sp., *Spirogyra* sp., *Zygnema* sp., Blue-green: *Stratonostoc verrucosum, Phormidium favosum.* Diatoms: *Diatoma, Synedra, Cocconeis, Navicula, Pinnularia, Cymbella, Gomphonema, Nitzschia* and others. Red algae: *Bangia purpurea.* On shores: sweetbrier, willow, honeysuckle, birch, poplar maple, rowan-tree, and juniper locally

Streams poor in algae in summertime

Kelte brook (affluent of the Kugart River, Ferghanskii Range)	Length above 10 km; width 1.5–2 m, sometimes up to 3 m; depth 10–30 cm; V 1–1.5; t 0.9°C (near snows, July), 4.5°C (downstream), 6–17°C (farther downstream). Turbid water	Very poor floristically owing to high water turbidity. (Another brook nearby with transparent water has mass development of *Hydrurus foetidus*, diatoms)
Kumtor stream (the upper Naryn River near Petrov Glacier, Eastern Tienshan)	Powerful turbid flow. Width 20–25 m at distance of 8 km from the glacier; depth down to 1 m; V 1–1.5; t 2–3°C (August); dissolved salts 28 mg/l. Transparency 3–8 cm (August).	Algae absent in August; films of blue-green, green, and diatomic algae are rare. In late August *Ulothrix zonata* and *Ceratoneis* occur.

Swift streams	Short description	Vegetation
Karakuldja River (Karadjar region, Ferghanskii Range)	Flowing in narrow gorge. Steep slopes, on peaks snow even in summertime. Width from 7–8 to 15–20 m; V 1.2–1.5; t 5°C (August); dry sediment 256 mg/l; large stones in channel. Rather transparent water	No macrophytes on stones, no algal films on stones. Coniferous belt, mainly of juniper. Willow, birch, rowan-tree and others, shrubs on shores
Karakuldja River (in Aktash region, Ferghanskii Range)	Width 16–30 m; depth 23–155 cm; V 0.74–5.0; Q mean 10.5–25 (max 46–49, min 3.3–3.5); t 0.5°C in winter, from 0 to 8.2°C in spring, 10–12°C in summer, 10.1, 5.6 and 1.9 in autumn; stony bed. Transparency 2–3 cm	In late August algae growths not observed on stones, which is due to extraordinarily high turbidity of water
Alaiku River (Kizyldjar region, Ferghanskii Range)	Width 300–500 m; has 4–5 arms; depth 20–50 cm; in gorge, width 50–60 m, depth down to 1 m; V 1.4–1.7; t 9–19°C (August)	Very poor in algal flora, single *Spirogyra*
Kugart River (Uyumbash region, Ferghanskii Range)	Width from 5–7 to 30–40 m, depth 15–30 cm; V over 1; stony bed. Turbid water with changing transparency during the day	No algae on stones ("due to turbidity")
Aksai River (the Shahimardansai River basin, Alaiskii Range). Downstream from Karashur to Kara-Kazyk region	Width 3–4 m, depth 30–70 cm; t 3°C (28 July, 0600), 4.1°C (28 July, 1400); O_2 222% (0600, t 3°C); pH 7.5 (28 July, 0600), 8.0 (the same day, 1400); large stones on bed. Transparency 10 cm decreasing towards midday	Floristically very poor river. High-mountain forms of algae only in springs and brooks: *Hydrurus foetidus*, diatoms. On shores, locally: juniper, rowan-tree, birch, sweetbrier, barberry, honeysuckle, dwarf cherry, mountain elm, meadowsweet
Aksai River (the Shahimardansai River basin, Alaiskii Range). Near Hamza-Abad village	V 3.8 (12 July); t 8°C (22 May, 1800), 17°C (22 May, 1500); O_2 146% (12 July at t = 9°C), 96% (30 August at 9.3°C); pH 7.5 (up to 7.8, rarely 8.0); pebbled bed. Transparency 3–4 cm in summer, up to 1 m in winter	Mosses. *Cladophora glomerata, Oedogonium* sp., *Phormidium favosum, Ph. autumnale.* Films of diatoms on stones: *Cymbella affinis, C. helvetica* var. *punctata, C. ventricosa, C. parva, Gomphonema constrictum, G. lanceolatum, Eucocconeis flexella* and others. In films: *Calothrix fusca* and other blue-greens. All periphyton disappears after mid-May with flood onset. In November, in addition: *Ulothrix zonata, Zygnema* sp., *Bangia*

River	Characteristics	Vegetation
	atropurpurea, Stratonostoc verrucosum occur. On shores: poplar, willow, mountain elm, walnut, apple-tree, apricot-tree etc. Wildgrowing sweetbrier, hawthorn, tamarisk, and sea buckthorn locally	
Naryn River (near Naryn city, Central Tienshan)	Width 40–100 m; mean depth 1.17 m (August), 2.36 m (July), max 3.20 m; V from 0.41 in winter to 4.58 in July, average 1.52–2.68; Q mean 83.6, max 421 in July	Thin films of diatoms on stones, very few of other algae ("high turbidity of water"), leafy mosses
Isfairamsai River (at Uch-Kurgan village, Alaiskii Range)	Width 20 m, depth 0.5–1.75 m; V from 0.6 to 3.71, average 1.5–1.6; Q 7–84; O_2 98–120%; pH 7.4–7.6, rarely 8.0; dense residue 231 mg/l (October–March), 493 mg/l (April–September)	Due to very high water turbidity algae do not grow in summertime. In late August, with increasing water transparency, films of green algae (*Hydrurus foetidus, Ulothrix zonata* and others) and diatoms occur on stones. Blue-green *Stratonostoc verrucosum* may be met during September–March
Shahimardansai River (at confluence of the Kuksai River and Aksai River, Alaiskii Range), point "A"	Width 12–30 m, depth 30–40 cm (down to 1 m); V 3.44 (12 July); t up to 15° in summer, 2–3° in winter; pH 7.5; cobbles on bed. Transparency 8–9 cm; 2–3 cm during snowmelt	Poor periphyton on stones in summer. In autumn, with increasing water transparency, green algae: *Ulothrix zonata, Cladophora glomerata*, more seldom *Spirogyra communis, Sp.sp., Zygnema sp.* Blue-green *Stratonostoc verrucosum, Phormidium favosum.* Diatoms: *Synedra, Eucocconeis, Achnanthes, Pinnularia, Cymbella, Rhoicosphenia, Diploneis, Navicula, Gomphonema, Hanthschia, Nitzschia* and others. On shores: willow, poplar, mountain elm, apricot, walnut, peach-tree. Wildgrowing hawthorn, barberry, sweetbrier, sea buckthorn and others.
Shahimardansai River (18 km downstream from point "A", at Kadamjai region, Alaiskii Range), point "B"		During summer flood in very turbid water, algae nearly absent. Only after mid-August, when transparency increases, green film of diatoms appear on stones. Diatom species are about five times more numerous than other algal groups. In October: *Hydrurus foetidus, Stratonostoc verrucosum, Ulothrix zonata* and others.
Tar River (piedment, in Chalma locality, Ferghanskii Range)	Width 31–37 m, depth 0.38–1.95 m; V 0.58–4.40; Q 46–57 in average, max 203 (June), min 6 (February); t below 0°C in winter, up to 23°C in summer. Transparency from 3–4 up to 10 cm in summer, over 1 m in winter	In summer algae and mucous film formations not observed. During winter, in large quantity: *Cladophora glomerata, Hydrurus foetidus* and mucous formations

Swift streams	Short description	Vegetation
Yassy River (piedmont, near Uzgen mountain, Ferghanskii Range)	In summer: width 7–50 m, mean 20–25 m; V 0.8–1.0; t 18°C (September). Transparency down to 50 cm. In winter: width 5–6 m, sometimes 10–12 m; depth 20–50 cm; t 7–9°C (February). Transparent water (over 1 m)	Very seldom *Cladophora golmerata*, small colonies of *Didimosphaenia geminata* on stones. *Rivularia dura, Ulothrix zonata, Hydrurus foetidus*
Karadarya River (in Uzgen mountain region, Ferhgana Valley)	Width 500–600 m, depth 0.5–3.7 m; V from 0.88 up to 5.75; Q 35.6–145.0 in average, max 500, min 26.8–55.0; t 2.9–6.2°C in winter, 7.8–16.1°C in spring, 18.6–20.7°C in summer, 17.4–6.9°C in autumn. Very turbid in summer, transparency below 5–6 cm, 38 cm in September	In summertime algal growth on stones not observed (high turbidity of water). In winter, stones are covered by mucous films of *Cladophora glomerata, Hydrurus foetidus, Ulothrix zonata* and other diatoms. In late February flora is three times richer than in September. Rice fields at study site

Streams with unshown quantity
of algae

Swift streams	Short description	Vegetation
Arabel River (affluent of the Kumtor River, Eastern Tien-shan)	Width at rips 10–12 m, depth 40–50 cm; V up to 1.5; t 5°C (August); dense residue of salts 86 mg/l. Rather high transparency	Coldwater and bottom algae: *Ulothrix, Scytonema, Calothrix parietina, Phormidium favosum*, diatoms. On shores in sazes: flowering plants, lichens, mosses. Among emergents: *Ranunculus natans*; among submerged: *Potamogeton filiformis, Batrachium divaricatum* and others
Zeravshan river (Sudjino village, Zeravshanskii Range)	Length of mountain section 300 km; t 12.5–16.5°C (May); O_2 88%; pH 7.4–7.6; dense residue 225 mg/l; stony bed. Turbid water, maximum transparency in January–February, minimum in June–August	On stony beds: *Stratonostoc commune, Lyngbya amplibasinata, Zygnema* sp., *Spirogyra* sp., *Microspora tumidula, Pleurotaenum minutum, Pinnularia* sp., *Achnanthes* sp. Near and on shores: *Equisetum arvense, E. ramosissimum, Equisetum arvense, E. ramosissimum, Typha angustifolia, Potamogeton perfoliatus, P. pectinatus, Schoenoplectus lacustris, Bolboshoenus maritimus, Veronica sp. Miriophyllum spicatum*

| Chirchik River (at confluence of the Chatkal, Pskem, and Ugam Rivers, Western Tienshan, Chatkalskii Range) | Snow-icy alimentation. Length 396 km; V 5 (mean velocity 3–4); Q max in July and August, min in December–January; t: 13.6°C Pskem, 17°C Chatkal, 18.8°C Ugam, 16°C Chirchik; O_2 71–108%; low-alkaline water; dense residue 143–480 mg/l; stony bed | On stony beds: *Amorphonostoc paludosus, Stratonostoc commune, Oscilatoria granulata, Diatoma elongatum, Synedra ulna, Cocconeis placentula, Cymbella proschkinae* and others. Totals: green algae – 39; blue-green algae 17; diatoms – 23 species. At shores: *Ecuisetum ramosissimum, E. arvense, Phragmites communis, Typha latifolia, Sagittaria trifolia, Calla palustris, Utricularia* sp., *Potamogeton pectinatus, P. perfoliatus* |

Note: V is flow velocity, in m/sec; Q is water discharge in m^3/sec; t is water temperature in centigrades; O_2 is dissolved oxygen in percent saturation; width and depth are given for the channel.

3000–3500 m, while the data relate only to the swift waters of the Tien Shan. These for the Pamirs and the adjacent regions to the west (data by Kiselev and Vozjennikova, and partly by Muzafarov) are not incorporated in the table.

If we look at this table carefully, we shall be able to make the conclusion about the relationship between algal abundance and different environmental factors. The first thing that attracts one's attention is the significance of water transparency. It should be interesting to cite some general conclusions from Muzafarov's monograph (1958, p. 94, 95): "In summer, it is the streams and brooks with transparent water that have more diverse flora. Streams and brooks with direct snowy, icy, and mixed alimentations, which do not have any reaches of slow current, are very poor floristically due to the excessive turbidity of water during the flood, *i.e.* in summertime. With increasing transparency in autumn, there is a noticeable increase in the algal abundance. In the piedmont regions, the algae grow rapidly in late autumn and during the first half of spring and, to a lesser degree, in winter. As the water turbidity increases in spring, algae begin to diminish and almost vanish during the flood". These points are repeated by Muzafarov in his "Conclusions", where he also gives examples of the algal genera and species emerging and disappearing throughout a year (*op. cit.*, p. 148): "Algae are almost absent in summer owing to the extraordinary water turbidity during the flood. After mid-August, blanket algae occur (*Cladophora glomerata*). The number of species increases in September and becomes maximum in November or, less frequently, after mid-December. In winter, cold-resistant species increase in number (*Hydrurus foetidus, Bangia, Chantrasia, Batrachospermum*, spp. *Ulothrix, Prasiola fluviatils, Stratonostoc verrucosum*, diatoms). The number of species somewhat decreases during January and February, then it again increases towards the spring with a second maximum in March. In April, many forms disappear owing to some increase in the water turbidity and temperature, especially the forms common to cold and clear waters. In May, the number of species reduces sharply and with the flood onset, algae almost vanish from the water."

As a general conclusion, the author makes the following statement: "Thus, a factor of the greatest importance in the inhibition of algal growth in mountain streams during summertime is the extraordinary water turbidity caused by the raised water level and the great power of current, and, as a result, a lower content of nutrients" (p. 99). On the next page (p. 100) it is pointed out that the flow velocity and, especially, its sharp changes due to cataracts in the river bed is a significant factor for algal life, which causes change in the species composition and contributes to the production of specific life forms (for the latter point, see Kiselev and Vozjennikova (1950)). Summarizing Muzafarov's assertions, one may conclude that, in spite of the great effect of water turbidity on the quantitative growth of algae, it is difficult to assess the role of this factor without its relation to other factors specific to mountain streams, such as the flow velocity and water discharge. We arrive inevitably at the conclusion about the necessity for a complex assessment of factors. However, we have found no clear position of the cited author as to the question of the interaction of specific ecological factors which affect the abundance and the qualitative composition of algae. For example, Muzafarov wrote in 1965 in the chapter entitled "Regularities in the algal distribution and growth in waters of Middle Asia": "... the distribution of algae is determined to a large degree by the water temperature as well by transparency, pH, quantitative and qualitative composition of the dissolved salts and gases, water movement and many other factors" (p. 196). This is nothing more than a listing of factors, from which it is hard (or rather, impossible) to imagine the whole complex of things determining the ecology of algae in swift waters. It appears that such a general treatment of the major ecological problem may stem from the lack of data on "algal" microbiotopes in the Middle Asian rivers, although investigations of such kind are already underway outside of the Soviet Union (see, for example, Douglas, 1958). When examining the lithorheophylous fauna in mountain streams of Middle Asia, we gave special attention to the biotopes, keeping in mind that different species inhabit different environments.

Taubaev (1970), who had investigated the flora and vegetation in all types of waterbodies in Middle Asia, expressed a more definite view on the necessity of a complex approach to the study of ecological environments of aquatic plants. He wrote: "... generally, these factors affect aquatic organisms as a complex rather than individually, and their effect can be understood only if they are examined in their relation to all biogeocenotic conditions. One thing is often neglected, however, that, depending on physiographic conditions of the milieu, some factors can be more distinct and have a stronger effect on biocomplexes than others" (p. 50). The same point of view we suggested in 1935 and are trying to follow it in this essay.

The above data on the essential role of water transparency and algal abundance, a factor dependent on flow velocity and water discharge, raise the question of seasonal variations in both the quantity and species composition of algae. This dynamics is pronounced so well in mountain streams that Muzafarov (1958, pp. 110, 111) could make the following remark: "There is an opinion, widespread in the literature, that mountain streams in Middle Asia are very poor in algae. Some works report even a complete absence of algae in the streams examined. But such statements are erroneous. The point is that the field investigations of the Middle Asian mountain streams were carried out mostly in summer, during the flood time, while there have been no year-round investigations."

A few data concerning the seasonal variations of algae in the Gulcha River (the upper Kurshab River on the southern slope of the Alaiskii Range; Karimova, 1972) may be appropriate. At the Gulcha headwaters, just after the flood, which sets in from the second half of July, there is an active growth of *Hydrurus foetidus* (a dominant form), *Ulothrix zonata*, *Oscillatoria terebriformis* f. *amphigranulata*, *O. rupicola*, *O. prolifica*, diatoms such as *Diatoma elongatum* var. *tenue*, *D. hiemale* var. *mesodon*, *Meridion*, *Synedra*, *Cymbella*, and others. In the nearshore zone, *Spirogyra* sp., *Zygnaema* sp. and others are abundant.

Algal growth is appreciable in autumn mainly owing to coldwater forms of

diatoms. The diatoms dominate in winter (80–85% of all algal species). At
the beginning of spring, in March, the algal composition is almost unchanged
although a thin layer of blue-green algae (*Oscillatoria, Phormidium*) occurs.
In April and the middle of May the abundance of the coldwater forms
reduces sharply. During the flood time, algal growth is not observed. There
are no algae in the middle Gulcha River during summer (*i.e.* in the flood
time). In the second half of August, as water transparency increases, single
filaments of *Cladophora* and algae of genera *Spirogyra* and *Mougeotia* occur
together with diatoms. At the end of October and during November,
colonies of *Hydrurus foetidus* are present. In winter, *Cladophora glomerata*
and some diatoms are observed.

It is impossible to express the seasonal variations of algae in mountain
streams in terms of biomass or individuals per square metre. We have not
found such data published for rivers in the Tien Shan. Only one illustrative
example may be given on the seasonal variation in algal species diversity in
the Isfairamsai and Shahimardansai Rivers (Fig. 24). In both rivers, the
number of species, decreasing in summer (under 10 species), has two
maxima: the first in November (more than 70–80 species) and the second in
March (70 and more than 100 species, respectively). Such seasonal distribution of the species number is typical for the Tienshanian rivers with an
insignificant shift of dates in the Northern and Southern Tien Shan.

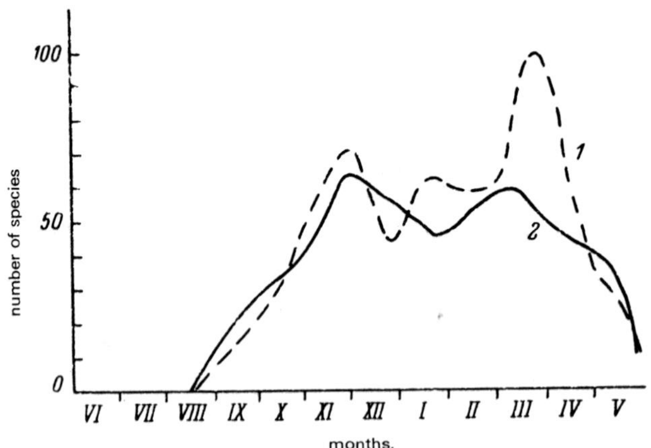

Fig. 24. Monthly variations in the number of algal species. (After Muzafarov, 1958).
1. the Isfairamsai River; 2. the Shahimardansai River (northern slope of the Alaiskii Range).

To return to Table 4, in which streams are grouped, as it was said,
depending on the algal abundance in summertime, we have attempted,
however, to arrange the streams according to the elevation of the study sites
(this principle however could not be maintained throughout). Analysis of
each group gives a general idea about the vertical zonality of algal species
composition. Muzafarov (1965, pp. 20–45) has suggested a scheme of
vertical distribution of algae.[1] (This scheme has been adopted by other
authors – botanists who study both aquatic and terrestrial plants.)

[1] Partly described in Chapter 1.

30

High-mountain belt, or *yailau* (from 2700 to 5000 m and higher, snowfields, rocks, syrty, alpine meadows).

Brooks and streams fed by ice and snow. Muddy water; transparency 3–4 cm in summer; water temperature from 1 to 5–6°C; flow velocity 2–2.5 m/sec; dissolved salts 40–60 mg/l; pH 7.5–8.0; O_2 100–230%. Algae are almost absent, only small number of *Hydrurus foetidus*, some diatoms.

Brooks and streams fed by lakes. Transparency down to the bottom (1 m); water temperature 8–14°C; dissolved salts 80–200 mg/l; pH 7.5–8.0 Aquatic mosses, in particular, *Hydrohypnum dilatatum*; *Hydrurus foetidus* is absent; diatoms.

Brooks and streams of mixed feeding. Transparency from 8–10 to 20–30 cm; water temperature from 5–6 to 10°C; dissolved salts 60–100 mg/l. Abundant algae. *Ulothrix*, diatoms, leafy mosses.

Mountain belt, or *tau* (from 1200 to 2700 m, the belt of hardwood and coniferous forests).

(Five groups are identified with respect to the alimentation type: brooks and streams of icy and snowy alimentation; lacustrine; mixed; groundwater, and snowy alimentations. Muzafarov's scheme includes also a general description of the belt, which is partly being cited here.)

Transparency varies with alimentation type of the different waterbodies. Flow velocity is higher than in the yailau belt, from 1.5 to 4.0 m/sec; water temperature 8–19°C; O_2 95–150%; suspended sediments 250–1600 mg/l; dissolved salts 130–300 mg/l; pH 7.5–8.0. In summer the following species are present in brooks and rivers: *Ulothrix aequalis, U. zonata, Prasiola fluviatilis, Stratonostoc verrucosum, Phormidium favosum, Ph. automnale, Diatoma hiemale, Eucocconeis flexella, Didimosphaenia geminata, Cymbella affinis, C. cistula, Bangia atropurpurea* and others. From late autumn to the next flood, we find forms characteristic of the high-mountain belt. Even in summer there is *Hydrurus foetidus* in some cold-water brooks.

Piedmont belt, or adyr (from 500 to 1200 m, adyr, debris cones, cultivated lands).

(Streams of this belt are divided into four types on the alimentation principle.)

Transparency from 3–6 to 30–40 and even to 100 cm; water temperature 10–20°C and higher; flow velocity 1.5–3.0 m/sec; dissolved salts 200–300 (450) mg/l (in summer).

With few exceptions, all species of the mountain belt and, the more so, of the high-mountain belt are absent in summer *e.g. Hydrurus, Ulothrix,* and *Stratonostoc* genera. Their growth is observed from autumn almost to May. Owing to high water turbidity, many rivers of icy and snowy alimentation have no algae during the first half of summer or the whole summer.

In Chapter 10 of this book we shall discuss the vertical zonality of mountain streams in terms of the distribution of the most representative faunal groups and ecological factors of the greatest importance to the groups assessed in their interrelationship. We have confined ourselves to a discussion of vertical subdivision of only one type of swift streams – the torrent.

Table 5. Distribution of algal groups in streams by species number (After Muzafarov, 1958, abridged).

Stream	Flagellatae	Green	Blue-green	Diatoms	Red	Totals
Shahimardansai	2	46	41	206	3	298
Isfairamsai	1	46	41	152	4	243
Tar	2	33	46	146	0	227
Yassy	1	26	18	135	0	180
Karakuldja	1	10	6	49	0	66
Karadarya, at Uzgen	1	10	7	33	0	51
Naryn, streams and brooks	1	30	43	164	4	238

The scheme of vertical zones in Chapter 10 is based mainly on changes (dynamics) of the factors and the specific fauna.

Some idea of the plant species composition in the Tienshanian swift waters can be obtained from Table 4, while Table 5 presents some general information on the distribution of algal groups in the Tienshanian rivers (Muzafarov, 1958, Table 32, p. 95).

Quantitative predominance of diatoms over other algal groups is quite evident. For all types of Middle Asian waterbodies Muzafarov (1958) gives the following proportions of the species, varieties and forms: Chrysophyta 4, Pyrrophyta 2, Euglenophyta 11, Chlorophyta 171, Charophyta 6, Xanthophyta 158, Rhodophyta 5, Baccillariophyta 454, *i.e.* more than a half are diatoms. In the swift waters the proportion in favour of diatoms is still more striking. In addition to the diatoms which forms mucous films on stones, there occur coenoses of rheophylic and coldwater forms: thickets of *Hydrurus foetidus* with admixture of *Ulothrix zonata* and *Callothrix fusca* or thickets of *Didimosphaena geminata, Cladophora glomerata* and *Phormidium favosum*. The representative forms are also *Bangia atropurpurea, Lemanea fluviatilis, Chantransia, Stratonostoc verrucosum, Prasiola fluviatilis* and others.

Phytogeographical analysis of the torrential flora is not discussed in this essay, but it should be noted that both A. M. Muzafarov and T. Taubaev argue against assigning the cosmopolitic distribution to many species of algae. Thus, Muzafarov (1958, p. 136) writes: "As far as the algal cosmopolitism is concerned, some investigators consider nearly all algal species as cosmopolites. Such an approach to this phenomenon seems to us mechanistic". The author points out an abundance of endemics in the algal flora of Middle Asia, the Tibet and the Himalaya. Taubaev speaks with greater certainty about the distribution of aquatic-mire plants: "It is quite clear from the geographical analysis of the Middle Asian flora that it possesses a high degree of endemism".

As will be seen from the section on torrential fauna, the present author, when presenting the data on the most specific groups of invertebrates in Middle Asian streams, also points out high endemism of the groups, sometimes on the generic level.

From the vast list of aquatic-mire plants given by Taubaev (1970), Table 6

Table 6. Mosses found in speedy streams of Tien Shan (After Taubaev, 1970).

Species	Habitat
Family *Grimiaceae* *Grimmia alpicola* Sw.	In water of streams, brooks, on watered stones (tau, yailau). Totally submerged plants in Zailiiskii Alatau
Family *Bryaceae* *Bryum Funckii* Schwaegr.	Sazy (tau, yailau) submerged in water. Syrty of Central Tien Shan
Family *Bartramiaceae* *Philonotis marchica* Brid	Sazy, springs (tau). Grows in water. Rivers of Western Tien Shan
Family *Amblystegiaceae* *Cratoneurum commutatum* (Hedw.) Broth.	Wetted sites on flood plains, near springs, on watered boulder surfaces or under water, in rocky caves under waterfalls (tau). Submerged in water. Western Tien Shan
Drepanocladus intermedius (Lindb.) Warnst.	Mires, sazes. Nearshore plant. Rivers: Upper Issyk, Bolshaya Almaatinka, Malaya Almaatinka, Turgen
Drepanocladus aduncus (Hedw.) Mönkem	In water of brooks, springs, mires (tau, yailau). Submerged in water. Zailiiskii Alatau

includes only those moss species which were found in swift waters of the Tien Shan. This list should be supplemented by the moss *Hydrohypnum dilatatum* mentioned by A. M. Mazafarov and found in the Djergalan River (Terskei-Alatau), and *Cynclidonus fontinaloides* and *Hygrablystegium fluviatile* both identified from samples of the Issyk River. In mountain streams, however, mosses do not occur so frequently as in mires and sazes, and they have a lower species diversity.

To summarize, emphasis should be placed on such characteristic features of the mountain stream vegetation in the Tien Shan which, so to speak, are "landscape-building" for these streams and will be commonly encountered when the streams are examined. First, it is the mucous films coating the stones, boulders and rocks formed by abundant growths of different diatoms and, to a lesser degree, of green and blue-green algae. Second, rather frequent are the colonies often shaped as complex balls or semispheres, and, finally, the moss formations on the stable stony substrate and on the banks – on rocks, boulders and scree debris. Flowering, submerged and emergent plants are not typical of streams and occur only in smaller flets, bays, *etc.*, being completely absent from strong current.

3 Description of the mountain torrent using two examples of watercourses, the Issyk and the Akbura Rivers, from the northern and southern Tien Shan

Both the Issyk and the Akbura Rivers are typical watercourses of the Tienshanian terminal ranges bordering on plains or on large valleys like the Ferghana. Streams in the intermontane, relatively narrow valleys are, with few exceptions, similar to them. For comparison, the author has investigated several dozen mountain streams and rivers in the Tien Shan in order to reveal the most characteristic flows (Fig. 25–30).

The swift waters in the Tien Shan (as well as in other mountain regions) undergo significant changes due to both natural phenomena and human activity. For example, a catastrophic mudflow in 1963 caused by forest felling at the Issyk River's affluents overfilled the Issyk Lake storage and partly destroyed an old dam (moraine) ponding the lake (Kazakhstan, 1970, p. 299, 300). There is a project to build a dam in order to make a large reservoir in the middle Akbura River, and then the middle river will change to a lake. Therefore, it is quite timely to make a thorough study of the mountain streams and to record their "undisturbed" or "natural" condition, otherwise it will be impossible to foresee the changes that may result from the power plant construction. In this case, the very considerable changes take place in the river's irrigative pattern being transformed from a primitive system into modern hydraulic structure.

The data presented in this and the next chapters pertain to the Issyk and the Akbura Rivers in their "undisturbed" condition, *i.e.* before the catastrophic mudflow and transformation of the Issyk River irrigative system and prior to the construction of the Akbura reservoir.

The shortness of the Issyk River (about 30–35 km) can however give a good idea about all zones of a mountain torrent which on the northern slopes of the Alaiskii Range, for instance, are much more extensive (the Akbura River is 195 km long).

V. V. Sapozhnikov (1907) gave the following picturesque description of the zonality on the terminal ranges: "The Zailiiskii Alatau (where the Issyk River runs) rises up as a sudden powerful fold on the southern side of the broad Iliiskaia Valley. At the foot there is a wide spread of villages and farms buried in the greenary of orchards and surrounded by fields. . . . Beyond the line of apple-trees, poplars and other hardwood plants is a stretch of Tienshanian spruce, still higher the lush alpine meadows and above them snow-covered silverlike peaks. . . . The Zailiiskii Alatau is a rare example of a range where, at a glance from the wide valley, one can see all the vertical belts from the cultivated hot steppe to permanent snows".

Fig. 25. The upmost course of a torrent. The bed is composed of boulders and rock fragments (Photo by the author).

Running down the northern slopes of the Zailiiskii Alatau, the Issyk River originates from a glacier adjacent to the Bogatyr Glacier feeding the Talgar River. In the same locality, at a mountain junction with the highest peak 4774 m above sea level, originates the Chilik River (Palgov, 1954). The glaciers feeding both the Issyk and the Talgar Rivers are not large (see Fig. 3); in daytime in summer, the ice is melted intensively giving rise to small brooks which run down the glacier surface (see Figs. 1 and 2). There is the Akkul Lake (White Lake) upstream, which has the shape of a regular oval with opaque opal-green water. The water turbidity is caused by the glacier genesis of the river heads carrying the debris of the glacier-destroyed rocks. The Akkul Lake is surrounded by steep rocky slopes without any vegetation. The lower slopes are covered very uniformly by talus deposits. In places, one can see snow patches melting in the sun and feeding numerous little brooks which cause a continuous slipping of stones down the slopes. The lake has been formed by an end moraine of large boulders. There is no visible outflow from the lake, since the water percolates through the moraine material. Downstream from the moraine, having received several

35

affluents (brooks), the Issyk River looks like a rather large torrent with a bed of stones and rock fragments nearly unpolished by water. At an elevation of about 2700 m lies another lake but it is of a different type than the Akkul Lake. The Boskul Lake represents a large flat depression surrounded by lofty steep slopes covered by well-developed talus deposition. The slopes have no vegetation. But in the lake region one may find stunted spruce (Tienshanian spruce) as lonely single trees. Of interest in the Boskul Lake is the water filling cycle. At dawn the lake contains no water and its silty bed is exposed, with the Issyk River meandering over it from the southern end of the lake. Then the river rests against the end moraine of considerable height and filters through it. At night, when the ice melting is slow and the water level in the river becomes low, all the water has enough time to percolate through the body of the dam so that the river bed becomes exposed, covered with glacier silt. In the afternoon, under the highest sunshine, the snow and ice, which contribute to the highest reaches, produce so much water that it cannot be in time filtered through the dam and by night covers the lake bottom by a layer of 1–2 m high.

There is no need to emphasize that moraine lakes are of great importance to torrential fauna, in particular, when they are periodically filled by water and, therefore, play the role of natural water-level regulators. The Boskul dam is covered by thick spruce forest (*Picea schrenkiana*) undergrown by rowan-tree, bird-cherry tree, honey-suckle, and birch. The stream escapes from under the Boskul dam and, joined by the western affluents, becomes so full that it is impossible to wade it, although one can easily do this farther upstream (Fig. 26). The transparency is high, and the water has a beautiful blue color. Farther downstream the river is full of rapids and covered by foam; the bed is paved by large coarse debris and boulders which create a serious obstacle to the water flow. The stones in the channel and rocks overhanging the banks are coated with moss (*Hygramblistegium fluviatile*). The banks of the stream are overgrown by thick stands of Tienshanian spruce with leafy underwoods and shrub.

The river retains the same character down to the point of confluence with the Djarsu (or Zharsu) stream, which flows into it roaring through a rocky gorge from the south-west. Then the river changes sharply its aspect: the water becomes turbid (especially in the afternoon, when the snow and ice are intensively melted) and its amount doubles. The Djarsu bed is crowded by large but less coarse boulders which are then dragged by the water downstream into the Issyk River.

Before it runs into the Issyk Lake, the river has a very variable character. The rivers gorge has rather gentle slopes covered by a dense canopy of spruce. Sometimes the gorge becomes narrower, reminiscent of a rocky corridor through which the river roars its way in foam. Almost at its whole length, the river bed consists of large pebbles and boulders, while at the sites where cliffs approach it, the river flows over even and rocky ledged grounds. They are covered locally by mosses giving shelter to an original population, whereas the bed ledges produce a number of cascades, small waterfalls, *etc.* (Fig. 27). In wider parts of the gorge, the river is divided into 3–4 branches

36

Fig. 26. The middle reach of a torrent. Laminar flow is replaced by turbulent flow (Photo by the author).

which run more quietly in the bouldered channel, leaving behind pebbly and sandy patches. Where feet of trees at the gorge bottom are buried by river into debris, one can encounter whole plots of dead spruce forest. On its way to the Issyk Lake, the river receives 4 or 5 affluents and a large number of small brooks and springs. The affluents fed by the snow on the watershed summits contribute during the whole summer. About 8 km downstream, after the Djarsu stream empties into the Issyk River, the latter runs into the Issyk Lake (elevation about 1760 m and area about 1×2 km) surrounded by steep slopes with Tienshanian spruce or rocks of felsite porphyry. The river runs very violently out of the lake, falls in cascades on the channel of large slabs and fragments, and arrives at a small lower lake (Fig. 28). From there, through a narrow rocky gate, the river runs in a series of rocky steps down an old moraine, creating innumerable cascades and small waterfalls. Half the water is smashed into spray so that a white lacy cloud rises high over the seething cascades, producing a number of rainbows intercepting each other. The Issyk Lake's outflow (water discharge about 2.7 m^3/sec) starts only in mid-June (or even early July at the highest water level) and lasts to the autumn decrease when melting of snow and ice at the river headwaters is reduced. We have emphasized this circumstance because in field studies one may observe scarcity or even complete absence of fauna in "normal" streams and rivers unaccountable without a knowledge of the stream dynamics.

37

Fig. 27. An affluent of a torrent with very steep slope of the bed ("cascade of steps") (Photo by the author).

Along the main moraine (lake dam) lies the lower limit of spruce forest, under which the spruce is absent (about 1600 km above sea level) giving way to hardwood forest of apricot, apple-tree, plum-tree, hawthorn, and the dense shrub of honeysuckle, sweetbrier, *etc.* Farther down, at the foot of the last moraine, the river flows more quietly. However, it is still rather impetuous down to Issyk village, sometimes dividing into 2–3 branches, sometimes gathering into one channel (Fig. 29). The bed in the lower river consists of smaller and well-rounded boulders and pebbles; there are sand and gravel deposits in quiet places near the banks. On large stones there are characteristic peryphyton formations of mosses and algae (*Nostoc,* green algae).

All the lower part of the river and its irrigative system are located on a large debris cone. The drainage ditches end up at an old, so-called Kuldjinskii route, where the debris cone changes into an almost plane steppe. Farther down, the river, still maintaining at the village the character of a mountain torrent, becomes progressively changed. Downstream from the village, it runs in a wide channel paved with fine or medium-sized pebbles (Fig. 30). The channel divides into numerous arms, highly "ephemerous" and continually wandering about the river bed. Ephemeral grasses and wormwood occur on the shores. The air and water temperatures have increased. Near the Kuldjinskii route, the Issyk River loses the character of a mountain torrent

38

Fig. 28. An affluent of a torrent. Sharp change of microbiotopes: flow velocity drops at the foot of cascade (Photo by the author).

Fig. 29. Middle reach of a torrent. Laminar flow dominates with velocity over 3 m/sec (Photo by the author).

Fig. 30. Middle reach of a torrent. The water diffluences over the bed of rounded pebbles (Photo by the author).

and becomes a small brook. Its floor becomes narrower and sandy intercolations appear among the pebbles on the bottom. The water temperature becomes increasingly higher, the flow velocity slower, and foam is entirely absent. The powerful crashing of the mountain torrent so audible in the upper or middle river changes into the weak purl of a brook. At a distance of 3–4 km downstream the view again becomes entirely different because the Issyk River runs dry into a small mire. The old pebbly bed is covered by sands and gravels and overgrown by reeds and osier shrub. Here the water amount is very scarce, the velocity is hardly noticeable, the bed and shores are covered by a thick layer of silt with occasional flattened pebbles scattered about. Now the water has acquired a rather strong odour of mire,[1] although farther upstream it had no smell.

Thus, along the distance of only 30–35 km of the Issyk River, one could

[1] Description of the drainage ditches is not given here because they suffer rapid changes owing to intensive hydraulic work. Even so, the data, obtained from detailed studies of the ditches, have considerable and lasting value if one wants to understand the ecology of fauna in swift waters. Hydrological and faunistic data on the irrigation ditches will be presented when reviewing the fauna ecology.

40

follow all the modifications of a Tienshanian mountain torrent from glacier headwaters to the complete disappearence of the river in the steppe and semidesert area of the Iliiskaya Valley. During our observations of numerous streams in the Tien Shan we became convinced that this picture is typical of swift waters in Middle Asian mountains.

The other mountain torrent, the Akbura River in the Southern Tien Shan, is very similar to the Issyk River. Some differences, however, are caused by aridity of the Akbura River area, which is revealed mainly in the riparian vegetation. As it was noted above, because of the longer fetch of the Akbura River, its vertical zones are more "stretched" than those of the Issyk River. However, the ecology of the aquatic fauna is very similar in both rivers (Brodsky and Omorov, 1972a, b, 1973). The Akbura River refers to the Syrdarya basin (The Issyk River belongs to the Ili River basin). The river runs down the northern slope of the Eastern Alaiskii Range, the drainage areas of two other rivers: the Aravan and the Kurshab. Like the Issyk River, the Akbura starts from glaciers and snow fields and is expended for irrigation farther downstream and only a small part of the water reaches the large water manifold – the Ferghana Canal. The Akbura River originates at an elevation of over 4500 m, but it takes its name only at the confluence of the Kichik-Alai and Hodja-Kolon streams. The headwaters result from the melting glaciers and permanent snows so that a maximum discharge in the Akbura River falls on July and August.

The Kichik-Alai stream flows over moraines made up of rock fragments and coarse morainic unsorted material of both old and well-developed recent glaciations. Downstream, a series of cascades and small waterfalls is observed, but sometimes the river runs in a wide valley, over the pebbly bed, where it divides into several arms. In the vicinity of Osh city the river comes into a narrow rocky canyon. Downstream from the canyon several irrigation canals run out of the river, which form an extensive network over a debris cone.

According to Vyhodtsev (1956), the vegetation zonality in the river valley looks as follows. Since the Akbura basin is located in a more arid zone than the Issyk River basin or than south-western areas of Kirghizia, steppe plant aggregations, which dominate both in lower and middle vertical belts of the Alaiskii Range, occur also within high-mountain forests. Associations of mountain xerophytes and juniper stands play an important role here. Thus, a belt of couch-grass steppe, juniper forest, and open woodland lies at 1800–3000 m of absolute height. In contrast to the Issyk River valley, the mountain forest does not occur as massives, but rather as separated kolki (small insular groves). Brakes of ocier, poplar, sweet-brier, and barberry are widespread in the river gorges. It should be noted that Tienshanian spruce, so common in the Issyk valley, is absent in the Akbura basin. Ephemeral-absinthy semidesert vegetation (fescue and feather-fescue steppes) dominates in the adyr piedmonts at a height of 1200–1800 m above sea level. In a lower belt (500–1200 m) there is an agricultural oasis with crops of alfalfa, cotton, and mais. Hardwood vegetation consists of mulberry, Bolle's poplar or, in orchards, apple-trees, pear-trees, apricots and peach-trees.

Data on the mountain rocks in the Issyk and the Akbura basins are of importance, since it is known that water chemistry of torrents largely depends on the composition of rocks in the bottom. The nature of mountain rocks determines the morphology of the river bed, as well as errodibility of the underlying rocks.

Piedmonts in the Issyk, Talgar and Turgen Rivers' valleys are built from porphyritic rocks which form so-called melkosopochnik (hills of circum-denudation). These rocks are so well developed that they have nearly entirely replaced the exomorphozed granite and reached the very mountain range. Outcrops of highly weathered granite with overlying porphyric rocks can be seen in the Issyk valley. The Issyk Lake lies in greenish and red porphyries. Above the lake, the valley passes through felsitic porphyries erupted, here and there, by granite. The head of the range (Zailiiskii Alatau) consists of granites (Sergeeva, 1925).

Granite dominates also in the upper Akbura River; granodiorite, porphyry and porphyrite are common as well. Metamorphic rocks occur frequently. In areas with moderate relief paleozoic deposits are widespread: limestones, shales and also Cretaceous and Tertiary deposits of sand and clay. In piedmonts, Neogenic and Quarternary deposits are widespread. Piedmonts at heights from 800 to 1600 m above sea level are formed by Tertiary conglomerates and sandstones with particoloured wedges of marl. These deposits are generally overlaid by loess.

The climate in the Issyk gorge and in the Issyk Lake region, is very mild and moist. Maximum temperature in summer is not over 22°C, but in winter air temperature does not drop below −6°C. The annual range of temperatures equals 28°C with daily amplitudes up to 11°C, in average 6–7°. Short-term temperature variations are common and depend on cloud cover, as well on daily breezes from the valley or mountains. In the Issyk River gorge, above and beneath the lake, the weather conditions are quite different. Below the lake the climate is more continental, with sharp temperature changes and with higher summer maximum and lower winter minimum. Precipitation below the lake is much less than at the lake level and above it; cloudiness is much less there, foggy days are very rare and the air humidity is much less. The climate in the upper river, at the headwaters, is more severe, with low temperature in winter and rather low temperature at night in summer months (about 0°C even in the coniferous belt); the precipitation is exceptionally high in the form of rain, snow, hail or snow pellets; fogs are frequent. Even at the lake height the fogs last for 3–4 days on end (in July and August), and they are so dense that one can see nothing at five paces distance. Clouds occur in the lake area with high regularity: dense cumulus clouds appear from the south almost every day at 11–12 a.m. and cover most of the sky at 3–4 p.m. Farther upstream, the cloudiness is still greater, hiding the mountain peaks for several days at a time. Thus in some years, there was no clear day in the Issyk River region in July, but there were 14 days with precipitation. August has only one clear day, while others are overcast days or days with continuous cloudiness. The number of rainy days in that month is over 10. The wind, like the cloud, is highly regular:

starting at 10–11 a.m., it becomes very strong (on cloudy days) blowing along the gorge from south to north.

Similar weather conditions characteristic of the upper Issyk River were also observed on the southern slopes of the Kungei-Alatau and in many other regions of the Tien Shan, which gives good reasons to consider the above picture as typical for both moderate and high relief during summertime in the Northern and Central Tien Shan, *i.e.* in the period of highest activity of the organisms inhabiting the mountain streams. Undoubtedly, these conditions affect both the water and temperature regimes of mountain streams and, therefore, ecological environments of their faunas.

Of course, there are differences between the climatic conditions in the Northern and Southern Tien Shan: they are determined by the different geographical position of the areas and largely by the precipitation or the degree of aridity. For comparison, we shall give some data for the Akbura River.

The climate in the valley and piedmont of the Akbura River, at heights of 500–1000 m, is semidesert. The winter is relatively warm, with mean temperature minus 3–4°C in January. The summer is hot and arid, with mean temperature 25°C in July. The mean annual temperature equals 11.2–11.5°C. Precipitation amounts to only 200–350 mm per year with a maximum in March or early April. At an elevation of about 1000–2000 m the climate becomes moderately warm: the winter is cold, while the summer is warm. Precipitation is 450–500 mm per year. Most of it falls out during the spring, in March or early April. At 2000–3000 m the climate is moderate with a rather cold and long winter. The summer is cool. Precipitation totals are over 600 mm per year. Farther upstream, just at the Kichik-Alai heads, at 3000 m above sea level, the climate is severe.

It is evident that a general description of the climate does not reflect its variations depending on the position of the meteorological station; much will depend on the topography, slope exposure, etc. As an example, we could consider the observational data obtained at the station on the Kichik-Alai stream (Akbura headwaters, Table 7). In spite of the high hypsometric position (the station lies at the absolute height of 2360 m), the region is characterized by appreciable dryness; the mean annual deficit of humidity equals 5.6 mm with the precipitation totals only 291 mm per year. Most precipitation occurs in the warm period of the year, while winter accounts for only 68 mm giving a thin snow cover. The winter period, from November to February, is relatively short; in March the mean monthly temperature becomes positive, and the snow cover entirely disappears. The number of

Table 7. Mean monthly temperatures (t) and air humidity (%) Kichik-Alai station at the Akbura headwaters.

						MONTHS							
	I	II	III	IV	V	VI	VII	VIII	IX	X	XI	XII	year
t......	−6.9	−5.1	0.2	5.6	9.1	12.1	15.0	14.6	10.5	4.8	−1.4	−6.4	4.3
Humidity	55	54	58	55	58	54	50	49	48	51	54	58	54

sunny days is large here – over 150 days of clear sky per year with 25 days
every summer month (in the Issyk River area the conditions are entirely
different: see above). The mean relative air humidity in summer months
amounts to 35–40% and in the daytime is less than 30%. Precipitation
during the year is mainly in the form of rain and wet snow. Mean annual air
temperature equals 4.3°, with −6.9°C for the coldest month, January. The
freeze-free period lasts 131 days. The mean annual amplitude of air temperatures is 21.9° and the absolute amplitude 61.5°C (absolute minimum
−30.4° and maximum +31.1°C).

To conclude this brief account of Tienshanian mountain torrents, one
from the northern and the other from the southern regions, we should like
to emphasize that in the present state of knowledge the ecological environments of aquatic organisms, which spent, at their pre-imaginal stages, most
of the time in the aquatic medium, seem to be relatively alike throughout
the Tien Shan. Variations in the climate aridity affect mainly the seasonal
dynamics of water discharge in a stream (the extent and date of floods, *etc.*).
Very essential to the ecology of fauna is orography since it determines the
flow velocity, turbulence, *etc.*[2] All regions of the Tien Shan are similar both
in their orography and ecological conditions for torrential organisms. From
the point of view of these conditions, torrents of the Western Tien Shan,
differing from the Central, Northern and especially Southern regions in
having much higher humidity, are no exception.

[2] In this respect, the rivers in the Eastern Pamir and in syrty of the Terskei-Alatau, with the
character of lowland rivers meandering on the relatively flat surface of high-mountain plateau,
differ from the typical mountain torrents of the Tien Shan and western piedmonts of the
Pamirs.

Hydrological and hydrochemical characteristics of mountain streams and their microbiotopes

As noted above, the Issyk and Akbura Rivers have been investigated up till now in more detail than any other streams of the Tien Shan, although not in similar degree, which can be understood if one bears in mind that the Akbura River is about ten times longer than the Issyk. In the literature we have not found any ecological or faunistic data on the Akbura irrigative system as a stream derivative. So here we outline the data on the Issyk River irrigation canals. We believe that these data are of particular interest for the understanding of variations in the composition, distribution, and ecological peculiarities of the fauna in a mountain torrent after it gets into similar, but still somewhat different conditions of irrigation canals originating from the torrent.

From these considerations, we present data in this and the next chapters mainly for the Issyk River and its irrigation network, as well for the Akbura River, using also comparative data for other Tienshanian streams. The sites in which the faunas have been sampled and investigated are shown in Figures 31 and 32.

Before we present the data, we think it necessary to point out that the hydrological and hydrochemical peculiarities and the velocity distribution in microbiotopes are discussed here from the point of view of habitats of the lithorheophylous fauna. This determined both the selection of parameters and their more or less detailed investigation. In assessing the ecological conditions of algal life emphasis should be placed on "turbidity", *i.e.* sediment discharge. Alternatively, for a genetic analysis of the torrent some other parameters should be considered, and so on. In other words, both the hydrology and water chemistry are "biological" here, or, more precisely, ecological for the lithorheophylous fauna, or rather for the typical inhabitants of torrents – hymarobionts.

In this context, when discussing the flow velocity and water discharge, we review shortly some specific methods of measuring these parameters in a torrent. Other methods, not used specifically for running waters as compared to standing waters, are not considered here.

Morphometry. The channel width only varies a little along the Issyk River. The main channel width ranges from 3.5 to 7–8 m, mostly 5–6 metres. Sometimes the width increases to 25–28 m, but then the channel represents, in fact, a system of small islands of boulders projecting above the water surface. In some places the river is 12–15 metres wide, but again with rocky islands. The branches running off the main channel have very different widths, 2–4 metres, *etc.* The pebbly sections of the channel bed are usually wider (7–10 m) than the rocky ones (about 3 m). The Akbura River,

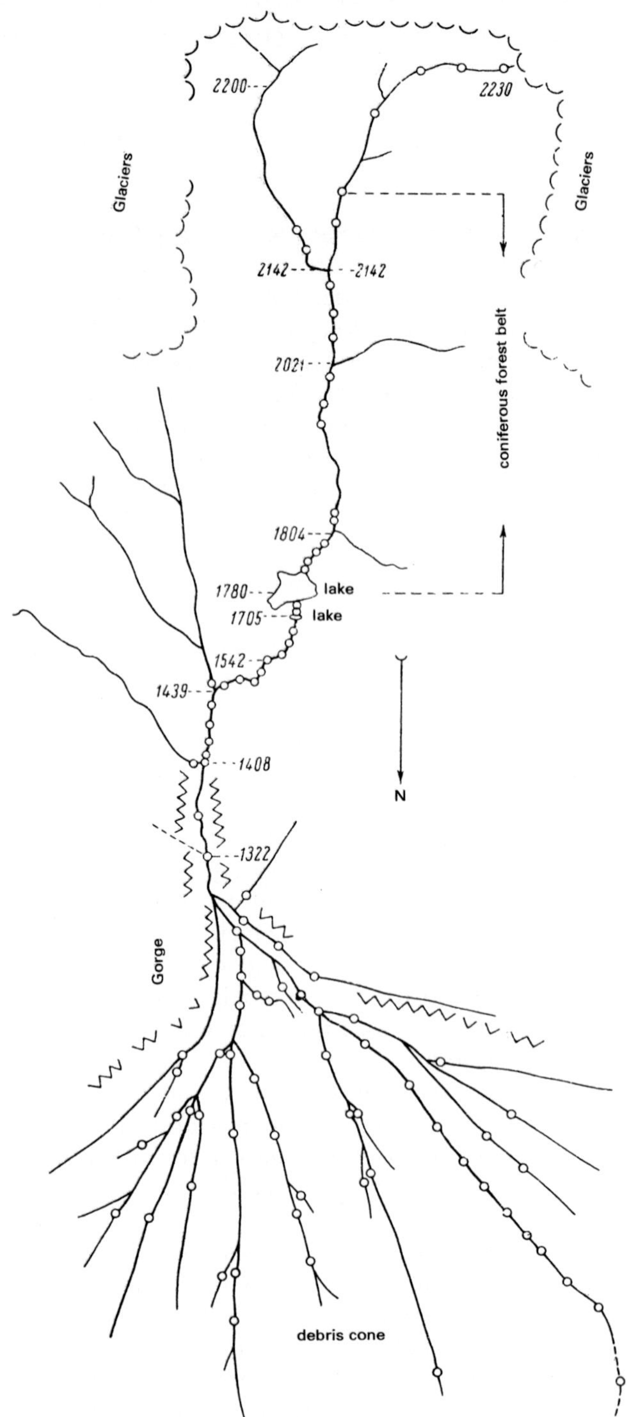

Fig. 31. A scheme of the Issyk River (northern slope of the Zailiiskii Alatau, Northern Tien Shan) with study sites (circles). Numbers are absolute heights in metres.

46

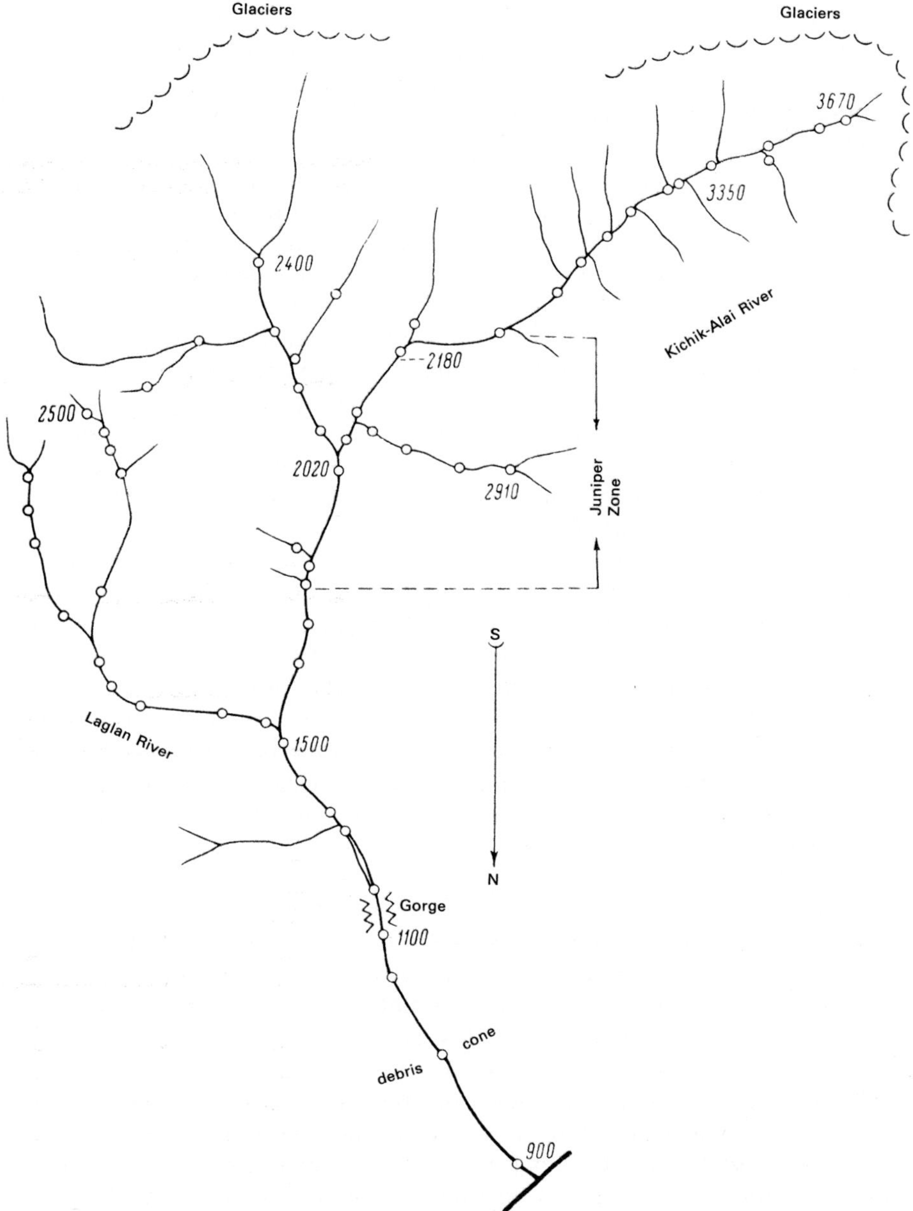

Fig. 32. A scheme of the Akbura River (northern slope of the Alaiskii Range, Southern Tien Shan) with study sites (circles). Numbers are absolute heights in metres. Irrigation network is not shown.

47

which is longer than the Issyk, has a width from 10–20 metres to 30–50 metres in the middle reach, but during the floods it may become as wide as 100–200 metres.

Both rivers have mostly random distribution of depths in the upper and middle reaches. Only in sections with beds of fine homogeneous material is there an increase of depth from the banks to the midstream, while in sections of coarse debris the depths are entirely irregular. The Issyk River depths vary from 25–30 cm to 1–1.5 m, in average 60–90 cm, and in the Akbura River from several centimetres to 1.5 m or up to 2 m and higher during the flood.

The above figures are nothing more than the order of magnitude of depth in the torrents, as it is well known that depth varies not only with bed roughness, but also with the water level which changes with the season and the time of day. Not only hydrobiologists, but anyone who visited the mountain regions of Middle Asia knows well that a river which can be waded early in the morning becomes unwadable later in the day. The reason is the increased discharge due to melting of permanent snow and ice in the light hours. In any case, it should be stressed that the lithorheophylous fauna of streams is living in conditions of continual and intensive fluctuation of water level and, therefore, of stream depth, so this circumstance is of particular importance to the attached (pupae) or slowly moving (larvae of some aquatic insects) stages.

The depth distribution is more regular in the lower Issyk river and its irrigation canals compared with the upper river. This may be due to the more graded slope of the river bed, less developed turbulence of flow, and absence of rocky slabs, large boulders, and rock fragments in the channel. The Issyk River depths near the village of the same name are about 90–100 cm and tend to decrease downstream to 60, 50, 35 and 10 cm, *i.e.* the depth is decreasing progressively from 1 m to 1 dm, changing the river farther downstream into a small mire which disappears in the reeds.

Most irrigation canals have horizontal and rectilinear beds. The bed profile in the affluents is more constricted and higher than in the canals. When one compares the profiles in the main channel, one notices their "flattening" towards the lower reach. Meanwhile in the irrigation canals we observe an inverse pattern: as the instantaneous water discharge decreased in the smaller canals, their profiles become higher and narrower. In contrast, in the lower Issyk River, near its end, the wetted perimeter of effective cross section increases with decreasing hydraulic radius, *i.e.* the profile become flatter. This phenomenon, we believe, can be explained as follows. In the lowermost Issyk River, almost at the base level of erosion, the eroding power of the stream is negligible or zero. Here, with the lower flow velocity the deposition of sediments takes place, which smooths out the bed roughness and causes the water to spread horizontally. In the irrigation canals which usually end up before the base level of the river and, therefore, also above their own base, the fluid force at an adequate flow velocity is still rather tractive, so the deposition of solids occurs only in the canals which approach the base level of erosion being not spent for irrigation. Of course, in the engineering type of canals the bed configuration is determined by human activity.

Flow velocity. In its direct and indirect effect on the torrential fauna the flow velocity is of paramount importance for aquatic organisms. There are various techniques of velocity measurement, but for description of daily and seasonal variations of velocity one should certainly use such records of gauging stations which present the flow velocity over the whole cross section rather than at a single point of the stream.

In route surveys the determination of mean surface flow velocity at single points of a stream can be made by means of the surface-float method, in which the float travels a certain distance measured along the river bank. It is a rather crude method, but it gives a possibility to compare velocities in various sections of the stream. The method has especially low accuracy in a violent torrent, nevertheless it provides "some idea" (Hynes, 1970) of the velocity. The only refined modification of the method suggested by an author consists in using an orange as a float. It is not the orange that matters, of course, but rather the gaudy, colored, rounded floats semi-submerged in the water.

A very important problem in the study of torrential fauna is the microbiotopes. This is particularly true of the violent streams with turbulent flow characterized by notable contrasts in the environmental conditions of aquatic organisms. When describing the upper and middle Issyk River (Brodsky, 1935), we gave special attention to it. Later, there were a number of works published on microbiotopes (Dorier, 1937; Linduska, 1942; Dorier and Vaillant, 1955; Ambühl, 1959; Macan, 1962; Pleskot, 1962). The significance of both boundary layer and so-called "dead water" (Das Totwasser, eau morte, eau calme) to the organisms was evaluated. However, no effective instruments were proposed for measurement of the flow velocity in microbiotopes. Quantitative data were mainly obtained from experimental investigations (Zimmermann, 1961; Bournaud, 1972).

Comparing various instruments for flow velocity measurements, we selected one which can give a detailed pattern of flow velocity in microbiotopes. This is the so-called Pitot-Darcy pocket tube (impact-pressure tube). Various types of current meters, Glushkov's bathometer-tacheometer and similar devices, require relatively large water depths and are unsuitable for velocity measurement in streams of centimetre or even millimetre depth over rocks and rocky slabs, or at the upper and side surfaces of stones, boulders, among plants, etc. The rod, control surfaces and the prolated form of the sensors in the Pitot-Darcy tube make it possible to set this device at the required point and depth in the torrent. Other advantages of the device are very small working area and the disposition of the tubes in a single horizontal plane. Velocity estimation is made from the formula $V = k\sqrt{h}$, where k is a calibration constant, h is the pressure difference in the manometer readings.[1]

[1] For impact-pressure tubes see Bykov and Vasiliev, 1972, p. 150–152. The authors give a different formula for estimation of local velocity by the tube: $V = \varphi\sqrt{2gh}$, where φ is a correction factor.

What devices have been proposed by ecologists in recent years to measure the velocity of flow? A number of current meters have been tested: the most suitable is the Edington-Molyneux meter with a revolving head of 1 cm in diameter (Edington and Molyneux, 1960). However, the large size and high cost make this instrument unpracticable.

The specifications of current meters made in the USSR are given in "Hydrometry" (Bykov and Vasiliev, 1972). The table on page 127 in this textbook shows the smallest diameter of the propeller for seven different meters (G-3; VMG-3; GR-21; GR-21M; GR-55; GR-11; GR-11M, each having two modifications) to be 60 mm, so they are unsuitable for measuring flow velocity at a single point or in a thin water layer (see also Burtzev and Baryshnikova, 1969).

Of some interest for ecological studies is the pygmy meter GR-96 developed by the State Hydrological Institute. This current meter is designed for velocity measurement in small streams: brooks, irrigation ditches, etc. The propeller diameter is 30 mm (rather too large), and as the authors pointed out, "the experience of using the meter is still insufficient" (p. 131). Also the Gessner meter (Gessner, 1955) has been proposed – similar to Gluchkov's bathometer – tacheometer. The inlet aperture is not more than 5–10 mm, but the meter, as a whole, is rather voluminous. The device consists of a metal cylinder with a rubber bag inside which is filled by water. By this, however, the meter creates a considerable obstacle to the flow, which distorts the velocity in thin water layer. Velocity measurements have been also carried out either by assessing the cooling time of a filament heated by current (Ruttner, 1953) or by comparing the temperatures of two current-heated thermistors: both were submerged in water so that one of them was pointed in the direction opposite to the flow (Kalmann, 1966). Another method of velocity measurement is to assess the solubility time of a standard tablet of salt (McConnell and Sigler, 1959). All these devices and techniques however, are less suitable for studying flow velocity in microbiotopes than the Pitot-Darcy tube. Among the existing modifications of the tube the best is the smaller model by Grenier (1949).

Some general principles are useful when mean flow velocities are measured. Thus, it is known that flow velocity is inversely proportional to the logarithmic depth of a channel, and velocity gradients are dependent on the bed roughness. The mean velocity is at a depth of about 0.6 of the total stream depth and is proportional to the square root of the hydraulic radius multiplied by the bed slope: $V = C\sqrt{Rs}$, where V is mean velocity, C is a constant, R is the hydraulic radius (cross section area of the wetted perimeter), S is the bed slope. The value of C is dependent on the component size of the debris in the river bed. Cummins has proposed a quantitative classification of the bed components (Cummins, 1962; Table 8).

As noted above, the bed roughness is significant for velocity evaluation. In mountain streams, the roughness coefficient has a rather large value; e.g. in a cascade type of channel with large boulders in the bed and great amount of foam over the water surface, it is equal of 0.190–0.200 (Sribnyi, 1960).

It is of interest to present the data on mean velocity of flow needed to

Table 8. Quantitative description of bed material (From Cummins, 1962).

Name of substrate	Particle size, mm
Boulder	>256
Cobble	64–256
Pebble	32–64
Gravel	16–32
	8–16
	4–8
	2–4
Very coarse sand	1–2
Coarse sand	0.5–1
Medium sand	0.25–0.5
Fine sand	0.125–0.25
Very fine sand	0.0625–0.125
Silt	0.0039–0.0625
Clay	<0.0039

transport the bed material along the channel of a stream (Nielsen, 1950; Table 9).

But the mean flow velocity, necessary for transportation of various types of bed material along the channel, is different for clear and muddy waters (Table 10; Schmitz, 1961, cited after: Hynes, 1970).

Therefore, the mean flow velocity in a stream depends on water discharge and on the width and depth of the stream, but even during the flood rarely exceeds 3–4 m/sec. Even in waterfalls the velocity is normal and does not exceed 6 m/sec, though at sites it may be higher. For example, in a rocky gorge of the Potomak River near Washington the velocity during a flood was equal to 810 cm/sec. In mountain torrents of the Tien Shan one usually deals with velocities about 3 m/sec.

This short review of methods for measuring flow velocity in streams enables one to conclude that the old and most efficient methods of float and Pitot-Darcy tube are best suited for biological aims.

The values obtainable by these methods are close enough to the real velocities to be used for field studies. It is better to gather a large amount of rough data than to obtain single precise measurements since the parameter itself – flow velocity in a torrent – is extremely variable.

Table 9. Mean flow velocity required to transport bed material (From Nielsen, 1950).

Velocity, cm/sec	Size of mineral particles, cm	Velocity, cm/sec	Size of mineral particles, cm
10	0.2	100	20.0
25	1.3	150	45.0
50	5.0	200	80.0
75	11.0	300	180.0

Table 10. The mean current velocity of clear and muddy water required to initiate movement along a stream bed of various types of bottom deposit (Schmitz, 1961; from Hynes 1970).

Type of substrate	Mean velocity, cm/sec.	
	Clear water	Muddy water
Fine-grained clay	30	50
Sandy clay	30	50
Hard clay	60	100
Fine sand	20	30
Coarse sand	30–50	45–70
Fine gravel	60	80
Medium gravel	60–80	80–100
Coarse gravel	100–140	140–190
Angular stones	170	180

The mean surface velocity in the upper and middle Issyk River was about 3.00–3.33 m/sec. In the uppermost section it was, however, lower, 1.84 m/sec, until a large affluent (the Djarsu River with 1.34 m/sec) runs into it. After the confluence of the Issyk and Djarsu Rivers, the velocity increases up to 3 m/sec and higher. The decrease of flow velocity in the upper part of a stream compared with its middle reach can be appropriately shown by the mean velocities of flow in various sections of the Akbura River from different vertical zones:

Absolute height, m...	4000–3500	3500–3000	3000–2500	2500–2000	2000–1500	1500–1000	1000–500
Flow velocity, m/sec.	1.7	3.0	3.0	3.2	2.6	1.9	1.0

It might seem that the flow velocity at the large bed slope in the headwaters of a mountain torrent should be much greater. However, there is a "humpering" effect on the flow exerted by the accidents of the bed, rough bed material, transverse and oblique vertical currents which arise from many obstacles in the way of the water jets, causing a stronger friction and disturbing the laminar flow which can be only measured by the float method.

At some places of the Akbura and Issyk Rivers the flow velocities were over 3 m/sec, sometimes 5–6 m/sec. Where the rivers are running out of the mountains (absolute heights of 1000–500 m), the velocity was lower, but larger than 2 and 3 m/sec. In the lower Akbura River the velocity was hardly over 1.5 m/sec.

For comparison, we present data on the mean flow velocity in the Turgen and the Kulsai Rivers (the latter an affluent of the Chilik River; Kurmangalieva, 1976). In the former, at absolute heights of 1150–1700 m, the velocity was 0.87–0.98 m/sec, and 4.9 m/sec at heights about 1800 m. In the second river, in its "upper reach" (about 1800 m above sea level), the velocity was 0.7–2.5 m/sec, in the "lower" 0.5–2.2 m/sec.

The flow velocity in the lower reach and towards the end of the Issyk River varies as follows (downstream): 3.33–3.66–3.57–3.33–2.50–2.20–2.00–2.17–2.00–0.50–0.31–0.06 m/sec. The analysis of the velocity variation in the river reveals a stepwise pattern of velocity in various reaches. At the site where it varies from 3.30 to 2.50 m/sec, which indicates a certain change in the character of the mountain torrent, two large irrigation canals run off the river. A still sharper change of velocity is observed at another site where the velocity of 2.00 m/sec drops down to 0.5 m/sec. This is the transition zone from a torrent to a brook.

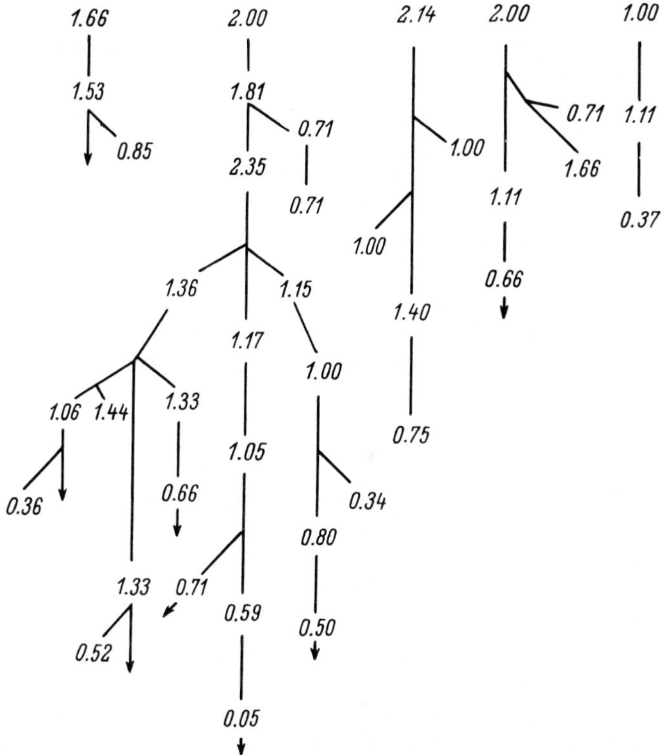

Fig. 33. Diagrams of velocity distribution (m/sec) in the Issyk irrigation network.
Here, as well as in other similar figures, on top are the ditches in the immediate vicinity to the river from which they outflow. At the bottom are the terminals of the ditches. Each scheme represents one of irrigation canals.

The velocities in the affluents are different, but due to their lesser power the velocity is not so large as in the main channel of the Issyk River: 0.40–0.56–1.11–1.80 m/sec.[2]

It will be recalled that for one and the same site in the channel the mean velocity depends on the following three factors: water surface slope, bed roughness and hydraulic radius. With the first two parameters unchanged, the increase of the mean velocity will depend on the hydraulic radius which in turn affects the water discharge variation in the stream ($R = F/P$).

Because of a lesser water discharge in irrigation canals compared to the main channel of the river, the velocities also decrease (Fig. 33). Velocities

[2] It is a well known phenomenon: increase of velocity with increase of discharge; thus the Volga River has an average flow velocity of 0.8–1.2 m/sec, but 2.35–3.20 m/sec during the flood (Behning, 1928).

up to 1.00–2.14 m/sec are observed only at the head of the canals decreasing in their middle parts down to 1.0 m/sec and less. In the terminal sections or in small branches the velocities are about 0.05–0.75 m/sec only. This fact is of paramount importance for the survival of typical torrential fauna in the conditions of an irrigation canal.

The above review of flow velocities is a more or less synchronous record for the summertime (July, August). Certainly, the seasonal dynamics of both velocity and discharge plays an important role in the fauna formation in a torrent, but, unfortunately, year-round observations are possible only in the lower section of the streams (because of inaccessibility of their upper, or even middle reaches during the winter), so they are very scarce in literature (Macan, 1962). The general pattern of seasonal dynamics of flow velocity in the Tienshanian torrents is as follows: in spring the increase of the water level in torrents is responsible for the increase in the flow velocity (approximately in April, as observed in the Akbura River); maximum velocities occur in June and July; then during October and November, the velocity decreases towards a minimum in winter. But in any season of the year, with rainfall, the water level rises sharply, sometimes increasing the flow velocity so much that bed deposits are washed away from the sayes (dry channels) and small affluents.

So far we have concerned ourselves with mean surface velocity in torrents. However, the great variety of velocities depending on the bed material, depth, large stones, boulders, rocks and distance from the bank causes a differential distribution of aquatic organisms, thus creating the specific uniqueness of a torrent as an ecological environment.

Leaving some considerations concerning the effect of microbiotopes on the distribution of organisms for the appropriate section of the essay (on the role of ecological factors), we present here the data on the flow velocity in microbiotopes of the Issyk River obtained from the original measurements by the Pitot-Darcy tube (Brodsky, 1935). Although the velocities at individual sites of the stream differ, one should keep in mind that they nevertheless largely depend on the mean velocity in a given section of torrent.

There are few data available in the literature on the velocity distribution at various points of a mountain torrent. We can cite only two or three authors who present such records in their publications, for example, Hubaut (1927) and Ambühl (1959, 1962). After a discussion of the physical nature of turbulent and laminar flows, Ambühl has concluded that the laminar flow hardly occurs in nature except for the surface layers of water, groundwaters and psammon.

So, what sites of the river bed and the torrent itself could be considered as microbiotopes? Since we speak about the distribution of aquatic organisms, it is clear that it is the difference in the composition of biocenoses will be indicative of the difference in the biotopes. These are the surfaces of a stone oriented differently with respect to the direction of the flow, various river depths, the boundary layer, various distances from the bank, *etc.* (see photos 19–23 and 25–30 of river sections). Table 11 and Figures 34 and 35a,b give

54

Table 11. Distribution of flow velocity (m/sec) in some points of the Issyk River near stones submerged into water.

At distance of 10–15 cm upstream off stone	5–6 cm downstream off stone	On upper surface of stone	75–80 cm downstream off stone	Near side surface of stone (relative to direction of flow)
7.74	0.94	1.93–2.03	2.09	1.74–2.09
0.82	0.46	Over 3.00	1.40–2.03	1.68–2.09
0.46–0.94	0.28–0.46	1.23–1.55		
0.66–0.96	Op.–0.46	1.23–1.55		
Op.–0.59	Op.–0.66	2.79–over 3.00		
1.42–1.56	0.84–1.25	2.03–over 3.00		
Op.–0.84	0.00–0.46	1.93		
		1.86–1.90		
		0.84–1.16		
		2.11–2.23		
Mean 1.00	0.48	2.13	1.84	1.90
Min and Max Op.–1.74	Op.–1.25	0.84–over 3.00	1.40–2.09	1.68–2.09

Note: Measurements were made for several stones, each about 1.75 m^3 in volume and total watered surface about 6 m^2. In this and next tables "Op." means "opposite" or weak current opposite to the general direction of flow.

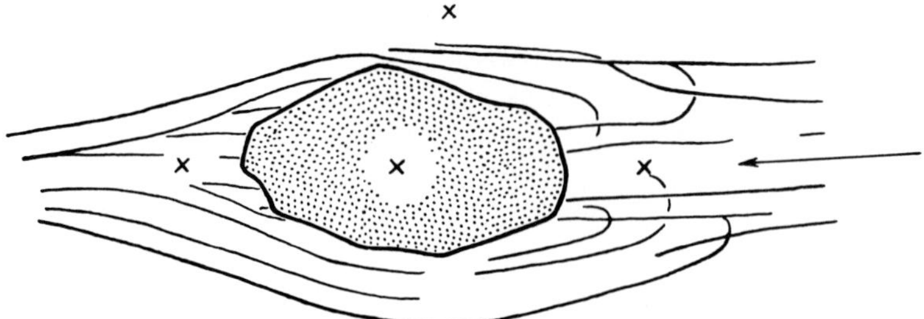

Fig. 34. Sites of velocity measurement (x) at a large stone (with surface area of $1.5 \times 1.0 \, \text{m}^2$) in the river channel. The arrow indicates the direction of the flow.

measurements of flow velocity near large stones and boulders in the Issyk River.

The values in this and the next tables are not single readings by the hydrometric tube, but the range of flow velocity observed with repeatedly sinking of the tube apertures down at the same point of the stream with 3–4 min duration of every measurement. The velocities in the tables are given in the order of observation, not by increasing values. The arithmetic means in the table show clearly enough their variety at the obstacles.

Measurements at different points have also been made in the water flowing around a thick log (spruce) fixed on stones: somewhat upstream, before the log, at a distance of a few centimetres, where the flow has lost some of its original velocity; then, after the log (downstream) where a strong vertical flow from under the log becomes mixed with the overall flow of the stream. Here the velocity was measured at two points: immediately upstream off the log and 75–80 cm downstream (Table 12 and Fig. 36).

Even greater differences were registered during velocity measurements between large, closely spaced stones, above small stones, at a depth of 10 cm, where some of the stones were partly emerged above the water surface; lastly, in the river arms between rocks completely submerged to a depth of 30–40 cm (Table 13).

It is worth noting the sharp decrease of flow velocity above the small stones and pebbles as a result of friction and formation of small irregular jets above the rough bed. In the gaps between the water-smoothed stones and above rocky slabs, the velocities are relatively high.

A detailed examination was also undertaken of the zone adjacent to the stream banks in the sites where the stream is flowing freely, unsuppressed by the rocky walls, but runs over a bed of pebbles and finely grained debris (Table 14). The water depth in the littoral zone is small (about 10–15 cm) compared to the typical depths in the midstream so that the pebbles and rock fragments on the bed emerge from under the water to form a favorable habitat of the abundant "nearshore" population (mainly, young stages of mayflies, stoneflies and caddisfly larvae).

A notable decrease of flow velocity is observed near the bank (at a distance of 15–20 cm from the water edge, at a depth of 25–40 cm) but only over a pebbly bed and in little bays where the depth is small. An entirely different pattern occurs when the velocity measurements are made at deeper

56

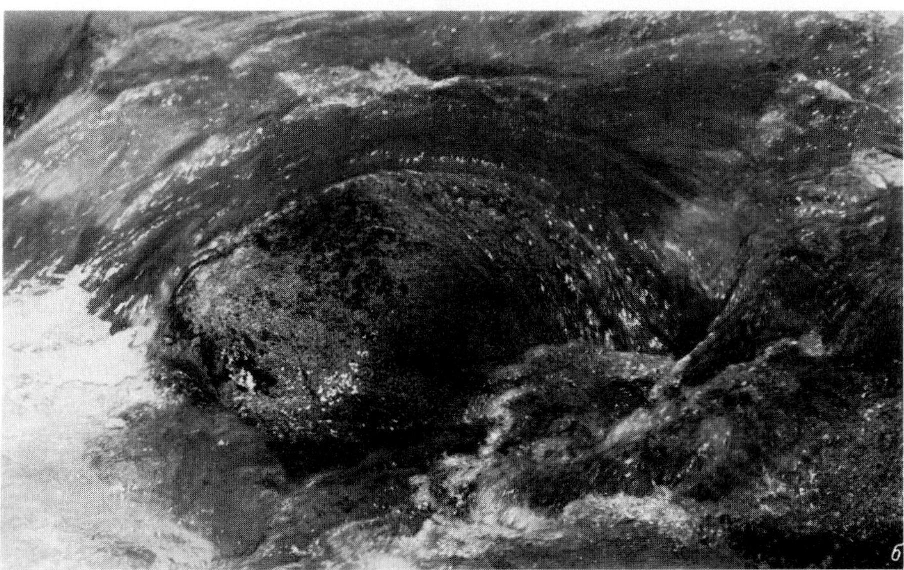

Fig. 35. Distribution of water jets on encounter with an obstacle (boulder). a. upper surface of the boulder is free of water; b. partly covered by water.

sites or near vertical rocky banks formed by slabs and large, although smoothed rock fragments.

A certain diversity of velocities is also observed within the river boils: from 0.46 to over 3 m/sec (Table 15).

There is a very specific velocity distribution in cascades (see Table 15) formed by the free water fall down from large slabs, rock ledges and even

Table 12. Distribution of flow velocity (m/s) near a log lying fixed on stones in a torrent.

50–60 cm up-stream off log	Near the log, just upstream of it	Near the log just downstream of it	75–80 cm down-stream of log
1.74	0.46	0.00	1.98
1.48	0.00–0.66		1.62
1.14			
1.74			
Mean 1.52	0.37		1.80
Min and max 1.14–1.74	0.00–0.66		1.62–1.98

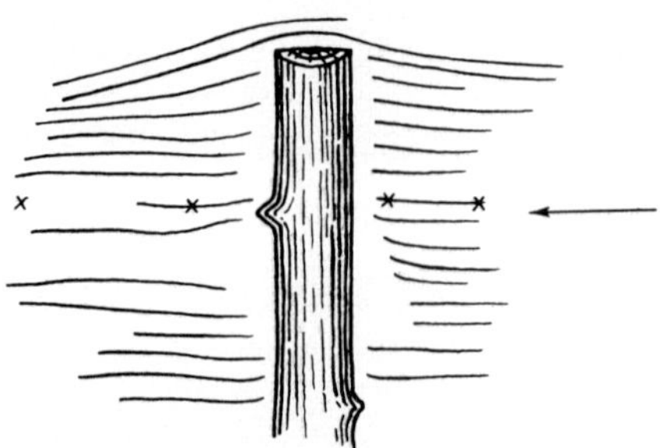

Fig. 36. Points of velocity measurement (×) near a log lying fixed on the river bottom (plan view). Arrow indicates the direction of the flow.

large cobbles. Here one should expect a regular (rather than random) velocity distribution since certain velocities are characteristic of certain sections of a cascade. The velocity distribution in a cascade is shown in Fig. 37. Maximum velocity is not observed over the whole length of the cascade but at the site where the angle of water fall is nearest to 90°. This conclusion seems to be a triviality; however it is of great importance for the ecologist to

Table 13. Distribution of flow velocity (m/s) between and above small stones and pebbles and in spaces between rocks and large boulders.

Between small stones	In spaces between large stones		Above small stones
0.66	Over 3.00	1.64–2.16	0.46
Op.	1.68–1.80	1.94–2.03	0.59–0.76
	1.77–2.00	1.44–1.77	
	2.06–2.25	1.32–1.74	
	1.28–2.28	2.09–over 3.00	
	1.95–2.16	1.74–1.93	
Mean 1.96			0.63
Min and max 2.25–over 3.00			0.46–0.76

Table 14. Distribution of flow velocity (m/s) near the banks and in little bays.

Near the banks, pebbles		Near rocky banks		In bays
0.00	0.46–0.82	0.19–0.46	2.28–2.47	0.56–0.59
0.66	0.66–0.82	0.19–0.63	1.05–1.23	0.00–0.46
Op.–0.46	0.46–0.72	0.28–0.41		Op.–0.51
0.66	0.66–0.82	0.70·		
0.00–0.28	0.00–0.79	0.00–0.46		
0.66–0.82	0.28–0.31	0.28–0.56		
0.00–0.46				
Mean.......	0.44		2.00	0.42
Min and max	0.00–0.82		1.05–2.46	Op.–0.59

Table 15. Distribution of flow velocity (m/s) in water eddies, in cascades and in "free" flow.

"Free" flow				In eddies	In cascades
1.62	1.55	Over 3.00	Over 3.00	Over 3.00	1.05–1.23
1.74	2.23–2.33	Over 3.00	Over 3.00	0.82–1.56	0.82–1.86
1.62–1.74	2.46–2.79	Over 3.00	Over 3.00	1.07–1.25	0.46–1.93
1.48	1.55–1.93	1.77–1.84	Over 3.00	0.46–1.55	1.55–1.80
Over 3.00	Over 3.00	Over 3.00	Over 3.00		2.33–2.28
1.74–2.19	Over 3.00	Over 3.00			Over 3.00
Mean.............2.85				1.35	1.86
Min and max.............1.48–over 3.00				0.46–over 3.00	0.46–over 3.00

Note. In calculation of mean values, "over 3.00" was taken to be 3.00 m/s, which somewhat reduces the mean velocities.

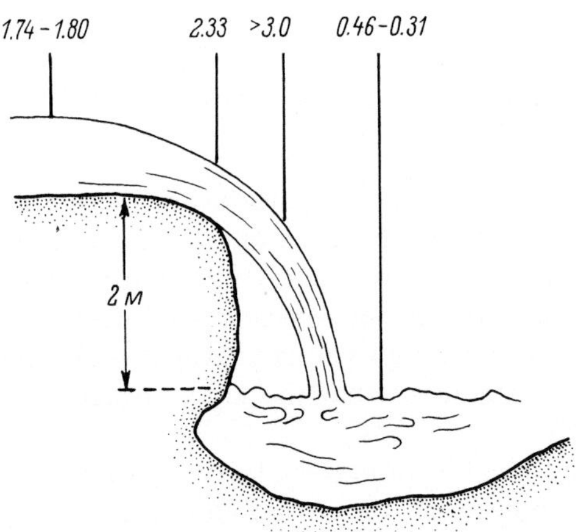

Fig. 37. Velocity distribution (m/sec) in a small waterfall.

assess the exact locality, in which any given living form occurs, rather than to simply detect its occurrence in a cascade.

Table 15 also gives data for the stream portion (at the same site where the velocity measurements shown in Tables 13 and 14 were made) 5–10 m away from any obstacles in the bed. This portion could be named conventionally as a "free-flowing" portion, and it has velocities of about 3 m/sec and higher. These data were obtained, like the above findings, not by means of the float method (which gives the average flow velocity), but with the Pitot-Darcy tube, *i.e.* they are the "point" velocities really observed over the area of the inlet aperture of the tube (about 2 mm^2).

The data on the velocity distribution show that for the natural state of a mountain torrent is indicative not the known predominance of a stable strong current, but a wide variety of velocities and even flow directions, which makes the living conditions for the organisms in a stream extreme not only because of the high velocity, but also due to the unusual variability of the velocity in space and time. Of course, it should be borne in mind (as we have pointed out above) that such variability could be possible only if the velocity in the stream *in toto* is high enough (Fig. 38).

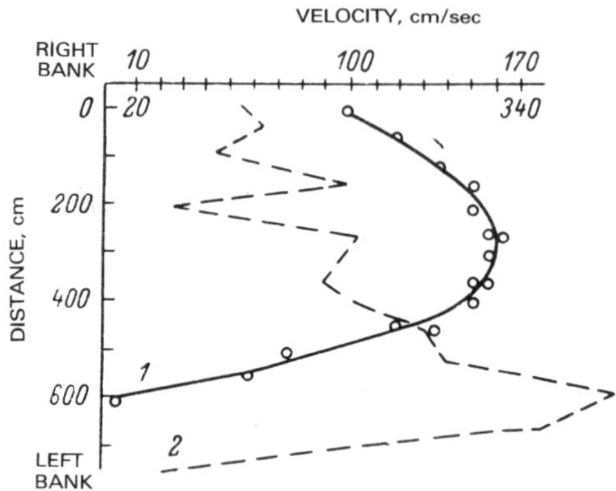

Fig. 38. Velocity distribution (cm/sec) at two points of a torrent.
1. slightly rough bed (small pebbles); velocity scale is above the horizontal axis; 2. very rough bed (shoulders and large stones); velocity scale is under the horizontal axis.

In connection with the extreme conditions in a mountain torrent it should be pointed out that this is not only a quantitative variability of the flow or its sign, but also the appearance of water-free spaces in the stream (in microbiotopes), *i.e.* generation of the so-called air cavities (Fig. 39). When a laminar flow is running at a high velocity over a rock surface it has no time to flow around the obstacle so that the side surface of the obstacle is left "dry", *i.e.* only moist.

Thus, not only the Prandtl boundary layer, but also many other sites in a stream serve as a temporary or permanent shelter for various forms of lithorheophylous life. However, this by no means implies that only sessile or substrate-attached stages (pupae or larvae of some aquatic insects) are

60

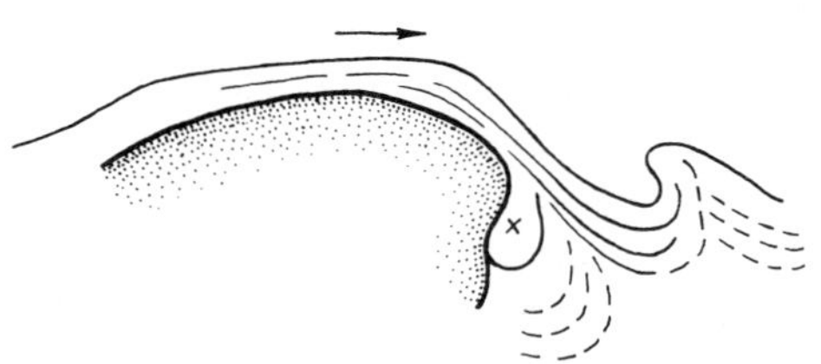

Fig. 39. Formation of "air" (water-free) cavities ($\times$). Arrow indicates the direction of the flow.

affected by strong currents: all inhabitants of the stream feel the effect of strong current some time or other. This happens because of their own movement in the torrent and because of the local and temporary variability of flow velocity. This accounts for the wide range of specific adaptations in rheophylous forms including those found sometimes under the conditions of slow current.

Discharge and level of water. The significance of water discharge as an ecological factor is most difficult to assess. It will be discussed later, but now it should be emphasized that discharge is closely related to other characteristics of the stream (as is water level). Discharge increase causes an increase in both the flow velocity and the water level; and, conversely, if these two change, the water level will also change, let alone the effect of discharge fluctuation on the water temperature, chemistry, bed material, *etc*. This happens because the abiotic and biotic environments in a mountain torrent are a well co-ordinated system, in spite of the fact that it is in a torrent that all the parameters can change very quickly within rather large limits.

For biological aims (and in route surveys) discharge may be evaluated from a simple formula $D = wdal/t$ (Davis, 1938), where D is discharge; w is the channel width; d is the mean depth; l is the distance of the float run during time t; a is a constant. Instead of D ("Discharge", in Davis) we use the designation Q. Coefficient "a" varies with roughness of the stream bed: from 0.8 for relatively smoothed beds (silt, sand or slabs of rock) to 0.9 at large bed roughness (fragments, boulders, pebbles). The discharge can be determined more correctly if a section of the stream is divided into short portions so as to average the discharges by all of them (Robins and Crawford, 1954).

To illustrate the ecological conditions of torrential fauna and in the related smaller watercourses, we present some original data on the zonal (within vertical zones of the torrent) and seasonal variations of water discharge in the Issyk River. A maximum mean monthly discharge in the river is observed at the end of its middle reach and at the beginning of the lower reach, *i.e.* after the confluence with all main affluents but before the water is expended in the irrigation canals. The discharge at this site was $18 \text{ m}^3/\text{sec}$ in a year of abundant water. Downstream discharges in the main channel decrease down to $13 \text{ m}^3/\text{sec}$ and farther downstream to $3 \text{ m}^3/\text{sec}$. The latter measurement was made nearly at the end of the river, at the lower debris cone. Farther downstream the water discharge decreases rapidly: 2.38–0.14–0.09–0.01 m³/sec. In the affluents the value of discharge ranges from 0.03 to 4.45 m³/sec. Figure 40 shows the discharge distribution in the irrigation canals. To understand the ecological situation in the irrigation canals it is essential to consider the discharge ratios in the canals in their different sections as compared to the main stream, rather than the discharge changes due to water level fluctuations. The rapid decrease of discharge in the Issyk River and in the irrigation network is accounted for the water outflow into small canals of the second and third order, by evaporation (it seems to be of little significance at high flow velocity) and by the percolation throughout the soil. The latter has a value of 0.031 m³ in the Issyk River and 0.01–0.02 m³ in its affluents per distance of 1 km. Certainly, the water level regulation causes an extremely variable ecological situation in the canals, particularly, in their terminal sections, affecting to a certain extent the canal fauna. There occur organisms typical of ephemerous waterbodies, whereas the fauna of swift waters becomes poorer.

The cited discharges in the Issyk River vary annually, which can be illustrated by comparing the discharges at a single site of the river during three consequent years (the 1930's; Table 10).

Water discharge during the year strongly depends on its genesis; the dates of flood also change. This question was discussed earlier, and we shall only

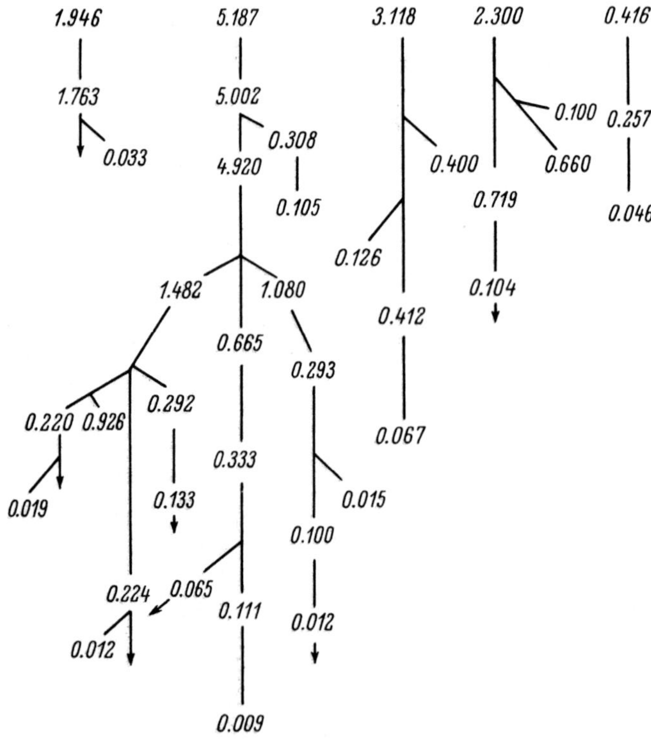

Fig. 40. Diagram of water discharge distribution (m³/sec) in the Issyk River irrigation canals. For designations see Fig. 33.

62

Table 16. Mean monthly discharges of water (m³/s) in the
Issyk River during summertime at a single point for a period
of three consecutive years.

	May	June	July	August
The first year	—	3.95	6.71	—
The second year	—	15.92	10.58	18.63
The third year	3.06	3.40	7.65	7.19

Note. "—" means absence of data.

recall that two floods are typical for the rivers starting above the snow line. The first is caused by the melting of seasonal snows in March and April and has a short duration in the upper and middle river, becoming longer in the lower reach. The second is caused by the melting of permanent snows and glaciers and lasts from mid-June to mid-September. Table 17 gives long-term ten-day discharges (averaged from 35 years of observation) throughout a whole year in the lower Akbura River where the two floods merge into one flood and the low stage between them disappears.

The Issyk River, owing to its short length and to snowy and glacial origin of its affluents, has the character of a glacial river throughout the whole of its length; therefore, the flood caused by the melting of permanent snows and glaciers is rather well-defined in the lower reach and occurs in summer months (Fig. 41). This typical annual discharge curve in mountain torrents during summertime is frequently disturbed by showers, rains or high insolation (increased melting of permanent snows and glaciers) producing catastrophic "non-seasonal" floods.

Variation in the mean monthly and daily discharges in streams leads to water level variation. For instance, the Issyk River, like other torrents, has a very unstable water level, a fact we have repeatedly pointed out. Large fluctuations can be observed even during a short period of one day (it is true also of the discharge). Figure 42 shows daily water levels for the summer months (July–September). The graph shows clearly the level increase in the

Table 17. Long-term mean discharges for ten-day periods in the
lower Akbura River (m³/s) (Hydrometeorological Station data).

Ten-day period	MONTHS					
	I	II	III	IV	V	VI
I	9.4	8.6	8.4	9.5	9.6	42.0
II	9.2	8.4	8.6	10.9	27.9	44.3
III	8.8	8.4	8.8	13.8	33.6	44.5

Ten-day period	VII	VIII	IX	X	XI	XII
I	48.7	43.0	26.0	16.3	13.3	10.6
II	48.8	38.5	22.0	15.1	12.3	10.1
III	46.8	32.1	18.5	14.0	11.4	9.7

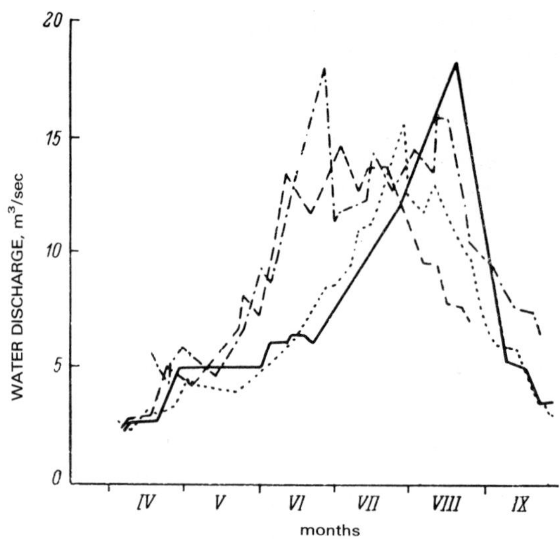

Fig. 41. Daily water discharges (m³/sec) in the Issyk River at 10 km upstream of Issyk village for 1912–1915 (one curve stands for one year).

summer months with a maximum in August. Yankovskaia (1948) gives valuable data on daily level fluctuations in addition to air and water temperatures for the Kugart River, Tar River and its affluent, Kaindy-Bular, and the Karakuldja River (Ferghana Range). The curves in Figs. 1–7 in the cited work give a vivid example of a sharp change of the water level

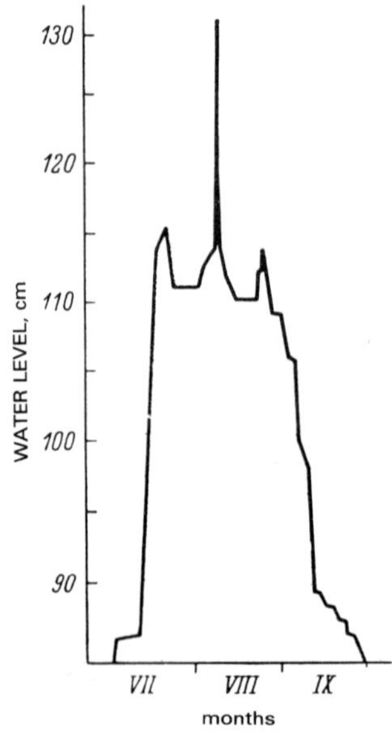

Fig. 42. Daily variation of water levels in the Issyk River (1926, 1927) for July–September at the same site as in Fig. 41.

64

throughout a day. Their specific feature is that the level increases towards the evening, and the lower the observation site above sea level, the later in the day the maximum level occurs. The level range is very wide. For example, in the Tar River the level varies from 23 to 85 cm, producing at its minimum a dry shoreline belt of 1.5–2.5 m wide (Fig. 43).

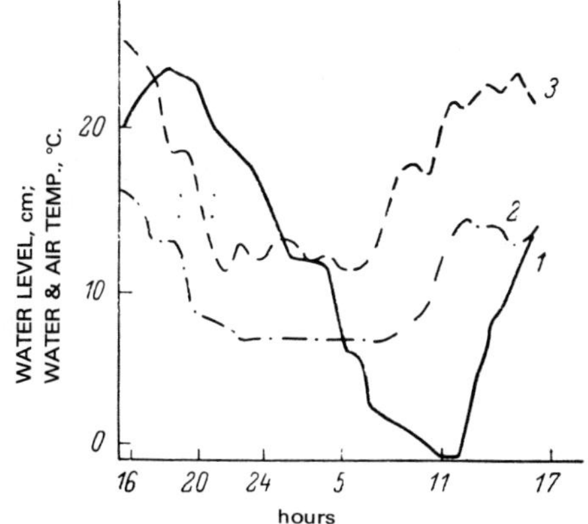

Fig. 43. Daily variation of water level (1), water temperature (2) and air temperature (3) in the Kugart River (Ferghana Range) near Djetymatu, 3 VII-1 VIII 1946. (After Yankovskaia, 1948.)

Although the hydrodynamic drag is directly related to the flow velocity, we consider it as an auxiliary hydrodynamic parameter and discuss it after the flow velocity, discharge and water level. The force acting on the organism in a torrent is little known so we have to confine ourselves only to general considerations.

The hydrodynamic drag is a force which acts in the direction opposite to the movement of a body, *i.e.* it is the resistance of the fluid to the movement, which can be estimated from: $X = C_x(PV^2/2)S$, P is the fluid density, V is the velocity and S is the body surface area. The dimensionless coefficient, C_x, is dependent on such factors as the body shape, position of the body with respect to the direction of the movement, and similarity numbers.

Because of the hydrodynamic drag, organisms are exposed to high impact pressure, which results in formation of some specific adaptive devices in these organisms. The organisms which inhabit the boundary layer (of 2–3 mm thick) or "dead water" zone are affected by the pressure to a lesser degree, but this "saves" them only temporarily, as we have pointed out above in connection with the microbiotopes. It is known, that the hydrodynamic drag increases with the flow velocity squared and reaches 0.5 g/mm^2 at 300 cm/sec. At such pressure many "small" animals which do not possess specialized organs of attachment will be washed off the substrate (Bournaud, 1963). Naturally, in order to understand the ecological situation

65

in which the torrential fauna exists it is necessary to make direct measurements of hydrodynamic drag in the stream in general, and in the microbiotopes, in particular, since the flow velocity data alone do not give an adequate idea of the drag (shape of the body, its surface area, and position of the animal to the flow are to be taken into account).

For crude empirical determination of hydrodynamic drag we have used an instrument with a spring balance movably joined to a square metallic plate (15×15 cm). The plate was plunged into water and the balance reading was taken quickly. Then, by subtracting the weight of the plate from the reading it was possible, although very approximately, to estimate the drag of the plate (in g/cm^2). Thus, values of drag determined for various microbiotopes were from 32 g/cm^2 to 80 g/cm^2 and higher. The highest value, 118 g/cm^2, was measured in a cascade.

Bed material. So far little attention has been given to a detailed examination of bed material in torrents. This question was discussed at some length in the works by Percival and Whitehead (1929, 1930). More recently Cummins (1962) proposed a classification of bed material by the size of components (for the purposes of biological study; see Table 8). It is known that river deposits are classified as authogenic or allogenic ground; the former is made up of individual stony lumps and boulders of different size and volume, the latter includes small stones, pebbles, gravel, sand, silt and conglomerates. Authogenic grounds dominate in many torrents making up 90% of the total volume. Because of high flow velocity and large tractive force of the water, the allogenic grounds are dragged away downstream. Bed material becomes relatively stable only in the middle and lower reaches. The stones, pebbles and gravels form an even, compact bottom which is covered by sand, clay and silt in the lower reach.

The upper and, partly, the middle river is characterized by materials which compose the channel base: rocks and stony slabs, fragments of mantle rocks, boulders and pebbles of different size and roundness. The pebbles are formed mainly from disintegrated granite which easily becomes weathered by running water, while plots of badrock base are, like in the Issyk River, formed by outcrops of felsitic porphyry.

In small bays and near the banks, among stones and pebbles, there are coarse sand deposits; then, also in bays, thin layers of glacial silt and mineral particles occur as a delicate skin on the pebbles. At the midstream this skin layer is absent because it is washed away by the fast current.

Again, in the lower reach of the torrent, the main material composing the bed will be rocks, boulders and pebbles. When current slows down (at the debris cone), the pebbles, even at the midstream, are usually covered by fine debris which consists chiefly of remnants of eroded soils. In the Southern Tien Shan they are mainly loess; sand is widespread and together with pebbles forms large pebbly deposits so typical for channels which during the non-flood period are watered only partly. Even in the lower reach the sand is easily transported by water and forms pure deposits near the banks and in small bays. Together with sand and pebbles one finds gravel, rounded or rough; it fills the gaps between the pebbles, while fine gravel serves as house building material for caddisfly larvae (*Rhyacophila, Agapetus, etc.*). Gravel is more common in the middle and lower sections of irrigation canals.

Moss which occurred in the upper and middle stream is absent from the lower reach and canals. In the Issyk River it disappears at about 1600 m above sea level, which seems to be due, among other things, to the changing bed material as there are no rocks and the material itself becomes more movable.

When examining the river bed consecutively from the glacial headwaters down to its disappearence in the steppe, one could reveal a pattern very characteristic of mountain torrents, namely, not a stringent sequence in material distribution by absolute heights (succesive replacement of rough material by pebbles and sand), but rather a "random" alternation of the substrates. A more regular succession of bed material could be traced only in the lowermost reach, near the erosion base level, although some patch-like distribution also occurs. The reason for this phenomenon can be understood if one analyses the bed slope. Where the slope is steep, erosion of bottom deposits is intensified so the bed rocks were outcropped here. Where the slope is small, erosion is slower and debris deposition takes place. This situation holds not only for the river *in toto*, from the headwaters to the lower reaches, but also for any section of the river down off its heads. Another reason for the "patchy" distribution of bed material is heterogeneity of mountain rocks in the outcrops. Table 18 summarizes various sections in the Issyk River, from the headwaters down to its disappearence; however, it must be noted that the general pattern does not reveal the abrupt change of the substrate in any section of the river.

Finally, it should be stressed that a significant feature of the torrent bed at most places is the mobility of deposits lining the bed. Downstream transport

Table 18. Distribution of bottom deposits in the Issyk River downstream from glacial headwaters to steppe.

Approximate distance from river heads, km	Bed material in river channel
0.75	Ice of glacier
2.75	Large, rough boulders; rock fragments
25	Large well-rounded pebbles, rocks, boulders
	Medium-size pebbles, boulders
	Large rock fragments, smoothed moss-covered cliffs, slabs, boulders
	Medium-size pebbles
	Boulders, coarse pebbles
27	Well-rounded, fine and medium-size pebbles
35	Fine pebbles, gravel, sand
	Fine pebbles
	Silt, solitary fine flatened pebbles

of pebbles and rock fragments depends on the flow velocity and discharge: as these parameters change, the substrate mobility also changes. This can be proved without any complicated instrumental methods. Towards evening, when the water discharge and level in torrents rises, we were able to hear clearly the noise of large pebbles and stones rolling over the bottom, which increased with the water level.

The mobility of bottom deposits, along with large variability of other parameters constitute a specific feature of torrents, creating highly extremal ecological conditions and making them so different from slowly running torrents or standing waters.

Water temperature. The great importance of water temperature to hydrobionts is well known. This ecological factor is also essential to inhabitants of swift waters. Already Thienemann (1912, 1926) and Steinmann (1907, 1915) gave much attention to water temperature in running waters. Thus, Thienemann noted large fluctuations of water temperature in the zone of springs throughout a year, however the water is still cold here. At the heads of a trout brook the annual temperature variation approaches 10°C; at the lower brook, where the grailing zone begins, the amplitude is largest, 17–18°C. As early as 1915, Steinmann defined various types of swift waters by means of the annual amplitude of water temperature. These data are meaningful enough to be cited here: glacial headwaters – 0–1°C; high-mountain springs – 1–6°C; limnocrenes (the Schwarzwald, the Jura) up to 12°; heads of the trout brook – about 8–10°C; its lower reaches – 15° and higher; a barbel river – about 19°C; a bream river – about 24°C.

In relation to temperature, inhabitants of mountain rivers and brooks had been invariably characterized as cold-water stenotherms. The same is true of the members of torrential fauna, but the temperature factor is not always the most significant among other factors characteristic of mountain torrents.

Considering water temperature as an ecological factor, we should point out a somewhat one-sided approach to the study of temperature in swift waters. Writers usually record the temperature at a certain study site without any generalization, or give annual temperature variation at a single point of torrent or brook. Little attention has been given to such an essential factor in the life of torrential fauna (especially for vertical distribution of organisms) as a simultaneous change of temperature in the longitudinal profile. One example is the distribution of water temperature in different sections of a torrent from its headwaters to the lower reaches. Tables 19 and 20 list data for the Akbura River (Southern Tien Shan) and Issyk River (Northern Tien Shan), respectively. For the Issyk River, up to a height of 1340 m above sea level, the table gives averaged temperatures (for detailed records on the Issyk River, including study sites, *i.e.* elevation, time of air and water temperature measurements and the temperatures, see our earlier work of 1935), while below 1340 m it shows temperatures at individual sites.

Water temperature in affluents varies greatly with water discharge, absolute height, *etc.* In August, in the afternoon, their temperatures were about 5, 7, 12 and 14°C. It is of interest to note the distribution of water

Table 19. Water temperature (in °C) in the Akbura River, August 1966.

Altitude above sea level, m	Time (hours)	Water temperature	Air temperature
3670	9.00	1.5	2.5
3550	10.00	3.5	7.0
3500	11.00	3.5	7.0
3420	13.30	6.5	12.0
3300	15.00	7.0	10.0
2860	16.30	7.5	14.0
2770	17.30	7.5	10.5
2710	19.00	7.0	10.0
2700	15.00	9.0	18.5
2600	14.00	9.5	19.0
2440	16.30	10.0	18.0
2180	18.00	9.5	17.5
2020	18.00	9.0	20.0
1890	16.00	10.0	20.0
1800	17.00	10.0	19.0
1700	18.00	10.0	17.0
1600	8.30	10.0	26.0
1550	10.00	10.8	27.0
1500	11.00	12.0	27.5
1480	12.00	12.0	31.0
1470	13.00	13.0	35.0
1450	15.15	13.5	31.0
1400	16.00	14.0	30.0
1100	14.30	15.4	32.0
1000	13.30	18.4	32.2
950	13.00	18.0	33.0
900	15.30	20.0	34.0

temperature in the Issyk irrigation canals measured also in August, afternoon (Fig. 44).

Water temperature in the upper and partly in the middle reaches is rather stable compared to the air temperature. Thus on June 27 air temperature was 23.4°C and water temperature 10.6°; on June 28 they were 18.0 and 10.5°, respectively. Thus, the difference for the air was 5.4°, but for the water only 0.1°. Another example, on July 21 and 22 air temperatures were 28.5 and 24.5°, but for water 11.5 and 11.5°, respectively. Then, the difference for the air is 4.0°, but for the water 0.0°. Such constancy of water temperature is typical of glacial torrents in their upper part, where water has not enough time to be warmed by the sun and, therefore, its temperature remains low. Comparison of water temperatures at closely spaced sites, but in different time intervals (about a month), gives very small differences (about 0.2°C). This points to a high constancy of water temperature in a torrent. Of course, some daily variations occur, but they are not significant in the upper reach, being about 1–2° (see below).

At the end of the upper stream and the beginning of the middle reach, where the river enters the coniferous forest belt, the water temperature still

69

Table 20. Water temperature (in °C) in the Issyk River, July–August, at midday.

Altitude above sea level, in m	Water temperature
Over or about 3000	0–0.6
2200–2000	5.0
1780–1730	7.8
1730–1700	8.6
1700–1440	10.4
1440–1340	11.0
1300	10.2
1280	11.5
1180	11.5
1020	11.8
980	14.5
950	15.8
860	15.8
830	16.4
800	16.7
800	17.7
795	18.8
790	23.0
770	25.5
720	29.0
690	29.3

varies very slowly.[3] Such rate of change can be traced down to the end of upper-one-third debris cone, where a steep rise of water temperature is observed. Along nearly the whole lower reach, the water temperature rises very rapidly, and some stabilization of daily temperature is observed only at the very end of the stream. For example, in the middle Issyk River the water temperature remains unchanged for many kilometres, whereas in the lower reach simultaneous readings differ by 2.5°C at a distance of 300 m (23.0° and 25.5°, respectively).

Simultaneous records of temperature for the whole length of mountain torrent, from the glacier to the steppe, enable us to outline the thermal zones differing in the rate of temperature variation (in space, not in time; Table 21).

The difference between the Central and Southern Tien Shanian torrents depends on the flow power, torrent length and on the presence or absence of spruce forest in the middle reaches, whereas the rates of water temperature change are rather similar. Thus in the Akbura River, along the "upper thousand" metres (from 3670 to 2600 m above sea level), the water temperature changed only by 2°C, while along the "lower thousand" metres (2000–1000 m) the change was 8°C.

---

[3] This is particularly true of torrents in the Central and Northern Tien Shan, with the Issyk River as one example.

70

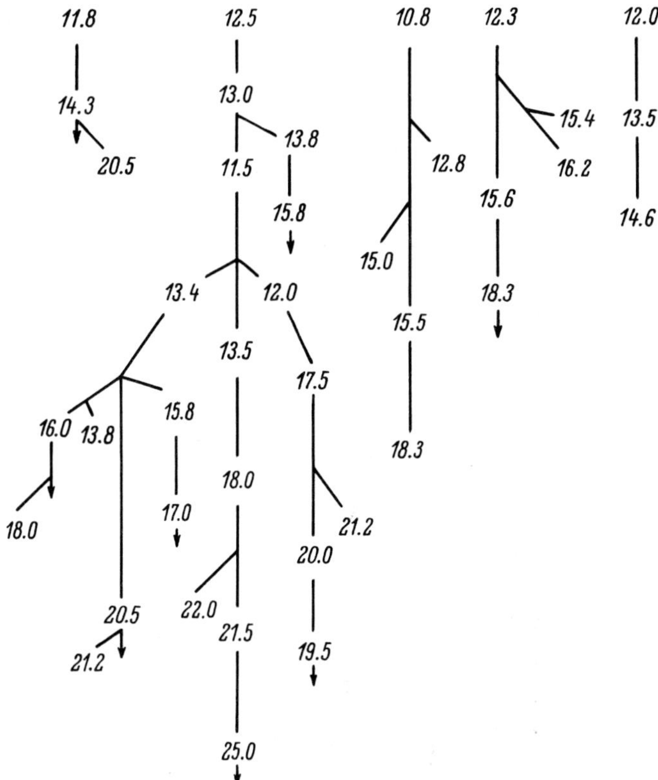

Fig. 44. Diagram of water temperature distribution in irrigation canals of the Issyk River (August, afternoon).

Unfortunately, there is little data on the annual range of water temperature over the whole length of the Issyk River, but it is close to that observed by Steinmann for brooks and streams of Switzerland. In winter, the Issyk River is frozen only in its lower part, whereas it is ice-free in the upper and middle reaches.

There are many factors determining the variation of water temperature in a mountain torrent. Mainly it is a function of air temperature which, in turn, is dependent on the elevation. However, as it was shown above, water temperature in a torrent is not immediately affected by the air and soil temperatures. The water temperature is an extremely "conservative" parameter. The water discharge in a torrent, its power, may be thought as a main reason of the temperature inertness mentioned, but of much importance are also the hydraulic radius and flow velocity. In addition to factors intrinsic to the mountain torrent, there are such important conditions as overshading of the torrent sections by vegetation in narrow rocky gorges ("kopchegai"), *etc.*

No doubt, the slow change of water temperature in the middle Issyk River, *i.e.* both time and space constancy of temperature, is associated not

71

Table 21. Thermal zones in the Issyk River.

Section of river	Geographical zone	Approximate size of zone in km	Rate of water temperature change	Distribution of isotherms in zone
Glacier	Permanent snow limit	Short, 0.75	Very slow	Widely spaced
Glacial headwater	Moraines, glacial region	Short, 1.2	Relatively fast	Relatively closely spaced
End of upper reach, middle reach	Coniferous forest belt	Long and very long, 24	Fairly slow	Widely spaced
Lower reach	Piedmonts, shrub zone	Shorter than previous zone, 8	Fairly fast	Very closely spaced, almost overlapping
Transition from mountain torrent to river	Steppe, lowland	Long, but may be short, 8	Slow	Widely spaced

only with the air temperature in the coniferous forest through which the river runs, but also with large discharge and fast current. Some influence on the water temperature in the torrent surely comes from the affluents which usually have either higher (if smaller discharge compared to the main stream) or lower temperature (springs). For example, at 2.40 p.m. on 15 July, at an air temperature of 22.5°C, the water temperature in the torrent was observed to change from 7.0 to 9.0°C after a tributary inflow. Therefore, along with such factors as distance from the glacier and lower elevation, the temperature in a torrent may become higher due to the affluents, whose number increases as the stream approaches the mountain foot.

Owing to the large inertness of water temperature, the zones in a mountain torrent are displaced with respect to the geographical and vegetational zones towards lower heights. The zones with fast change of temperature may be considered as intermediate, or transient zones lying between the zones of constant temperatures. Sometimes, the zone shift determined by the power of a torrent exerts a large ecological effect on the fauna. Thus we could frequently observe two neighbouring streams with different power, which had entirely different water temperatures.

Water temperature distribution in irrigation canals is also affected by the stream power and flow velocity. The water temperature at the end of one such canal was 15.8°, while in another, located much lower but having larger discharge, it was 15.4°. Temperature at the point of canal outflow is dependent, naturally, on the water temperature in the main channel and varies from 10.8 to 12.5°. Downstream, this dependence becomes weaker and in the middle sections of the canals we find 13–16°C, whereas in the river at the same absolute height and at the same time 11–14°C. The terminal sections of canals have water temperature of 18–25°C against 15–16°C in the river at the same absolute height (see Fig. 44).

a. *Annual variations of water temperature.* The data presented above on water temperature in torrents are characteristic of the warm, summertime period of the year and differ in other seasons. These changes are of great ecological importance, since many preimaginal stages of aquatic insects and other organisms of a torrent live in the same environment all year round, sometimes, two or three years.

Figure 45 shows annual water temperature curves for the Akbura River at three sites: the upper reach – at the Mantynek River mouth; the middle reach – at Papan village; the lower reach – at Tuleken site. The temperature curves for all three sites are very similar, but they lie at different levels, particularly, during the warm season; in winter they differ very little. The highest temperatures at all the sites falls on June–August; however, upstream the water temperature maximum tends to delay towards late August–September.

b. *Daily variations of water temperature.* Large daily variations are also observed in the water temperature of Tienshanian torrents (circadian cycles), but these variations change with the season; they are very small, 1.0–1.5°, or entirely disappear in winter. The range begins to increase in spring and reaches a maximum in summer. But in summertime the daily

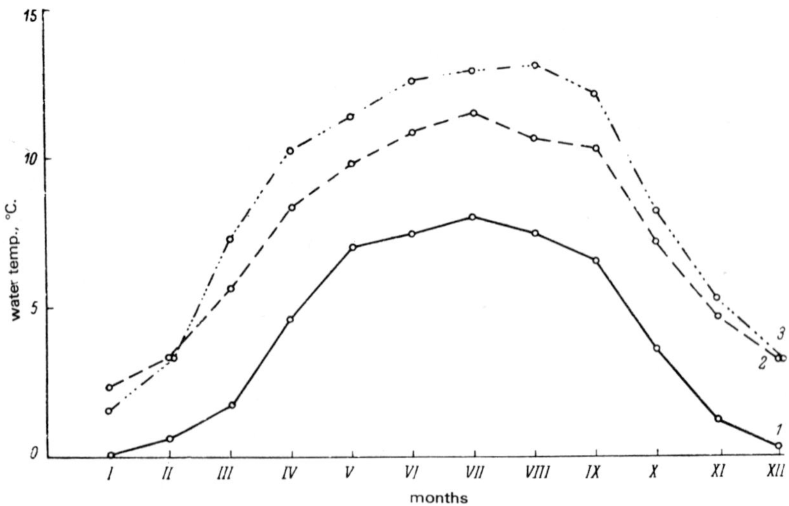

Fig. 45. Annual variation of mean monthly water temperature in the Akbura River. (Hydrometeorological Station data for 1967.)
1. The Mynteke River mouth (2700 m above sea level); 2. Papan village (1500 m above sea level); 3. Tuleken locality (1000 m above sea level).

variations of water temperature are different at different heights: the larger the absolute height, the smaller the variation. For example, in the upper Kichik-Alai River (about 3500 m above sea level) it does not exceed 4.5°, while downstream, at the Akbura mouth (about 900 m), it equals 10.5°C. Figure 46 illustrates daily variation of water temperature at four sites of the Akbura River at different heights, including the river heads, i.e. the Kichik-Alai River with an affluent, the Chogom river. For comparison we give also the daily air temperature variation. Some conclusions can be drawn from the graphs.

First, what attracts one's attention is the large inertness of the water temperature as compared to the air temperature. The inertness is particularly large at absolute heights less than 2000 m (with the same water discharge in the torrent). Both minimum and maximum water temperatures, observed early in the morning (5–7 hours) or in the afternoon (15–18 hours) respectively, do not correlate with the air temperatures. But the main feature, of course, consists in the smaller range of water temperature against that of the air. Although at the Kizyl-tuu site (the middle Akbura River) the air warms up to 32°C, the water temperature in the same hour does not rise over 15°C, which is surprising in view of the shallow depth of the stream.

Similar data were also obtained from air and water temperature measurements in the Kugart River (near Djetym-ashy, 31 VII-1 VIII 1946), the Tar River (near Kizyl-djar, 13–14 VIII 1946), the Yassy River (at Kara-Tiube, 1–2 IX 1946) and in the Kara-Tiube brook (an affluent of the Yassy River, 1–2 IX 1946).

All these data obtained by Yankovskaia (1948) in the streams of the Ferghana Range demonstrate well the "levelling" of temperatures in the

74

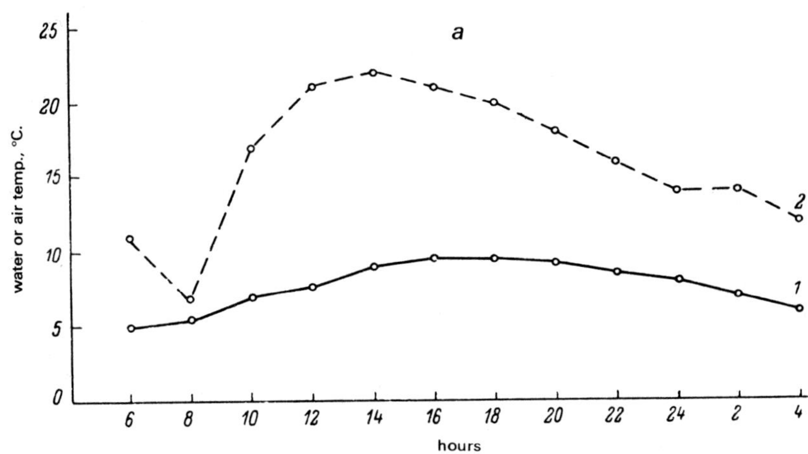
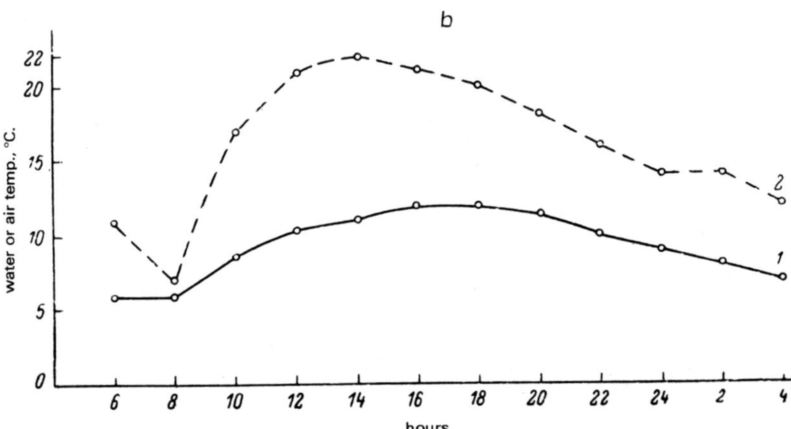

Fig. 46. Daily variation of water and air temperatures in the Akbura River. Data by E. O. Omorov.
a. the Kichik-Alai River (2300 m above sea level, 23–24 VIII 1968); b. at the offing of the Chogom River (about 2500 m above sea level, 23–24 VIII 1968). 1. water temperature; 2. air temperature.

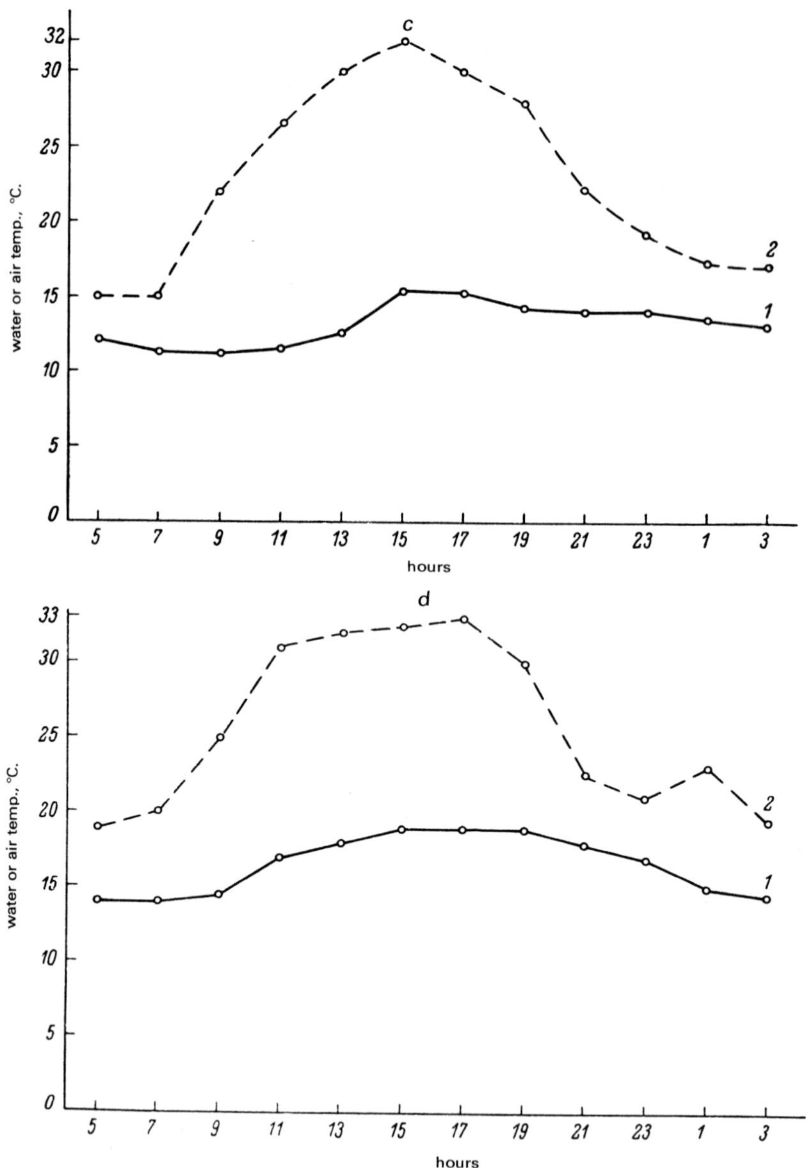

Fig. 46 (continued).
c. Kizyl-tuu (1600 m above sea level, 24–25 VIII 1968); d. Osh city (Botanical Garden, 950 m above sea level, 24–25 VIII 1968).

torrents as compared with the air temperatures (see Fig. 43). The difference between the water temperature in a stream and the air temperature may be as large as 7–12°C in the warmest time of day (3–4 hours p.m.). At the cold time of day, when the air temperature drops down to 5, 6, 10 and 12°, the water temperature differs but little from it.

Every waterbody represents a kind of thermostat with much more uniform and lower water temperature than the temperature of the surrounding

76

air. But a peculiarity of the thermal regime in a mountain torrent consists in the fact that, with a shallow depth, any standing or slow-running waterbody would have the same temperature variation in the water and the air, while the torrent has a lower and more stable temperature of the water than that of the air. This phenomenon is due to the large flow velocity in a torrent which constantly carries new portions of cold water from the glaciers and snow fields. Aquatic organisms live in a mountain torrent as if it were a perfect thermostat protecting them both from sharp changes and from high air temperature.

Dissolved oxygen. For organisms living in torrents the dissolved oxygen is a factor which does not limit their distribution and is not so vital as for the inhabitants of standing waters. Owing to the fast and turbulent flow, absorption from the air ensures nearly 100% saturation of water by oxygen, sometimes even higher. A certain role in this process belongs to the air pressure (the absolute height) and the temperature regime, but in normal torrents the oxygen amount does not fully correlate with temperature (Creaser, 1930). However, torrential organisms are always well supplied with oxygen (even in summer). In swift rivers and torrents vegetation contributes very little to the oxygen balance.

Dissolved oxygen contents in the upper and middle Issyk River observed in the light hours of July and August are listed in Table 22.

The oxygen data for the lower section of the Issyk River and irrigation network are presented in Table 23 in the order of increasing water temperature. The dependence of dissolved oxygen on the air pressure is well known: the lower the pressure, the lower the oxygen solubility. Therefore, in mountain torrents, the dissolved oxygen amount will be under the normal saturation at 760 mm Hg, which is explained by the lower oxygen solubility at a lower air pressure in mountains. If the data are reduced to the normal pressure at a given temperature, the picture will change: there will be an appreciable supersaturation. Thus, a factor is involved here which exerts a large enough effect to give the supersaturation. If one imagines a rough, foamy torrent, one can easily conceive what brings about the saturation and supersaturation by oxygen. The oxygen content varies with the character of the stream section and with its position with respect to the lakes. So, in the waterfall of the Issyk river where half of water is converted into foam, the oxygen content is not very large ($7.18\,\mathrm{cm^3/1}$), since here the water runs out directly from the lake with oxygen content of $7.02\,\mathrm{cm^3/l}$ at air pressure 613.4 mm and water temperature 14.6°C ($7.12–7.24\,\mathrm{cm^3/l}$ at other points of the lake).

A south-Tienshanian stream, the Akbura River, is also characterized by saturation and supersaturation of water with oxygen. Thus in the main channel the oxygen content is 7.5–10.9 mg/l, *i.e.* over 100% saturation; the same values were observed in the affluents. Oxygen saturation in the torrents of South-Eastern Tien Shan was near 100% (Yankovskaia, 1948); in the upper Zeravshan River it did not decrease below 80% (Sibirtzeva *et al.*, 1961); in the Chirchik River the oxygen content varies from 74.8 to

Table 22. Dissolved Oxygen in the upper and middle Issyk River.

Measurement site	Air pressure mm Hg	Water temperature	O_2 at a given pressure	O_2 at 760 mm	O_2 saturated at a given water temperature	O_2, %
Lower section of upper reach and a part of middle reach	607.0	6.3	7.82	9.28	8.68	From 95 and over 100
Before inflow into the large lake	616.8	7.0	7.05	8.67	8.47	
In cascades, downstream off the small lake	614.0	10.6	7.18	8.90	7.87	
Middle reach	627.0	10.4	7.61	9.20	7.87	
At debris cone, before exit from the gorge	638.0	11.0	7.50	8.92	7.69	

Table 23. Dissolved oxygen in the lower Issyk River and in irrigation canals.

Water temperature, °C	Air pressure, mm Hg	O_2 at a given pressure	O_2 at 760 mm Hg	O_2 saturated at a given water temperature	O_2, % %
11.5	647	7.65	8.95	7.60	>100
11.5	667	7.43	8.47	7.60	98
11.5	663	7.16	8.16	7.60	94
13.5	668	7.30	8.32	7.25	>100
13.8	672	6.98	7.68	7.19	97
15.5	647	6.30	7.30	6.95	91
17.0	694	6.48	7.13	6.75	96
17.3	686	6.42	7.12	6.75	95
20.5	683	5.90	6.49	6.30	93
24.5	686	5.67	6.23	5.83	97
29.3	698	5.36	5.84	5.36	100

108.5% (Sibirtzeva, 1964); and in streams of the Northern Tien Shan it was between 9.7–13.5 mg/l (Kurmangalieva, 1976).

The comparison of dissolved oxygen in the Issyk, Akbura and other rivers of the Tien Shan with that in mountain streams, rivers and brooks of Europe and North America shows that the oxygen regime in the Tienshanian ones, at least during summertime, is typical of watercourses with fast and extremely turbulent flow. The rich oxygen content in torrents is independent of organic life, because it is mainly determined by oxygen absorption from the air. This is the main point of difference of swift waters from slow-running or standing waters, where the contribution of organic life to gas exchange is very great.

Salt composition and hydrogen ion concentration (pH). The waters of mountain torrents are not so similar in their mineralization as in oxygen saturation. However, the torrents of snowy and snowy-glacial origin have usually low salt content, their water being almost distilled at the heads. Some increase of mineralization is observed only in the lower reaches (Uklonskii, 1925). Depending on the type of mountain rocks which make up the river bed and on the inflowing highly mineralized affluents, both the torrents and their affluents may have a varying degree of mineralization.

Analysis of water samples from the Issyk River and some of its affluents has shown that the content of dissolved salts is very small (Tables 24 and 25). Water sampling in the Issyk River was made only for the middle reach, but there are no grounds to expect a great difference in the salt composition between the lower reach and the irrigation network. One may assume only a slight increase in the salt content. Tables 24 and 25 give a possibility for the comparison of data to be made for the Issyk River, Issyk Lake, an affluent of the Issyk River, and the Arys River near the point of its inflow into the Syrdarya River, *i.e.* in the desert-steppe zone. The comparison shows clearly that the waters in the Issyk River and Issyk Lake are unusually poor in dissolved salts. The difference in salt content between these waterbodies is

Table 24. Salt composition (mg/l) in the Issyk River, Issyk Lake, the affluent Northern Trench, and Arys River.

| | | | | Alkalinity | | | | | | | | | | |
Rivers and lakes	Dry residue	Ignited residue	Loss on ignition	total in HCO_3	undeter- minable in HCO_3	NaCO in HCO_3	$Ca(HCO_3)$ in HCO_3	$NaHCO_3$ in HCO_3	SO_3/SO	Cl	CaO	MgO	pH	Hardness in Germane degrees
Issyk R.	0.072	0.052	0.020	0.047	0.016	—	0.031	—	0.012	0.04	0.01	0.050	7.2	<1
Issyk L.	0.071	0.053	0.018	0.051	0.016	—	0.035	—	—	—	0.01	—	7.2	<1
N. Trench	0.250	0.225	0.025	—	0.042	—	0.128	—	0.043	—	0.084	0.019	—	11.1
Arys R.	0.338	0.227	0.100	—	—	—	—	—	0.057	0.011	0.070	0.044	—	13.2

Note. In this table and tables 25–27 the sign "—" means that the compound was not found in water.

Table 25. Salt composition (mg/l) in the Issyk River and in the Northern Trench (expressed in salts).

River	$CaCO_3$	$MgCO_3$	$CaSO_4$	$MgSO_4$	$MgCl_2$	Na_2SO_4	NaCl	KCl	Total	Mineral residue
Issyk R.	0.025	—	—	0.015	—	0.003	0.007	—	0.050	0.052
N. Trench	0.105	0.033	0.063	0.009	—	—	—	—	0.210	0.225

Table 26. Salt composition (mg/l) in torrents of the Tien Shan, 1946 (Yankovskaia, 1948; data on Issyk R. from Brodsky, 1935).

River and observation date	Dry residue	HCO_3	Cl	CaO	MgO	Hardness in German degrees
Yassy R., upper reach (IX)	0.138	0.013	0.021	0.096	0.018	10.9
Yassy R., downstream of Uzgen (IX)	0.353	0.268	0.026	0.168	0.028	18.8
Karadarya R. in Uzgen region, before inflow of Yassy R. (IX)	0.171	0.177	0.013	0.132	0.035	15.3
Karakuldja R. (VIII)	0.256	0.182	0.015	0.112	0.038	13.9
Tar R. at confluence with Kara-kuldja R. (VIII)	0.139	0.195	0.012	0.106	0.028	12.6
Issyk R. (VII–VIII)	0.072	0.047	—	0.01	—	less than 1°

Table 27. Salt composition (mg/l) in rivers of the Zeravshan River basin (Sibirtzeva *et al.*, 1961).

Sampling site	Date	Time	Water temperature	Dissolved gases CO_2	O_2 mg/l	O_2 % %	pH	Ca″	Mg″	Na+K	HCO_3	SO_4''	Cl′	Mineralization
Zeravshan R., at Sudjino Vill.	16 V 1958	15	14.5	—	9.2	88.2	7.4	45.2	8.580	—	136.0	32.7	2.09	224.970
Kshtutdarya R. (affluent of Zeravshan R., at Rudaki village)	26 VI 1958	14	13.0	—	9.4	87.3	7.6	45.2	10.808	—	157.3	9.25	—	222.550

so negligible that we can say with confidence that the lake water is very slightly changed water of the river. The mineral residue in the river water is unusually small, 0.052 mg/l. In the Arys River it amounts to 0.388 mg/l; the Turgen River, according to Kurmangalieva (1976), has the "highest mineralization", 181.2 mg/l, whereas in its affluents it does not exceed 122 mg/l. At the same time, the mineralization in the brooks of Westfalien and in the Rügen springs, according to Thienemann (1926), amounts to 0.116 mg/l and 0.484 mg/l, respectively.

Even when compared with other streams of the Tien Shan, the Issyk River is very poor in salts. For comparison, Table 26 gives data on salt composition in some torrents of the Ferghana Range (south-eastern Tien Shan; Yankovskaia, 1948) and Table 27 presents similar data for streams of the Zeravshan basin (south-western Tien Shan; Sibirtzeva *et al.*, 1961). One can easily notice a higher mineralization and lower content of dissolved oxygen in the Zeravshan River and its affluent, the Kshtutdarya. This is the degree of mineralization which is more typical of mountain rivers (in the narrow sense of the word!) rather than for mountain torrents of the Issyk or Akbura type.

The hydrogen ion concentration (pH 7.2) indicates almost perfect neutrality of water in the Issyk River. In general, the unusual scarcity of dissolved salts in this river is characteristic of the upper and middle reaches of a mountain torrent of glacial origin which flows over igneous rocks, granites and porphyries.

When comparing the faunas in various affluents and irrigation canals which belong to the Issyk River system, we paid attention to a very specific fauna in one affluent, the Severnaya Shchel ("Northern Trench"). A careful examination of the substrate of the affluent bottom revealed an amorphous coating. When treated with the muriatic acid, the samples of the coating and even of the algae taken from the torrent evolved numerous gas bubbles, thus indicating a large amount of $CaCO_3$ in the water. This fact was confirmed by analysis (see Tables 24 and 25). Comparison of various types of rivers shows that calcium content in the water of Northern Trench is large enough to be a significant factor to life and distribution of the fauna in this affluent. Thienemann (1946) described a spring with a high calcium content and noted that this results in rapid formation of a thick calcareous layer on submerged twigs and stones so that the moss growing on them becomes covered by a calcareous crust. No calcareous tufa can be formed in the fast current and only calcareous sinters can be observed.

Similar phenomena, but a little less pronounced, were noted also in the Northern Trench. As is known, some authors have used the calcium content as a base for classification of swift waters (Ohle, 1937; Dittmar, 1955) in the same way as others used hydrogen ion concentration (Harrison and Agnew, 1961). However, it should be recalled that the Issyk River must be classified as deficient in calcium, according to Ohle.

It is essential to trace on a factual basis the changes in salt composition of water from the headwaters to the lower reaches. A general idea can be obtained from comparison of analyses made for two points of the Yassy

River located at different absolute heights (Table 25). However, a more complete picture is available for the Akbura River (Table 28) where a number of water analyses for salt composition was made from the headwaters to the lower reaches (we are indebted to E. O. Omorov who kindly made these data available to us). Some variation in the cation and anion concentrations, total hardness, and the solid residue is certainly associated with the season when the samples were taken and with the effect of affluents. But the mean values for consecutive sites from the heads to the lower reach show a progressive increase of mineralization.

The hydrogen ion concentration (pH) in mountain torrents is characteristic of the waters with neutral reaction. Tables 24 and 28 give such data for torrents of the Northern and Southern Tien Shan (see also Table 4). In the north pH = 7.2, but in the south the same value, pH = 7.2, in the upper reach increases towards the middle reach to 7.7–7.9. Similar concentrations were observed in the lower reach. Therefore, although the dominant value of pH in mountain torrents is indicative of the neutral water, but, as a rule, pH in the lower reach shifts to alkaline water.

Since the effect of hydrogen ion concentration on the distribution of organisms in mountain torrents of the Tien Shan has not been studied, it will not be discussed as an ecological factor in the chapter about the influence of environmental factors on the organisms.

Colour and transparency of water. The distribution of water colour and transparency is different for different sections of a torrent and for its affluents. If the transparency is high owing to poor erodibility of rocks or to capture of the glacial silt (or "rock flour") by some filters (by morainic materials, in landslips like dams, in lakes, *etc.*), the water of mountain torrents acquires an astonishingly beautiful blue or greenish colour. It is clear why the common local name for torrents in the Tien Shan – "koksu" – means light or dark blue water. However, washout of banks, affluents, or heavy showers alter the blue colour so that the water turns to greyish-brown and its turbidity increases. An abrupt change of water colour and transparency can be observed frequently as a result of showers in summertime or other seasons of the year. For example, the Akbura River carries bluish transparent water in autumn, but reddish-brown in April (maximum of rainfalls), which turns the river into a peculiar red stream. Most commonly in summertime, during the period of flood (see the chapter on flora and Table 4 therein,), the mountain rivers and torrents in Middle Asia carry greyish-yellow or brownish waters, turbid and nearly opaque. This is especially typical of rivers of glacial origin (such as the Chilik, Kebyn, Naryn Rivers, *etc.*) except for the mentioned above streams which have some filters at the headwaters (for example, the Koksu River in the Chatkalskii Range).

Now we shall present the colour and transparency distribution data for the Issyk River. Upstream from its confluence with a large affluent, the Djarsu River, the Issyk is very transparent, with a beautiful blue colour turning dark blue as the depth increases. Daily variation of transparency cannot be noticed by eye; the water remains absolutely transparent during the whole

Table 28. Salt composition (mg/l) in the Akbura River, 1968.

Locality	Date	Cations					Anions						Total hardness	Oxidizability (mg MgO/l)	pH	Dry residue
		Ca''	Mg''	$Na+K'$	NH_4'	Fe''	HCO_3'	SO_4''	Cl'	CO_3''	NO_2'	NO_3'				
Upper reach:																
Kichik-Alai	23 IX	42.1	14.0	1.4	—	—	143.4	41.6	3.6	—	—	0.8	3.30	0.5	7.2	208
Chal-Kuiruk	23 IX	46.1	12.2	4.6	—	0.3	140.3	53.9	3.6	—	—	—	3.25	3.3	7.4	190
Middle reach:																
Akbura	14 II	48.1	13.4	12.4	1.0	—	158.6	61.7	7.0	—	—	0.4	3.50	1.0	7.8	222
Papan	21 IV	44.0	18.0	25.0	—	—	134.2	49.9	3.6	3.0	0.02	0.8	2.35	1.0	7.8	234
	18 XI	48.1	13.9	13.5	—	0.05	158.6	61.3	8.8	—	—	0.8	3.55	1.2	7.7	234
	28 II	53.1	14.6	18.8	—	2.05	167.8	72.4	14.2	—	0.06	1.5	3.85	1.1	7.9	288
Gorge	14 V	44.0	15.2	10.6	—	—	146.5	52.2	3.6	20	—	2.0	3.45	1.0	7.9	212
	6 XII	48.10	17.6	14.7	—	0.10	176.0	66.6	7.0	—	0.06	1.5	3.85	0.8	7.9	276
	28 II	53.1	14.6	13.5	—	0.1	167.8	69.1	10.6	—	—	1.5	3.91	1.6	7.7	280
Kurpa village	14 V	80.2	34.0	37.2	—	—	265.4	173.2	14.2	—	0.30	4.0	6.80	0.8	7.9	500
	6 XII	50.1	24.3	2.3	—	2.4	183.0	67.5	10.6	—	0.1	1.5	4.50	2.0	7.7	258

day. An affluent, such as the Djarsu River carries a great amount of suspended mineral particles, glacier silt, and products of bank weathering when it runs over badly disintegrated grey and red granites which form the large rounded boulders and pebbles. The amount of fine suspended solids is so large that they sink down in quiet places of the torrent, near the banks and on the bottom in the form of a thin mucous-like layer; the water is very turbid, nearly opaque.

Downstream of the confluence with the Djarsu River, the Issyk River becomes turbid, greyish-yellow as a result of the mixing with the affluent waters. This picture persists down to the Issyk Lake. Small affluents and springs upstream from the lake, although having almost absolutely transparent waters, can hardly change the grey-yellowish colour and low transparency of the Issyk River. At the site of inflow into the transparent greenish lake, the river waters make a large yellow-brown spot which spreads radially like a fan and is easily noticeable from adjacent slopes. The lake plays the role of a settling basin by absorbing most of the solid particles carried by the river: the water running out of the lake is again almost absolutely transparent and bluish. Its transparency and bluish colour the water preserves farther downstream, but before coming out of the gorge onto the debris cone, it acquires a grey colour and its transparency becomes much lower, but is still higher than before the inflow into the lake.

The colour and transparency of water in the lower river do not remain unchanged during the day. As in most mountain torrents, the water level in the river rises sharply between twelve and one p.m. (the greatest insolation); the bank washout and the supply of suspended particles start to increase causing a sharp reduction of water transparency. After three o'clock it becomes almost non-transparent.

To summarize the review of abiotic ecological factors acting in mountain torrents of the Tien Shan, we should like once again to emphasize the interrelation of these factors, as well as the high variability of those factors which brings about changes in the ecosystem – the mountain torrent as a whole.

Sharp changes, periodic and non-periodic, of water discharge and, consequently, of water level, flow velocity, and even transportation of bottom deposits provide for the high adaptation of the organisms to such unstable conditions. Another consequence of such a variable ecological situation is the alternation of conditions in any microbiotopes. Thus, any shelter in a mountain torrent ceases to serve as such under sharp changes of water discharge. An extreme example of such a change is the mudflow which modifies the river channel and sometimes completely sweeps away the lithorheophylous fauna in a given site of the torrent.

We have not discussed the biotic conditions in the torrent, which play a great role in the ecosystem, particularly, the biocenotic factor. Unfortunately, this aspect of the torrential life is still poorly understood and only a few examples can be given in the chapters which follow.

5 The Issyk and Akbura Rivers as Tienshanian examples of the mountain torrent

Before we can identify the Issyk and Akbura Rivers as Tienshanian examples of the mountain torrent, we should discuss properly the conception of "mountain torrent". So far, the words "torrent", "stream", "river" or even "brook" have been used in a general sense, but now we shall try to refine these terms, *i.e.* to specify these watercourses from such additional criteria which would help distinguish one watercourse from another.

Trying to place the Issyk and Akbura Rivers in the existing classificational schemes described in the first chapter and based on the hydrological and geographical parameters, beginning from Oldecop's scheme (1917) to the schemes by Kuzin (1960), Shultz (1965), Shadin (1950c, 1951) and Muzafarov (1958), we concluded that these classifications can hardly provide a type of watercourse which could be fitted to the swift waters named in the heading of this chapter. The same conclusion was drawn from analyses of a number of other Tienshanian watercourses similar to the Issyk and Akbura Rivers.

However, the hydrological schemes have a certain value for revealing the genesis of a watercourse and this should be discussed at some length.

A comparison of discharge curves for the Issyk River with those for other mountain streams of Middle Asia, such as the Chirchik, Soh, Arys, and Bugun Rivers, *etc.* (Davydov, 1929) shows that the maximum water discharge in the Issyk River, which occurs during the warmest months, is in phase with that of other rivers of glacial origin. For example, the Soh River has a maximum discharge in June as do the Kekemeren, Isfairamsai, Shahimardansai and Isfara Rivers. In addition to the characteristic maximum during the warmest months, the discharge curves for these rivers, including the Issyk, have a specific feature – a steep, peak-shaped form. The discharge curves for other types of river are smoother and more stretched at their peaks. All rivers of non-glacial (snowy, spring or mixed) genesis have an earlier date of the maximum discharge. This is particularly characteristic of the rivers of spring or snowy origin, like the Bugun River. At the heads, a river of the glacial type may become of the mixed type in the middle or lower reach. The Issyk River maintains the regime of the glacial river from the heads down to the irrigation network, like the Akbura River does in its upper and middle reaches.

Such rivers of Middle Asia as the Issyk, Talgar, Almaatinka, Shaty, Turgen Rivers and the like deserve to be placed in a special group of "mountain torrents". When identifying this type of swift waters one should take into consideration, besides the thermal and chemical factors, bed material, *etc.*, those which describe the power of watercourse: its water discharge, width and depth.

As for its alimentation sources, the Issyk River may be referred to the glacial-snowy type of rivers. However, the weighted mean altitude estimated for such rivers is too large for the Issyk River. Judging by the intra-annual pattern of discharge (Kuzin, 1960), it should be referred with respect to the flood time to the arctic mountain zone and the Tienshanian hydrological region. But the Issyk River does not exactly fit this scheme, as can be easily seen from its altitudinal position above sea level. We abandon further attempts to classify the Issyk River on the basis of hydrological schemes, since all of them classify by different characters (mainly, by the regime), only the large rivers. Like many Middle Asian watercourses of relatively short length, the Issyk River seems to be quite an original phenomenon – a special type of "mountain torrent". To this type we should also refer the Akbura River, however not all of it, because it considerably differs in its lower reach. The Issyk River is a "mountain torrent" almost throughout its entire length, and the River forfeits this name when it disappears as the mountain torrent.

Turning to the ecological classifications of watercourses, which consider not only the hydrological and chemical characteristics, but also the fauna, it should be noted that a variety of such schemes have been suggested; they are rather similar and have been analysed in detail by Illies and Botosaneanu. A discussion of these classifications is outside the scope of this book, but all of them are largely of local significance. The authors themselves do not claim for their geographical versatility. The schemes have a wide range of application: from the purely local significance for the British Isles (Carpenter, 1927, 1928) to applicability to most mountain regions of Central Europe (Steinmann, 1907; Thienemann, 1912). Some of the more recent classifications are of the local character too: for the Carpathians, by Botosaneanu (1959); for the Balkans and the Tatra Mountains, by Kownacka and Kownacki (1972). Thus, these ecological classifications have no room for the Tienshanian mountain torrents.

The European and American schemes for subdivision of swift waters consider rivers, brooks and mountain streams, but with all the similarity in terminology, the characteristics used to describe these waters are different from those for the Tienshanian watercourses. This is exactly what we intend to show, partly in this chapter, but mainly when examining representative groups of organisms in Tienshanian torrents.

The basic distinction of our proposed scheme for the Tienshanian swift waters is that we do not classify these into two kinds of scheme, one of which deals with the types, *i.e.* dimensions of watercourses, and the other with vertical section. These are the principles underlying all the schemes available in the literature, which was emphasized by Illies and Botoseneanu who considered separately the "Classification of running waters as a whole" and "Classification of running waters by zones". In our conception, different types of watercourse, namely: brooks, torrents and rivers, may be vertical zones of other types.

Our objective is not a classification of Tienshanian swift waters. In conformity with the subject-matter of this essay, our aim is more modest: to find a place for the mountain watercourses of the Tien Shan in the general scheme and to render their specifications. The latter we expect to do taking into account the physical and chemical features of watercourses described earlier, especially, the unique features of the fauna, to which the following chapters will be devoted. Therefore, we classify only the speedy flowing watercourses which have a specific fauna. So, our classification is an ecological, rather than a hydrological one.

The following scheme can be proposed for subdivision of Tienshanian speedy flowing watercourses. Types or categories, of watercourse in this scheme are allowed to be both ecosystems in their own right and vertical sections of a "supertype", or of the total river system, but in the latter case, too, the categories are not deprived of a certain specificity through their abiotic and biotic features.

1. The flowing springs (an azonal type, but, within the mountains, sometimes forming a self-dependent ecosystem or entering into an alternative type, say, the torrent).
2. The brooks (sometimes a self-dependent type or entering into an alternative type such as the torrent or river).
3. The glacial headwaters (either a self-dependent type, if they flow into a moraine lake, or belonging to the torrent category).
4. The mountain torrent (a self-dependent type or part of a more general system composed of the glacial headwaters, the mountain torrent proper and the river, but in the latter case the torrent retains its specificity, *i.e.* it will be a "type").
5. The mountain river (part of the "glacial headwaters – torrent – river" system; may disappear into the desert or turn to large waterway like the Syrdarya, Amudarya, Ili Rivers, *etc.*; rarely occurs as a self-dependent type).
6. The steppe or desert river (in fact, this type does not belong to the speedy flowing watercourses proper, but is included in our scheme only because it may be the terminal section of a mountain river or torrent).

It goes without saying that no sharp boundaries can be drawn between these types, or categories. There may be either gradual transitions or different variations between the types, making it impossible to create a rigid and formal classification of such a complex natural phenomenon as the speedy flowing watercourse of the Tien Shan and, probably, of other mountain regions. One can speak only about the typical or specific to each type (having in mind the hydrological and hydrochemical parameters and, especially, the ecological ones).

In connection with the suggested classification, we should like to point out the following. All types other than the mountain torrent are only mentioned without description or differentiation. The vertical location of the Tienshanian torrent is well defined: from the snowy summits to the debris cone (piedmont). If a torrent is not expended for irrigation at the debris cone, it turns into river. By the glacial headwaters, the brook, the mountain torrent and the river we mean not only the vertical sections of a river system, but also well-defined self-dependent ecosystems. If a section of a river (in the general sense of the word) lies at the same altitude as a section of another river but differs from it in the size and other essential characteristics, then these sections must be considered as different categories, or types.

The flowing springs possesses the smallest water mass, the river has the largest mass, whereas the brook and the torrent occupy in this respect the intermediate position. All these categories, as pointed out above, are considered not only as vertical subdivisions of the river system, but they may exist parallel to one another.

The Issyk River, almost over its whole length, is identified as a mountain torrent, and the same is the Akbura River except for its lowermost section.

Let us consider in some detail the conception of mountain torrent as an objective system. What conditions (for the time being, only the physical and chemical; the fauna will be considered later) characterize the mountain torrent? These are the fast turbulent current (3–4 m/sec in average, at times,

(a)

Fig. 47. Typical sections of a mountain torrent characterized by well-defined changes of the bed slope and roughness.
a. a typical section of the middle and lower reaches with alternating laminar and turbulent flows at a small slope of the bed (Photo by the author).

(b)

Fig. 47 (continued). b and c. an affluent of a torrent with considerable slope and roughness of the bed. The step-like pattern: cascades alternating with "platforms" (Photo by the author).

faster), nearly always violent with a large amount of foam on the water surface: in other words, a torrent represents a series of cascades with alternating more or less violent rapids. The water is saturated or supersaturated by oxygen; the water temperature is low and stable both in space and time; the bed substrate is nearly exclusively stony and consists of rocks, rock fragments, boulders and pebbles. Rarely, only near the banks and in pools, are there deposits of gravel, sand and glacial silt. Especially specific for the mountain torrent is its large power, which is greater than in the brook but smaller than in the mountain river (Fig. 47).

Why is powerfulness considered as a specific feature of the torrent as opposed to the brook which possesses many of the torrent's characteristic features? The powerfulness affects the distribution and composition of the substrate. This may be assumed to be due to the high flow velocity, but it is known that an increase of discharge in one and the same channel results in an increase of flow velocity. The shallower the water depth in a channel, the

90

(c)

larger the effect of friction of the water against the bottom; the friction in mountain watercourses is known to be closely related to the flow velocity. The water mass affects the water temperature in a torrent. If the mass is large, the water in the torrent and the stony bed will warm up slowly. It follows that in a torrent with a large discharge the water temperature during summertime is lower than in a brook with a small amount of water, so the variation in the water temperature will be much slower.

Two other features of the torrent are to be emphasized: the glacial or glacial-snowy origin and the relatively high position above sea level (from 1000 to 4000 m in the Tien Shan; Kemerih, 1958; Ilyasov, 1959). The mountain torrent as a complex system is non-uniform and diversified, which results in a patchy distribution of depth, velocity, substrate, *etc.* However, this diversity is more or less permanent quantitatively and does not disturb the general integral pattern. In the annual regime of a torrent one may find features of other types. In winter, the discharge in a mountain torrent drops tremendously, sometimes down to that of the brook type, then it rises many-fold in summer and approaches the values for a mountain river. But

91

(d)

Fig. 47 (continued). d. the flight of cascades at a very steep slope of the bed, water completely in foam; e. a small freely dropping waterfall (Photo by E. O. Konurbaev).

this is not enough to assert that a mountain torrent with a small discharge is a brook in winter but a river in summer. One can only state that such regime of discharge is typical of glacial mountain torrents in the Tien Shan. With a different dynamics all other conditions would be different. In intermittent and perennial watercourses, the characteristics of the substrate, the bed itself, the water chemistry and transparency and others will be different. The spring brook changes but little its debit during the year, but the snowy brook commonly ceases to exist in winter.

The mountain torrent may be either a section of a river system or a self-dependent type, *i.e.* it will be a torrent over its whole length, from the heads down to the end. In the first case, the different types of watercourse should be classified as zonal (vertical) sections, or zones. The length ratio of the zones (or types) may differ in the Tienshanian mountain torrents. A typical pattern is: (i) a very short zone of the glacial headwaters followed by

92

(e)

(ii) the mountain torrent zone which dominates over the whole river system and (iii) a very short zone of the mountain river proper. When the torrent runs between two mountain ranges (the Chilik, Kebyn, Naryn, Aksai, Chatkal, Zeravshan Rivers and others), the type of "mountain torrent" will have a greater extent.

The schemes available for the speedy flowing watercourses of Europe, North America or, even, the Eastern Pamirs are unsuitable for classification of the Issyk and Akbura Rivers and many similar watercourses of the Tien Shan. All of them are based on the local findings and, therefore, are applicable only to the regions where the ecological conditions and the aquatic fauna have been studied.

The Issyk and Akbura Rivers are similar to many watercourses of the Tien Shan and represent a specific type – the "Middle Asian mountain torrent". Both in Europe and North America, the mountain stream (a term generally used in the literature) is only a part of a river system (its "upper section"), whereas in the Tien Shan, for the most part, the mountain torrent represents a self-dependent, closed system, from the glaciers and permanent snows to the semi-desert and desert areas. This is especially characteristic of the marginal ranges. The dynamics of the Middle Asian mountain torrent is much affected by the climate aridity (in this it differs from the torrent of the Altai). The fauna of Tienshanian torrents is specific with respect to species, genera and dominance or absence of various animalistic groups (thus the *Mollusca* are absent from the Tienshanian torrents).

One may ask for what mountain regions is the "Middle Asian" type of torrent representative? Perhaps (the data are still very scarce), for the Eastern Tien Shan (within China), the Western Pamirs, the mountains of the arid Central Asia, the Karakoram, the Hindu-Kush, partly the Himalaya and the Tibet and other regions with a similar climate. For instance, the Caucasian torrents have a different morphology and a quite different fauna! The same is true of the Far East (Levanidova, 1968). Unfortunately, there is nothing much to say about the watercourses of the Kopet-Dagh, but the available data, however scarce, cannot refer them to the "Middle Asian type" the way we have described it above.

Our scheme is only a first approximation to the classification of Middle Asian mountain watercourses, but there is no doubt that further investigations of the streams will allow differentiation of the "type" into subtypes, *etc.*

Major groups in lithorheophilous fauna of torrents

The first statement with which we must begin this chapter, unfortunately, is that knowledge of the systematics of mountain stream fauna in all regions of the Earth in general remains poor (including even Europe: the Alps, the Vosges, the Caucasus, *etc.*), and in the Asian mountains in particular. The systematics and phylogeny of some most representative groups of lithorheophilous fauna (dwellers of stony substrates in swift waters) are developed very unevenly. It is out of place here to analyze the reasons for the serious gap in systematic treatment of aquatic fauna in torrents, but two points should be emphasized. First, it was not until recently that we became cognizant of the great significance of studying mountain stream faunas, which is important from many points of view: for sizing up the way of life of organisms in extreme conditions; for bionics; for acclimatization, as a source of fauna for man-made lakes; lastly, for nature conservation and solution of the problem of "clean" water. Second, the dominance of larval and, in general, or preimaginal stages within the lithorheophilous fauna presents a serious problem for systematical treatment of the fauna under study.

It was not until recently that torrents began to attract the attention of specialists in ecology and systematics, who were faced with the fact that faunas of speedy flowing watercourses are almost *terra incognita.* As evidence, we refer the reader to a statement made in 1963 by Illies and Botosaneanu, the authors of the comprehensive work on swift waters, that the lack of specialists (taxonomists) and keys for species identification is the main difficulty in producing an adequate species list. This difficulty could be partly overcome by using *nomenclatura aperta, i.e.* designation of species by numbers or letters, which the authors believed was more valuable than just a mere indication "div. spec. indet.", "spec. spec.", *etc.* Indeed, the species lists for groups of lithorheophilous fauna, even from the swift waters of Europe, are literally stuffed with signs like "sp.", "sp.1", "sp.2", *etc.* Why do we say "even"? Because the study of lithorheophilous fauna from this region started as early as 1907 with the work by Steinmann! Although (i) some keys have become available for Central Europe and England, in particular, on mayfly larvae (Schoenemund, 1930; Kimmins, 1942; Bogoescu and Tabacaru, 1957; Macan, 1961b), on stoneflies (Hynes, 1967) and caddisflies; (ii) quite a lot is known about the stages of dipteran *Blepharoceridae,* a family typical of running waters; and (iii) the data on these groups are available now in Russian keys and guides for the European part of the USSR, we have nevertheless no such publications for Middle Asia.

The inadequate knowledge of the fauna from the Tienshanian torrents made Mani cite our works only for the 1940's when he discussed the systematics and ecology of aquatic fauna from watercourses of the Tien Shan. The matter is not only that there are still no regional guides on fauna of lotic waters, but that the systematics of some groups of torrential fauna is simply not yet developed, and much effort will be needed to revise these groups and make possible just the "identification" of the species.

A still worse situation is in the study of systematics of aquatic insects from Iran, Afghanistan, Pakistan, India, China, Tibet and Mongolia. Investigations by Hora for India in the 1920's were abandoned and there have been only random papers on mayfly systematics. The experience of revision and analysis of some groups in Tienshanian torrential fauna (mayflies, stoneflies, net-winged midges, chironomid midges and caddisflies) has shown, at least, high species endemism of this fauna (probably, about 80–90%). Thus, introduction of "European species" in Tienshanian faunistic lists, as observed in some works, is more than doubtful.

Nevertheless, in spite of the dismal picture presented above, it is possible to characterize the fauna of Middle Asian torrents in terms of the groups which are representative and specific for these streams. Among the inhabitants present in large numbers over the whole lengths of torrents which have a distinct adaptation to life in mountain watercourses are aquatic insects of such groups as mayflies, caddisflies, stoneflies and some families of *Diptera*. Mayflies, *Blepharoceridae* and *Deuterophlebiidae* were a subject of systematical treatment by Brodsky and more recently the mayflies were also treated by Tshernova, Sinichenkova and Kustareva; caddisflies by Lepneva, Kachalova and Sibirtzeva; stoneflies by Zhiltzova; blackflies by Rubzov and Konurbaev; chironomid midges by Pankratova.

This chapter contains data on the systematics and ecology of groups of animals which, in Martynov's terminology (1929), should be assigned to hymarobionts (after $\chi\epsilon\iota\mu\alpha\rho\rho os$), *i.e.* to the genuine inhabitants of torrents. They are present in large proportions among other groups and are common to all Tienshanian watercourses, a fact which we shall try to show in the following chapters (in particular Chapter VIII). Here we give only a short account of the state of knowledge on this group and its systematics, some comments about the adaptation, if studied, of different groups, and, lastly, data on the distribution in torrents and ecology of every species. The sequence of groups in our discussion is determined by their significance in torrential fauna: mayflies, caddisflies, stoneflies and *Diptera*; among the latter the most specific families are *Blepharoceridae*, *Deuterophlebiidae* and *Simuliidae*, then *Chironomidae* and other dipteran families. The chapter is concluded with comments on other groups of invertebrates, but, as it was pointed out earlier, these are poorly studied components of torrential fauna for the time being and they play a relatively insignificant role in torrents.

The technique for collection of fauna in mountain streams has some specificity. This is, naturally, because of the impossibility of using any techniques developed for lakes and large rivers with slow current in rapid streams. In recent years there has been a large number of publications on techniques for faunistic investigations of speedy flowing watercourses and their respective ecological conditions (Shelford and Eddy, 1929; Sadowskii,

1948; Robins and Crawford, 1954; Shadin, 1956; Harris, 1957; Longhurst, 1959; Edington and Molyneux, 1960; Würtz, 1960; Albrecht, 1961, 1966; Macan, 1961a, 1962, 1963; Cummins, 1962; Waters, 1962a; Müller, 1962–1963; Cushing, 1964; Mundie, 1964; Kalmann, 1966; Hynes, 1970; Edington and Hildrew, 1971 and others).

In recent years several instruments have been designed for fauna sampling in swift waters, which make it possible to record bottom invertebrates quantitatively. One can name a few: Surber's sampler,[1] Cushing's "drift" net, a spherical box for collecting fauna by Waters and Knapp, Usinger's drag, and, lastly, planktonic nets which were successfully used by Brundin (1966) to collect larvae of *Chironomidae*. However, all these devices (and those not mentioned) are designed for sampling mobile benthic organisms in running waters like mayflies, stoneflies, *Chironomidae, etc., i.e.* the so-called "organic drift" which has lately attracted much attention (Müller, 1954, 1962–1963; Waters, 1962a, b, 1964; Bishop and Hynes, 1969; Reisen and Prins, 1972, and others). Organisms which poorly trapped or not trapped at all are those which firmly attach themselves to the substrate or sedentary forms, which must be collected therefore only by hand, with the tweezers, from every stone or boulder. The cascades, which are widespread in the Tienshanian torrents and have extremely irregular and fast current, sometimes preclude any quantitative sampling with catching nets because the organisms are washed away by the water when a stone is lifted up out of a stream. All the content of even a large-mesh net is washed off by water at once, and even a proper setting of the net is usually impossible. As the stream bed consists of large boulders or rocks rather than of fine pebbles and stones, it rules out kicking up the substrate (Macan, 1958; Allan, 1975).

Torrents on the Tien Shan (as well as in other mountain regions) commonly have bed plots of large stony slabs, rocks and boulders. These are known to differ in their specific fauna which, probably, is more typical of a torrent than the fauna observed over movable substrates of pebbles or fine debris. The water layer above the slabs, rocks, *etc.* is usually shallow, which makes it possible to use a "barrage", *i.e.* withdrawal of water from the surface of a rock or boulder (the use of the above devices is ineffective). We use an elbow bend which consists of two metallic strips with hinged joint and a regulating rod (Fig. 48). The strip size is 15×30 cm. The bend, extended to a 50–60° angle, is pressed firmly to the rock, the vortex of the angle being oriented against the current. The water flows around the bend setting aside a waterless space between the strips. It makes possible the collection of larvae and pupae of *Blepharoceridae*, blackflies, caddisflies, *etc.* The sampled area may be determined rather precisely.

Some comments should be made about insectaries for torrential fauna. As is known, correct species identification for many groups of aquatic insects can be, unfortunately, made only from the mature, winged stages. Trapping of adults is very difficult and not always possible. It is better to deal with the winged stage through growing it from nymphs, pupae, *etc.* in the insectary.

[1] As Allan (1975) reports, the Surber sampler cannot determine the real number of species in the torrent under study.

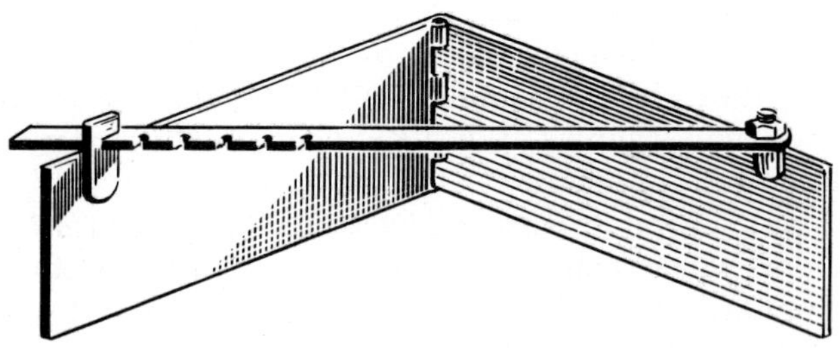

Fig. 48. A device for fauna sampling from smooth rocks, stony slabs and large boulders covered with thin water layer.

Numerous types of insectary have been proposed (Biarnov and Thorup, 1970), including even a simple "envelope" from the wire net. Drawings and description of various insectaries can be found in publications by Lepneva (1927; 1964, p. 182, Fig. 175).

For stationary study we used a collapsible insectary consisting of a copper zinc-coated steel net soldered to a frame of brass (Fig. 49). Very voluminous, the insectary, if collapsed, is easily transported which is essential for surveying in mountain regions. The insectary is dipped in a stream or brook near the bank to about one third of its height. On its bottom the stones with larvae or pupae of insects are placed.

The quantitative sampling is the most complex problem in surveying fauna in streams, as it requires measurement of the area or volume from which the sample is taken. The use of a measuring frame is not always possible, especially if the stream bed is composed of large stones, boulders and fragments, as usually is the case. To obtain approximate quantitative data and, chiefly, for comparison of different samples, we used two methods. One was the "timed" sampling, *i.e.* sampling for the same period of time at any point, which is especially convenient in route surveying. For this, preliminary experimental determination was made of the time needed for survey of stones with the total area of 1 square metre (or $10 \, m^2$).[2] A better data comparison can be made from determining the volume of stones under study by the balanced method, *i.e.* by plunging stones or pebbles into a large calibrated can (bucket!). This method is also suitable for measuring the biomass with a volumeno-meter. However, the tables in Chapter IX, which characterize abundances and biomass include data largely obtained by a measuring frame with an attached net. Of course, they underestimate the significance of hymarobionts in the total biomass.[3]

[2] Other investigations of fauna in mountain streams also used sampling "by hand" during a certain time interval (see, *e.g.* Mecom, 1972). Much detail on the methods of quantitative sampling for river benthos can be found in Bishop (1973), but for specific biotopes (cascades, boulders) he applied only qualitative sampling by hands! For designation of numbers of species in his tables he used such symbols: * casual, (x) single, x rare, +present, ++ common, +++ abundant. On page 278, he again emphasized that quantitative sampling on pebbly beds and in biotopes with boulders is impossible.

[3] For a comprehensive description of techniques for river benthos, see a paper by Shadin in the "Freshwater Life of the Soviet Union" series, vol. IV, part 1, 1956.

98

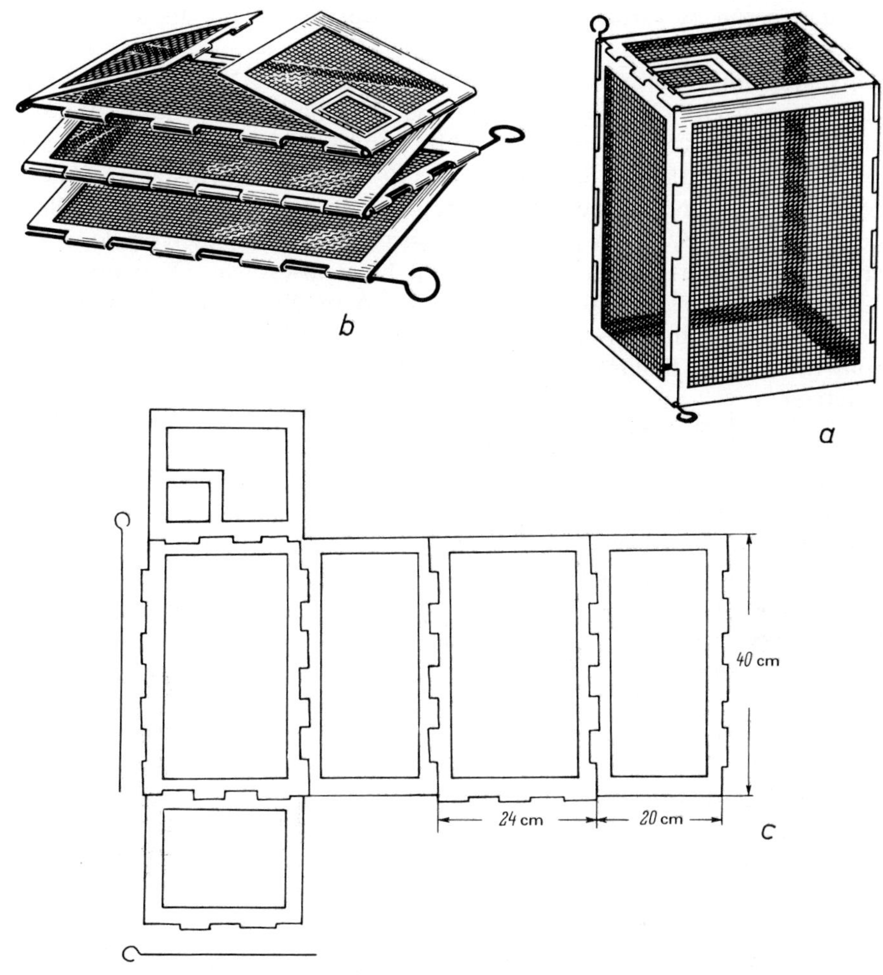

Fig. 49. The collapsible insectary. a. general view when prepared for installation in river; b. collapsing for transportation; c. full view.

1. *Ephemeroptera* (mayflies)

a. Species composition

As a constituent of the lithorheophilous fauna in the Tienshanian torrents, this order is of much greater significance than caddisflies or stoneflies, let alone other groups of invertebrates. In mountain streams, brooks and rivers, one can frequently see that all stones covered by mayfly larvae, whereas caddisflies, and particularly stoneflies, occur either in small quantities or are absent at all. This order also plays a great role in so-called "organic drift" in streams. A dry channel watered due to heavy rains or on the water level reduction teams with larvae and nymphs of mayflies. During certain periods of the year, when mayfly larvae are very small in size, their unusual

99

abundance can be estimated only if they are washed off the stones into a white-enamel cuvette.

Investigation of the systematics of this order of the Middle Asiatic fauna has only started. On mayflies from the Tienshanian mountain streams a number of papers has been published by Brodsky (1930) and, more recently, by Tshernova (1972), Sinichenkova (1973a, b) and Kustareva (1976b, 1978). These papers describe the major, abundant species of mayflies in mountain torrents, which provides a possibility to study their ecology, biology, their proportions in the torrential biocenoses, *etc.* Of course, these studies do not cover all the mayfly fauna. A number of forms, particularly those inhabiting the lower reaches of torrents and mountain rivers, chiefly from the orders *Baetis, Ecdyonurus, Cynigmula, etc.* still await their systematizer.

In the Issyk River the following mayfly species were found:

Iron rheophilus Brod.,[4]
Ir. montanus Brod.,
Ir. nigromaculatus Brod.,
Rhithrogena tianshanica Brod.,
Rhithrogena sp.1,
Ecdyonurus sp.1,
Ecdyonurus sp.2,
Ecdyonurus sp.3,
Ephemerella submontana Brod.,
Ephemerella sp.1,
Ameletus alexandrae Brod.,
Baetis issyksuvensis Brod.,
B. heptopotamicus Brod.,
B. transiliensis Brod.,
Baetis sp.1,
Baetis sp.2,
Caenis sp.1.

Now we have grounds to suggest that "*Rhitrogena* sp." is probably *Rh. brodskii* Kust, 1976; one of the species "*Ecdyonurus* sp." is *Notacanthurus zhiltzovae* Tshern., 1972, and another species of the same genus is *Cinygmula oreophila* Kust, 1978.

All forms that could be identified precisely proved to be new species endemic for the Tien Shan.

In the Akbura River (in samples collected by Omorov in 1966–1970) mayflies were represented by species of the same genera as in the Issyk River: *Rhithrogena, Iron, Ephemerella, Ecdyonurus* and *Ameletus*. The genus *Baetis* was found in the form of numerous nymphs of different species. The only species found in the Akbura River but not found in the Issyk was

[4] In the key for genera of *Heptageniidae* (Tshernova, 1974) the genus *Iron* is shown as a subgenus *Epeorus* Eaton, 1881. Here Tshernova follows Edmunds and Allen (1964). Yet the question about taxonomical status of the genus *Iron* is unclear: "In general, the problem of species interrelations in the *Iron-Epeorus* group remains unsolved" (Tshernova, 1976, p. 340).

100

Ecdyonurus rubrofasciatus Brod., earlier described from the Agalyk River and the neighbourhood of Samarkand city.

Although the mayfly fauna of torrents is known incompletely, we may insist on the great biogeographical identity in respect of the species and generic composition of this faunistic group in the Northern and Central Tien Shan and in the Alai Mountains which, at least its northern slopes, must be assigned to the Tien Shan rather than to the Alai-Pamirs, as suggested by some authors.

According to the samples collected by Sibirtzeva (1961; Sibirtzeva *et al.*, 1961), mayfly composition in the Chirchik River (the Chatkalskii Range) and in the Zeravshan River (the Zeravshanskii Range) is similar to that in the Issyk River. Thus in the Chirchik River were found:

Iron montanus Brod.,
Rhithrogena tianshanica Brod.,
Ephemerella submontana Brod.,
Ordella macrura Stern.,
Baetis sp.

and in the Zeravshan River:

Iron sp.,
Rhithrogena tianshanica Brod.,
Ephemerella submontana Brod.,
Ecdyonurus sp.,
Cinygma sp.,
Heptagenia sp.,
Ameletus alexandrae Brod.,
Cloeon sp.,
Baetis sp.,
Acentrella sp.,
Ordella macrura Stern.

In her first account of the Issyk-kul Valley, Kustareva lists the following mayfly species:

Ephemerella karasuensis Kust.,
Eph. submontana Brod.,
Iron montanus Brod.,
I. rheophilus Brod.,
Rhithrogena tianshanica Brod.,
Rh. brodskii Kust.

Sinichenkova (1973b) described nymphs from different rivers on the Tien Shan:

Rhithrogena tienshanica Brod.,
Rh. asiatica Sinitshenkova.
Rh. minima Sinitshenkova.

Taking into account numerous samples from different torrents and mountain rivers on the Tien Shan and the above cited lists, the following species of mayflies may be considered as the most representative for a typical mountain torrent: *Iron montanus, Ir. nigromaculatus, Ir. intermedius, Ir. rheophilus, Rhithrogena tianshanica, Ephemerella submontana, Ameletus alexandrae,* species of the genus *Baetis* (*B. issyksuvensis, B. transiliensus, B. heptopotamicus*) and *Cinygmula oreophila.* It should be borne in mind, of course, that all these forms occur in different sites of the torrent in respect to both its vertical sections and ecological conditions.

In addition to what has been said about the occurrence of different forms under a single name, another example can be given. A comparison of the description and drawings of *Rhithrogena tianshanica* made by Sinichenkova (1973b) and Kustareva (1976) shows a certain difference between them. As Kustareva writes, the differences might point to either variability of a species or that we deal with two different species.

As pointed out earlier, much work needs to be done on identification of species composition of mayflies from Middle Asian mountain regions, but one thing is certain, that all forms precisely identified are new species endemic for the Tien Shan.

A more complete list includes the following 22 species:

Iron montanus Brodsky
Iron rheophilus Brod.,
Iron nigromaculatus Brod.,
Iron intermedius Brod.,
Rhithrogena tianshanica Brod.,
Rh. brodskii Kustareva
Rh. minima Sinitchenkova
Rhithrogena asiatica Sinitchenkova,
Rh. stackelbergi Sin.,
Notacanthurus zhiltzovae Tschernova,
(= *Ecdyonurus zhiltzovae* Tshern.)
Ecdyonurus rubrofasciatus Brod.,
Cynigma asiaticum Ulmer,
Cynigmula oreophila Kust.,
Heptagenia tadzhikorum Tshern.,
Heptagenia perflava Brod.,
Ephemerella submontana Brod.,
Eph. karasuensis Kust.,
Baetis heptopotamicus Brod.,
B. issyksuvensis Brod.,
B. transiliensis Brod.,
Ameletus alexandrae Brod.

One can estimate the distribution of the above mayfly genera in torrents in Middle and Central Asia, in the Himalaya and adjacent areas from the data of the few expeditions that have been undertaken. For the Alai-Pamirs or, to be more exact, for the Western Pamirs, for the Kondara River, an incomplete preliminary list of mayfly nymphs was prepared (Shadin *et al.,*

1951) where the same genera are included: *Rhithrogena, Ecdyonurus, Heptagenia, Ephemerella, Ordella, Baetis, Pseudocloeon* (?).

According to the records of the 1960 Japanese expedition to the Pamirs (the Afghanian section) and the Hindu-Kush, the following mayfly forms were found:

Ameletus alexandrae Brod.,
Ecdyonurus sp. a,
Ecdyonurus sp. b,
Epeorus (*Iron*) sp.,
Rhithrogena sp.,
Baetis noshagqensis Uéno,
Baetis sp.,
Cloeon zimini Tschern.,
Ephemerella sp.

The mayflies were sampled in the form of adults or nymphs; the latter described by Uéno (1966).

The data by the Japanese expedition of 1952–1953 to the Nepal Himalaya include, as in the former case, forms not only from streams, but from lakes and small standing waterbodies:

Ephemerella sp.,
Baetis sp.1,
Baetis sp.2,
Baetiella sp.,
Ecdyonurus sp.1,
Ecdyonurus sp.2,
Ecdyonurus sp.3,
Rhithrogena sp.,
Epeorus sp.

All these forms were found and described as nymphs (Uéno, 1955).

A paper published by Kapur and Kripalani (1961) gives records on mayflies from the north-western Himalaya, including the following forms:

Adults	Nymphs
Ephemerella indica Kap. et Krip.,	*Paraleptophlebia* sp.1.,
Baetis chandra Kap. et Krip.,	*Paraleptophlebia* sp.2,
B. simplex Kap. et Krip.,	*Ephemerella* sp.,
B. punjabensis Kap. et Krip.,	*Baetis chandra* Kap. et Krip.,
B. hymalayana Kap. et Krip.,	*B. simplex* Kap. et Krip.,
B. bifurcatus Kap. et Krip.,	*Ecdyonurus* sp.,
B. festivus Kap. et Krip.,	*Iron* sp.,
Epeorus lahaulensis Kap. et Krip.,	*Ironopsis* sp.1,
	Ironopsis sp.2.

Judging by these data, except for some forms found in standing waters, the following six genera, *viz. Iron, Rhithrogena, Ecdyonurus, Ephemerella, Baetis* and *Ameletus*, appear to be typical for mountain torrents in Middle

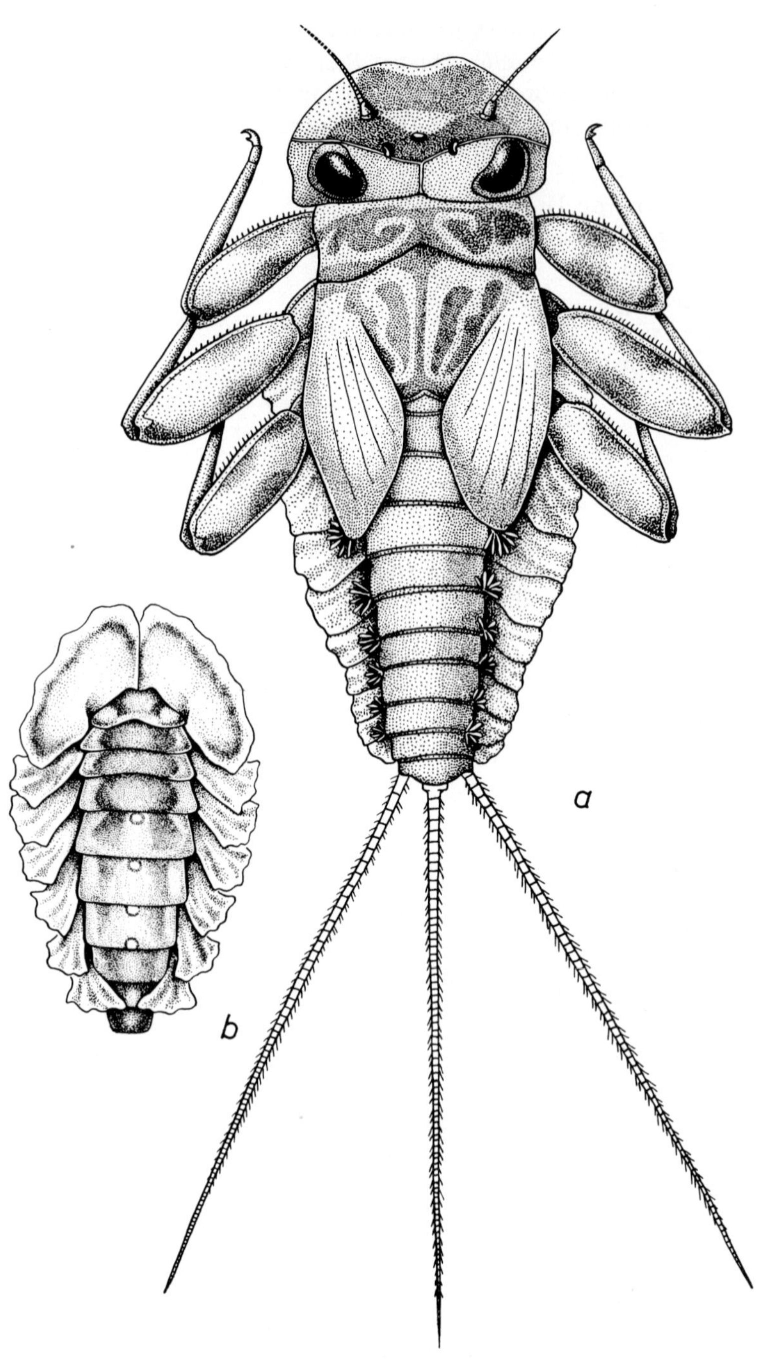

Fig. 50. Mayfly nymph *Rhithrogena tianschanica* BRODSKY (a, dorsal view and b, ventral view in limits of the gill system), *Acanthurus zhiltzovae* TSCHERNOVA (c, dorsal view), *Rhithrogena brodskii* KUSTAREVA (d, dorsal view). (a, d after Kustareva, 1976; b, original; c, after Chernova, 1972.)

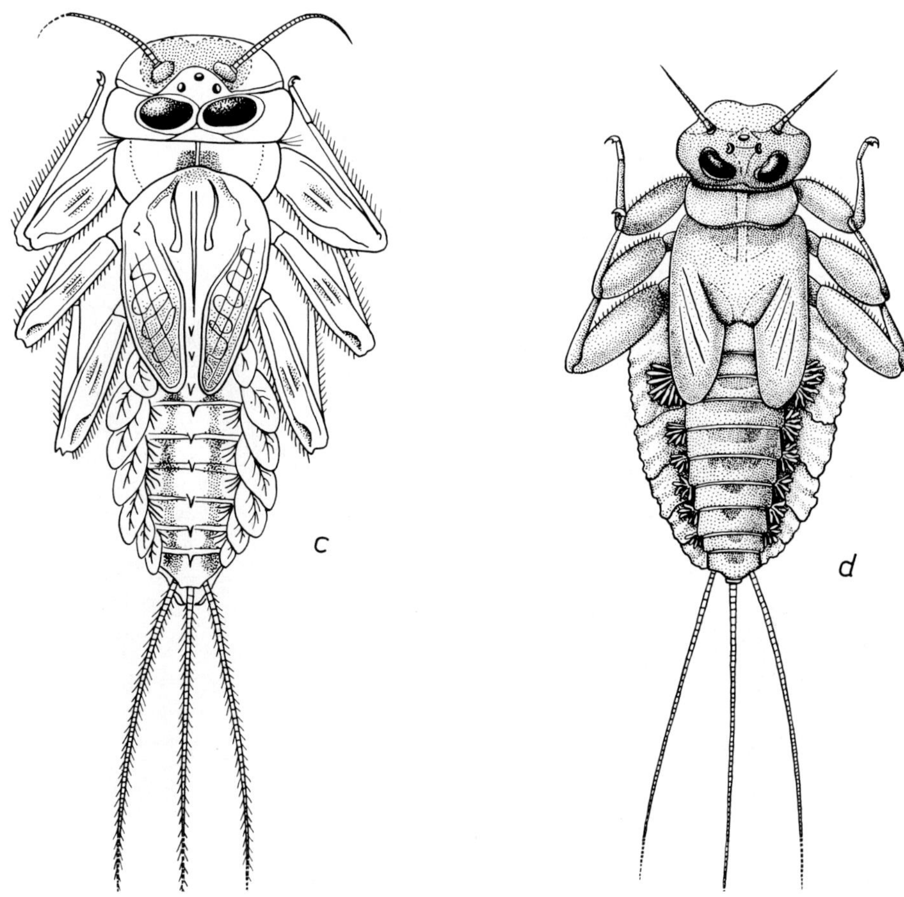

Fig. 50. (continued)

Asia, the Karakoram, Hindu-Kush, the Himalaya and the adjacent regions. It is very likely that further investigations will reveal species, at least some of them, previously described from the Tien Shan. This is evidenced by the examples which follow. *Ephemerella borakensis* Uéno, for instance, which is described from the Afghanian Pamirs and the Hindu-Kush, represents a junior synonym for *Eph. submontana* Brodsky (Allen, 1974). And *Sigmoneura anseli* Demolin and *Afghanurus vicinus* Dem. also from Afghanistan (Demolin, 1964) represent, respectively, junior synonyms for *Heptagenia perflava* Brodsky and *Ecdyonurus rubrofasciatus* Brod. (Tshernova, 1974).

Here we give a short description of nymphs commonly found in the Tienshanian torrents which is intended to help their identification in samples.

Rhithrogena tianshanica.[5] Large nymphs with a beautiful "annulated" reddish abdomen. The colour variations in the nymphs are from light-brownish to dark brown-reddish (Fig. 50a, b).

[5] Description and drawings see also in Kustareva (1976).

105

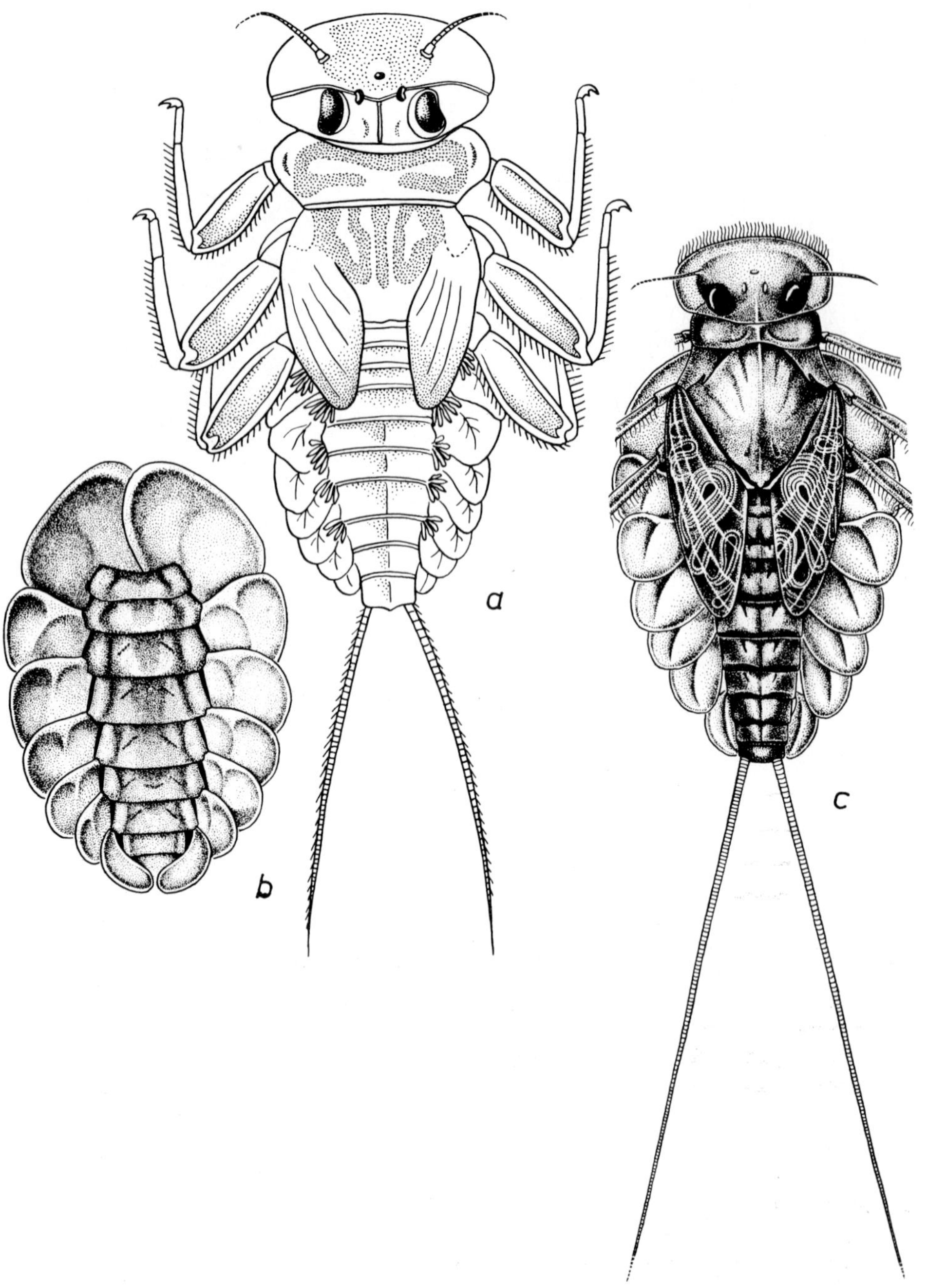

Fig. 51. Mayfly nymph *Iron rheophilus* BRODSKY. a. dorsal view; b. ventral view, within the gill system; c. more mature nymph, schematically. (a, after Kustareva, 1976; b and c, original.)

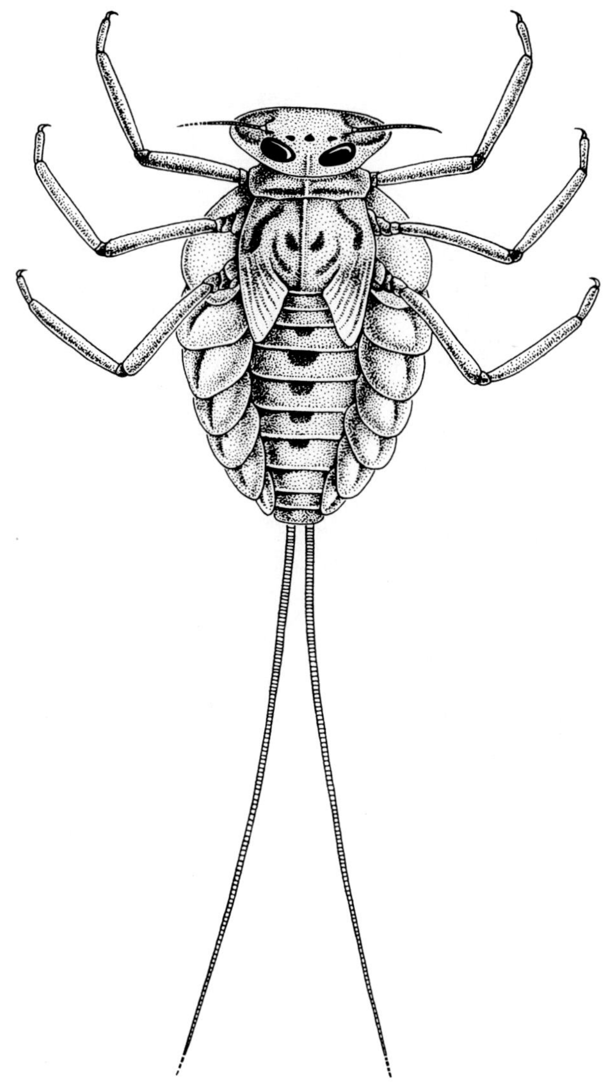

Fig. 52. Mayfly nymph *Ironopsis* sp., dorsal view. (After Kapur and Kripalani, 1961.)

Iron rheophilus.[5] A good diagnostic feature of nymphs of this and related species is the absence of spines on the abdominal tergites (Fig. 51a-c). This feature is also characteristic of nymphs of another genus, the *Ironopsis*, close to the *Iron* and found by Kapur and Kripalani in the north-western Himalaya. Foreseeing a possible discovery of this genus in Middle Asia, we give here the drawing from the authors cited (Fig. 52).

Iron montanus.[5] Sharp and long spines on tergites of abdomen. (Fig. 53a, b).

Iron nigromaculatus. Short and blunt spines on tergites.

107

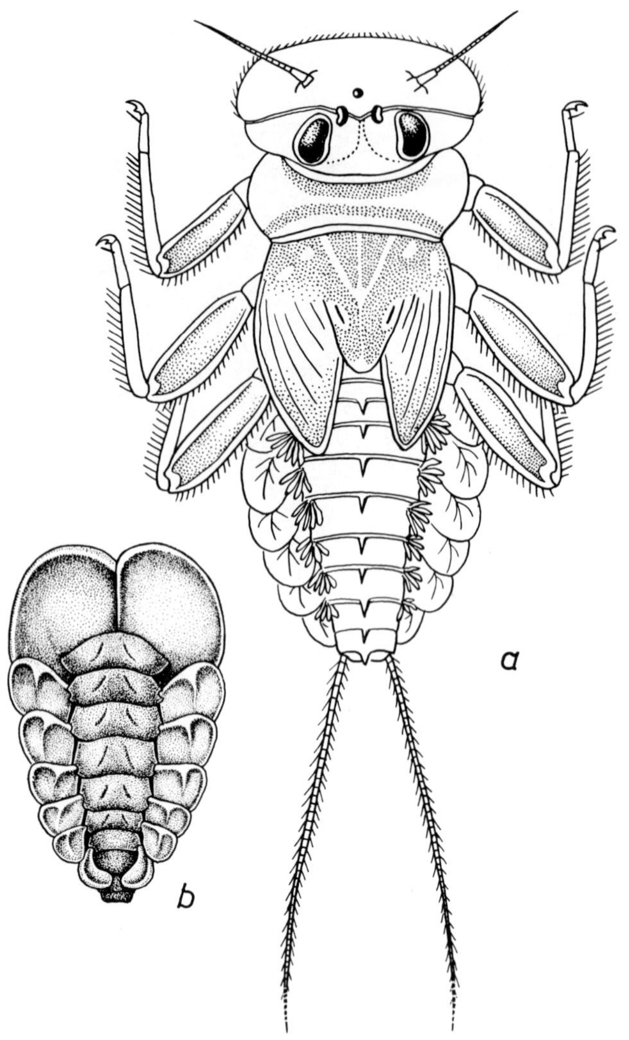

Fig. 53. Mayfly nymph *Iron montanus* BRODSKY. a. dorsal view; b. ventral view (partly, within the gill system). (a, after Kustareva, 1976; b, original.)

Iron intermedius. On tergites has sharp and long spines dilated at the base.

Ephemerella submontana.[6] A characteristic feature of nymphs of this species is powerful armament on the femur of the first pair of legs (as clearly seen in Figs. 54a, b). Similar formations are shown by Uéno in a drawing of the nymph *Ephemerella* sp. in accordance with the data of his expedition to the Karakoram and the Hindu Kush (Uéno, 1966, Fig. 15) and by Tshernova (1974).

[6] Tshernova assignes this species to the genus *Drunella*, but according to Needham it is a subgenus of the genus *Ephemerella*. Tshernova considers subgenera as genera. In the same paper she describes another species, the *Ecdyonurus zhiltzovae* Tshern., from the Aksu-Djobogly River (the southern Kazakhstan).

108

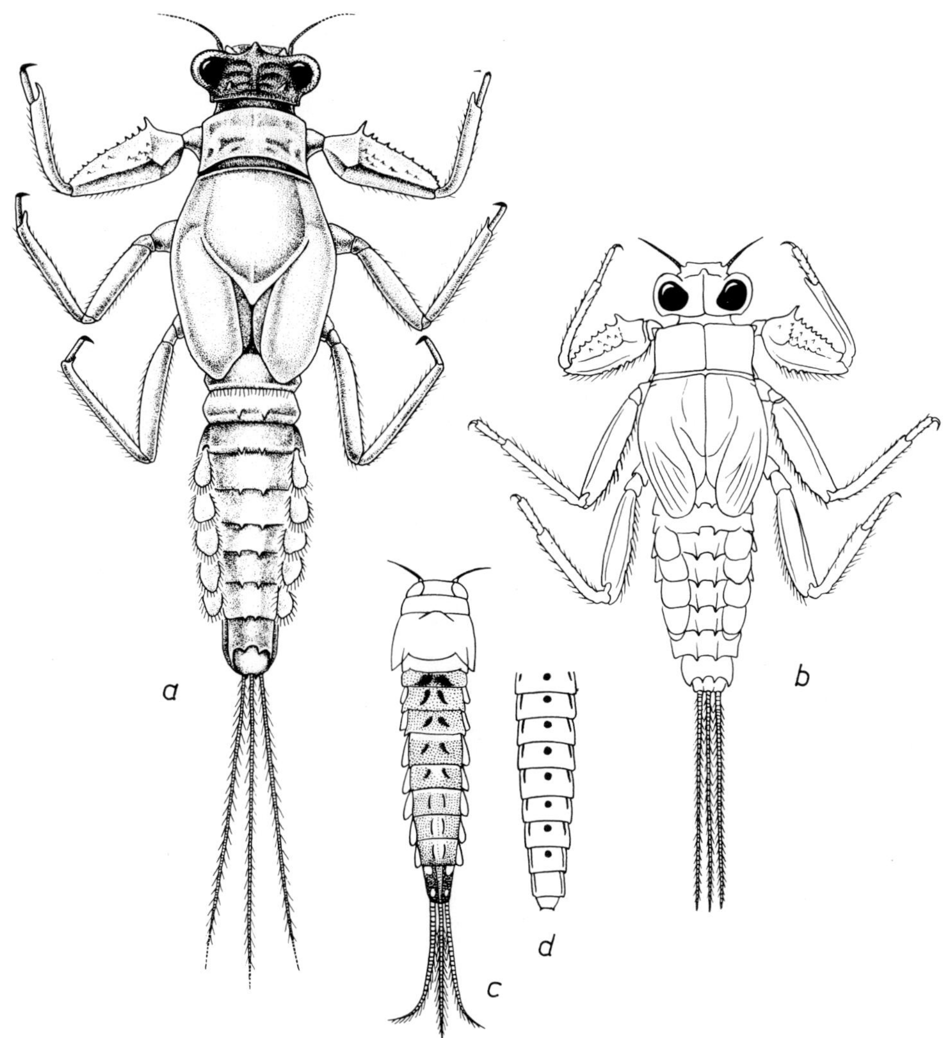

Fig. 54. Mayfly nymph *Ephemerella submontana* BRODSKY (dorsal view: a, female, b, male) and *Ameletus alexandrae* BRODSKY (c, dorsal view, the legs not shown; d, the abdomen, ventral view). e, *Cinygmula oreophila* Kustareva (a, original; b, after Chernova, 1972; c and d, after Uéno, 1966; e, after Kustareva, 1978.)

Ameletus alexandrae. For the first time we found this species in Issyk Lake. The nymph was also described and pictured by the Japanese expedition (Uéno, 1966, Fig. 2–4). And again, on the Hindu Kush it was found in a lake (the Shiva Lake), but more recent investigations (collections from the Akbura River and other streams on the Tien Shan) have shown that this species is a "permanent" member of the torrential biocenoses and, due to this fact, should be assigned to the torrential fauna.

Cinygmula oreophila (Fig. 54e). The nymph of this species has been described recently (Kustareva, 1978). "*Cynygma* sp." and, perhaps, some

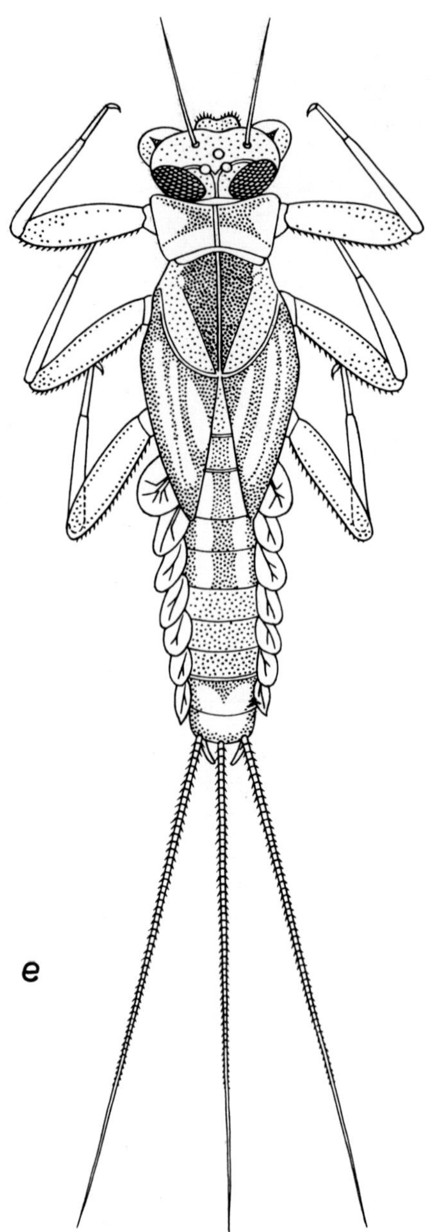

Fig. 54 (continued)

species from "*Ecdyonurus* spp." may be assumed to belong to this species, since it is widely distributed in the Tienshanian torrents. This species occurs at absolute elevation of 1700–3000 m and flow velocity from 1.5 to 2.0 m/sec. Judging by the constitution of the nymph which has no devices described below for other mayfly species and considering its preferable microbiotopes (the lower surface of stones, pools and quiet near-shore sites), we should think *Cinygmula oreophila*, the species of the genus *Baetis* and, evidently, *Ameletus alexandrae* to be of similar ecology and ethology.

b. Notes on adaptations in mayfly nymphs

The above listed mayflies, being typical inhabitants of the Tienshanian mountain torrents, have fairly perfect devices to live on stony substrates under conditions of fast current. The problem of adaptations in mayfly nymphs has been discussed by many investigators (Steinmann, 1915; Dodds and Hisaw, 1924a, b; Hora, 1930; Berner, 1950; Nielsen, 1951, 1955; Strenger, 1953; Macan, 1962, 1963; and others) and general information on this question can be also found in the monograph by Hynes (1970).

Here we shall briefly consider only such adaptive properties as are typical in mayfly nymphs from the Tienshanian mountain torrents.

It is known that adaptation to life in rapid waters leaves many marks in the body constitution, on the ecology and ethology of both larval forms and adults. In nymphs, the body shape, the head and even the total size of the animal, the number and structure of cerci, gill fringes, *etc.* are modified.

The species of the genus *Iron* should be regarded as the most typical dwellers of mountain torrents – hymarobionts. Bodies of larvae and nymphs of this genus are flattened and their dorsal side represents a smooth, perfectly streamlined surface. The head shield is very broad, covering all the mouth parts and closely attaching its frontal edge to the surface of a stone or rock. They do not have three filaments, as is common to mayfly larvae, but only two, the middle filament being reduced. The filaments are little felted, much less than in dwellers of standing or slow waters, which are good swimmers (*Centroptilum, Cloeon, etc.*). The filaments in live specimens are not arranged in parallel, but at a wide angle, and cling to the stone surface (Fig. 55). Most highly modified are the tracheal gills. Their outer edge is thickened and the fringes of the fore and the hind pairs are turned inwards, slightly overlapping each other so that something like a false sucker is formed (see Figs. 51–53).[7] In these animalistic forms the selection was directed not towards the retaining of the thin and delicate gill fringes (which helps oxygen absorption from the water), but towards their transformation into a mechanism capable of attaching the larvae to the stone surface. Such modification of the gill fringes is possible only in the waters saturated and supersaturated with oxygen, as it is the case in mountain torrents. The legs in the larvae of the *Iron* species are widely sprawled and bear rigid claws.

Thus, the general appearance of the preimaginal stage of all species of this genus, as well as the genus *Ironopsis*, indicates that in this instance we deal with true dwellers of violent mountain torrent. Their distribution in the torrential biotopes testifies that they are able at times, for a short or long period, to leave microbiotopes with relatively slow current and remain immobile or move over the surface exposed to water currents about 3 m/sec

[7] It is not a true hydraulic sucker, like in *Blepharoceridae*, but a false one (Pennak, 1953; p. 514). As Hynes observes, there is no negative pressure under it. The main function of this "sucker" is to increase the friction by its large area (Hynes, 1970, p. 134).

Fig. 55. Mayfly nymph and the *Blepharoceridae* larvae and pupae in natural pose on the stone. a. mayfly at the edge of stone with its headshield pressed to it, near the *Blepharoceridae* pupae. b. mayfly on flat granite boulder, near the *Blepharoceridae* pupae, *Blepharocera* asiatica and *Tianschanella monstruosa*) (Photo by V. Bukin).

and higher. Another feature common to the species of this genus consists in that the larvae, *e.g.* of the species *Iron montanus*, bear large straight or bent spines on their abdominal tergites. Their function is unknown, but similar formations occur also in larvae of *Blepharoceridae* and in some mayflies and, in all probability, it is a form of adaptation to life in conditions of high flow velocity. Perhaps these spines, which may be arranged in one, two or more rows, in some way break the water into separate jets.

112

In nymphs of species of the genus *Rhithrogena* and, in particular, of the *Rh. tianshanica* widespread in the Tienshanian torrents, the adaptations described for the genus *Iron* are less pronounced, although they retain their streamlined body shape and a broad head shield (Fig. 56), but they have three filaments, not two. The gill fringes are less thickened at the outer end and do not form such a perfect false sucker as in species of the genus *Iron* (Fig. 50). Still, when compared with other species of the genus *Rhithrogena*, the *Rh. tianshanica* seems to be a true inhabitant of mountain torrents, although it probably prefers less rapid places in a torrent and microbiotopes with slower current than larvae and nymphs of the *Iron* species.

The larvae and nymphs of the genus *Ephemerella* differ in their appearance from the above genera *Iron* and *Rhithrogena*. Their bodies are not flattened, the gill fringes are highly reduced and shifted onto the dorsal surface, their head shield is not broad and they have three filaments which are short and weak. The spines on the abdominal tergites are very short. Yet, the *Ephemerella* inhabits mountain torrents, although farther downstream from the headwaters than the *Iron* and *Rhithrogena*; they seek for quieter microbiotopes. They differ from these genera in their adaptations to life in torrents, namely, they have very robust legs, especially the fore legs, with powerful thickened femora (Fig. 54). The legs bear rigid claws. Here, the adaptation seems to have taken a different way, *viz.* at the cost of developing the strong claws. In spite of a different body shape, the nymphs of the *Ephemerella* species closely press themselves to the stone surface or,

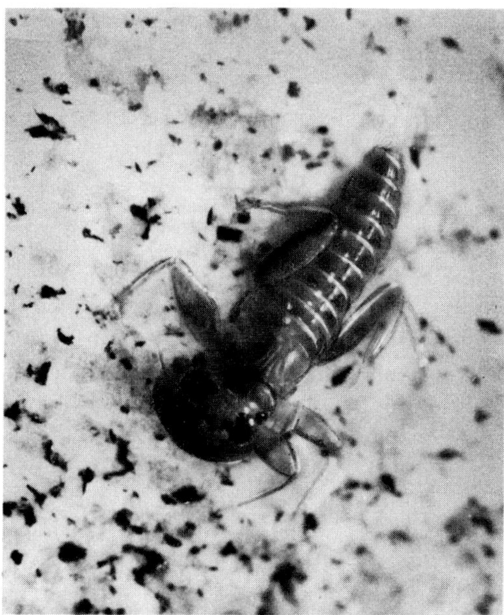

Fig. 56. Mayfly nymph *Rhithrogena tianschanica* on granite bounder. The head shield is clearly visible. Its filaments are not seen against the light background (Photo by V. Bukin).

more correctly, cling to it. In these species the *Eph. submontana* should be considered as the most typical dweller of mountain torrents. Nymphs of another species found in the lower reaches of torrents (*Ephemerella* sp.1) have much less robust legs, with weaker femora of the forelegs and they occur in relatively slow water.

Of much interest was the discovery of larvae and nymphs of some species of the *Baetis* genus in torrents. Nymphs of this genus are not flattened, but rounded in the cross-section, they have three well-developed and fairly felted filaments and thin delicate gills. The head shield is absent. All their appearance indicates that they are good swimmers and well-adapted to locomotion in slack waters. The *Baetis* larvae are low-resistant because of their streamlined bodies but the same streamlining is also advantageous when they remain motionless in swift waters.

In nymphs of the *Baetis* genus which are typical dwellers of mountain torrents one can observe modifications in comparison with species of the same genus living in slack or standing waters. They have lost the third (central) filament, the other two being felted only on their inner side. The gills are reduced and give very slight resistance to the water. The same features are found in nymphs of the *Baetis* species from the Tien Shan, the Himalaya and other mountain regions. Larvae and nymphs of the *Baetis* species cannot, of course, withstand such strong currents as the *Iron* species, but using suitable microbiotopes they are numerous both in species composition and in individuals living in torrents.

Similar peculiarities in body constitution are found also in nymphs of the *Ameletus alexandrae*, a species which occurs both in lakes and in torrents on the Tien Shan, Karakoram, Hindu Kush and the Himalaya. Owing to the streamlined body shape the nymphs of this species are capable of swimming in standing waters.

As to the behaviour of mayfly nymphs and adults, we can only say that they have adapted themselves to the complex ecological conditions of mountain torrents. In particular, the problem of imago hatching in conditions of rough and rapid currents has been solved. The hatching of mayfly adults is known to take place in the water, rather than on the nearshore stones or vegetation as in the stoneflies. When the time comes for emergence, the mature mayfly nymphs move closer to the water surface or towards the shore, but stay in the water. The sub-imagoes emerge in an air bubble, leaving its skin shed on a stone under the water. Interesting observations of the hatching of mayfly subimago from nymphs were made in a mountain torrent of glacial origin (the Alni River) in the north-western Himalaya (Dubey and Kaul, 1971). Unfortunately, the authors have not reported the exact names of the species they dealt with, but only the families *Baetidae* and *Ecdyonuridae*. The hatching occurred during a short time in the afternoon, close to the evening. Before hatching, as the authors describe, the nymphs leave their shelters under stones, swim in zigzag towards the water edge where they crawl out on stones clinging tightly to them. The hatching does take place here (the *Baetidae*?). When hatched, the subimagoes spread their wings at once and fly away within a few seconds. In fast turbulent

114

currents the hatching occurred under the water. The imago rises rapidly up
to the surface and twisting its body flies out of the water vertically up to 5–7
metres (the *Ecdyonuridae?*). The eggs deposited by females on submerged
stones or, in some species, on nearshore stones, have an adhesive secretion, by
means of which they stick themselves at once to the stones.[8] Of course, both
eggs and hatched larvae are swept downstream but this is compensated by
the fact that the imagoes, before oviposition, fly upstream.

c. Mayfly nymph distribution along the torrent and some other ecological data

The ecological data on the mayflies from the Tienshanian torrents are still
rather scarce. More or less detailed distribution of species of this order over
the whole length of a torrent (from its glacial headwaters to its disappearance in a desert or a lake) is known for Central and Middle Asia, the Hindu
Kush, Karakoram and the Himalaya from very few investigations: only in
the Tien Shan and only for two rivers, the Issyk and Akbura. The Issyk was
studied only in summertime and the Akbura throughout the whole year.
Although limited, the records allow to compare, for example, the vertical
distribution of mayflies in torrents in the Central and Southern Tien Shan.
Year-round observations in the Akbura River have revealed the seasonal
variation in the mayfly fauna and its ethology during winter time. The data
on mayfly distribution were obtained during August 1968 both in the river
proper (its total length is 190 km), in its heads (the Kichik-Alai River) and
affluents (at absolute heights 900 to 4000 m, *i.e.* over the range of 3000 m).
Other ecological data (number of generations per year, etc.) are based on
observations in the same river during summer and the other seasons of the
year (Omorov, 1973).

Genus *Iron*

Nymphs of this genus are true hymarobionts and very numerous in a
torrent over its whole length wherever it preserves its typical characteristics.
   In the Issyk River this genus occurs from the headwaters down to the
initial and middle sections of irrigative canals (Fig. 57). Differentiation by
species allows us to consider separately the ecology of nymphs of different
species. Thus in the Issyk River the typical form *Ir. montanus* is found at the
uppermost absolute elevations from 2500 to 1250 m at water temperature of
3.8–11.8°; another species, the *Ir. rheophilus*, occurs within the altitudinal
range of 1440–1050 m at water temperature from 11.5 to 15.5°; *Ir. nigromaculatus* lives farther downstream and chiefly occupies the lower river
and irrigative canals at 1020 to 900 m at water temperatures from 11.2 to
21.2°. The first two species, as true torrential dwellers and coldwater
stenotherms, keep close to the snowy and glacial sections of the river.
Nymphal abundance of these species greatly diminishes at 1330 m where
water temperature exceeds 7.0–11.0°.

[8] When depositing eggs, some mayflies "dive" into the water, *i.e.* drop down and then take off
upwards at such speed that the current actually has no time to engulf the mayfly female.

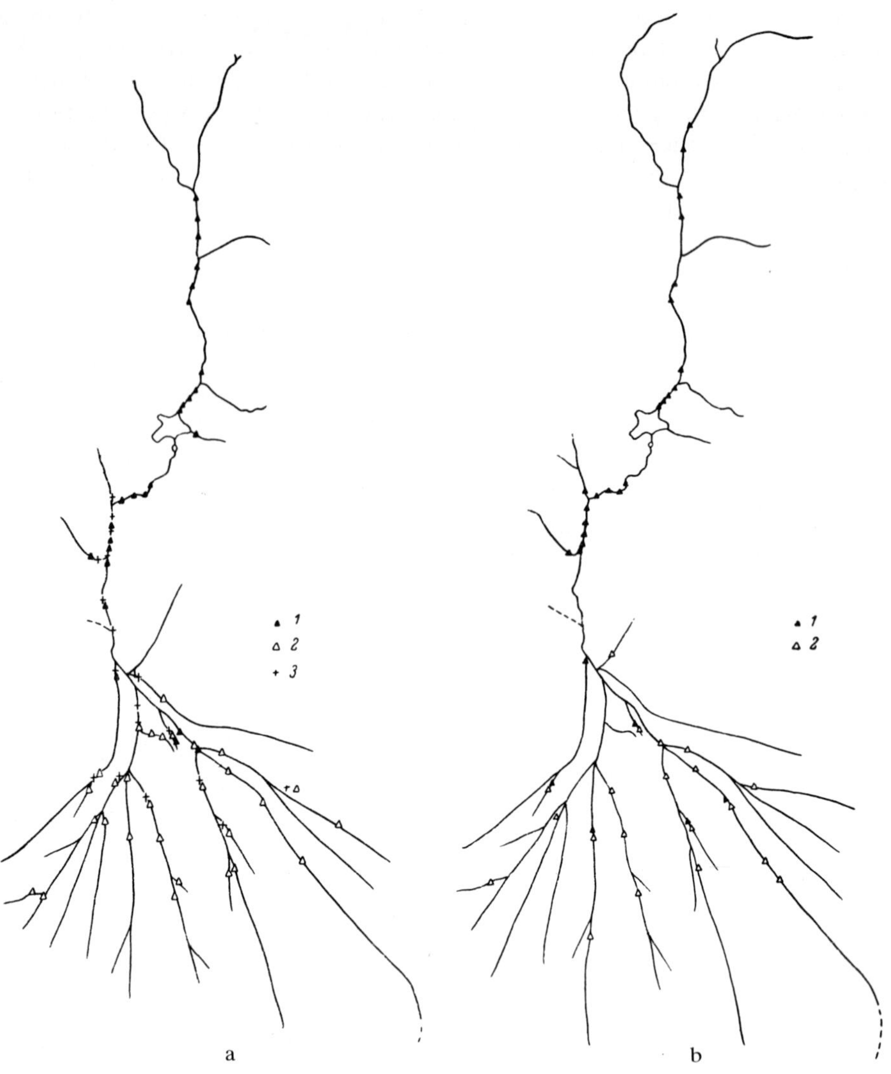

Fig. 57. Mayfly longitudinal distribution profile in Issyk River.
a. 1, *Iron montanus*, 2, *Ir. nigromaculatus*, 3, *Ir. rheophilus*;
b. 1, *Rhithrogena tianschanica*, 2, *Rhithrogena* sp. 1;

Distribution of different species of the *Iron* genus in the Akbura River is as follows (Figs. 58, 59 and 60). The *Ir. nigromaculatus* is most abundant in the lower half of the middle reach and, progressively diminishing in numbers, it disappears at the point of transition of the river from the middle to the upper reach. In contrast, the *Ir. montanus*, which is rather rare at the lower limit of the upper reach, increases considerably, reaching its maximum abundance (becoming the only species of the *Iron* genus) in the upper section of the upper reach. *Ir. rheophilus* seems to occupy an intermediate position, since it occurs in the upper middle and the lower upper stream.

116

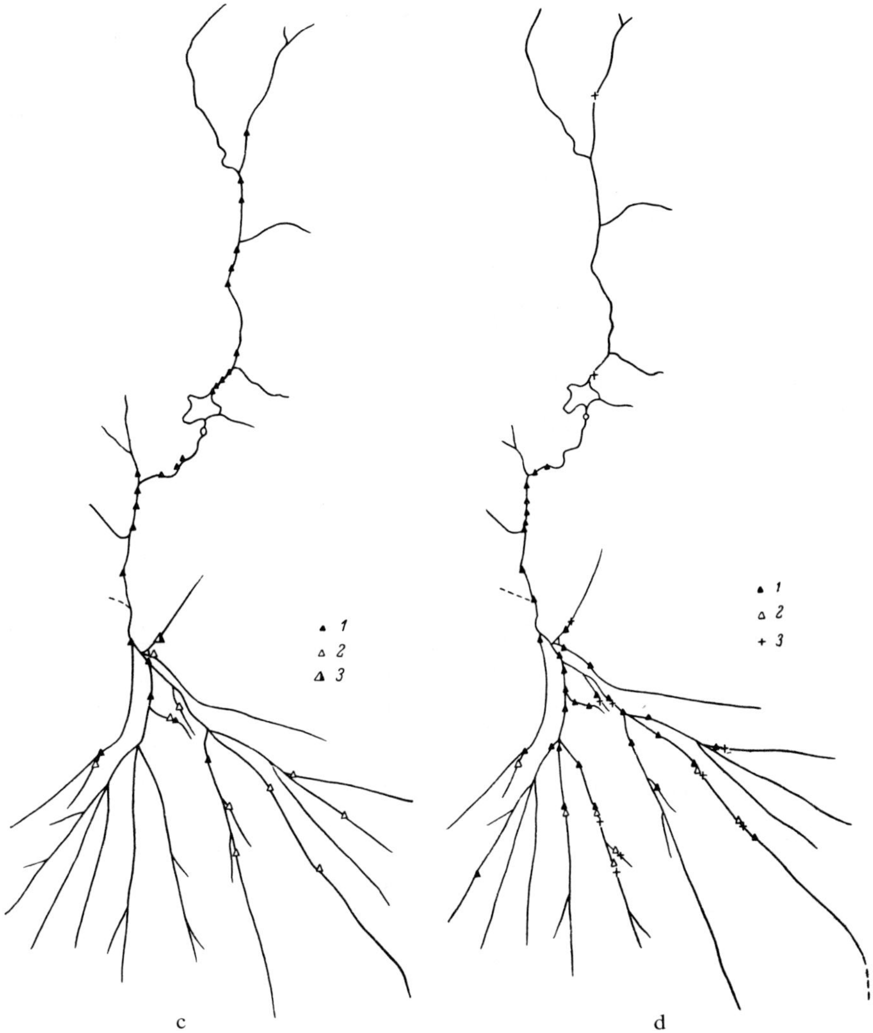

Fig. 57. (continued)
c. 1, *Cinygma* sp., 2, *Ecdyonurus* sp. 1, 3, *Edcyonurus* sp. 2;
d. 1, *Ephemerella submontana*, 2, *Ephemerella* sp. 1, 3, *Ameletus alexandrae*.

In view of such distribution in the torrent, the *Ir. montanus* should be characterized as a high-mountain species, the *Ir. rheophilus* as a mountain one and the *Ir. nigromaculatus* as a submontane species.

The *Ir. montanus* and *Ir. rheophilus* stations are the most typical parts of a mountain torrent; they live on stones, rocks, pebbles, being exposed to very fast and rough currents; they are absent from still bays and near to the banks, where the current is notably slower. They are abundant in cascades, waterfalls and in the channel line. The *Iron* nymphs inhabit places almost inaccessible for other mayfly species. They prefer the upper surface of

117

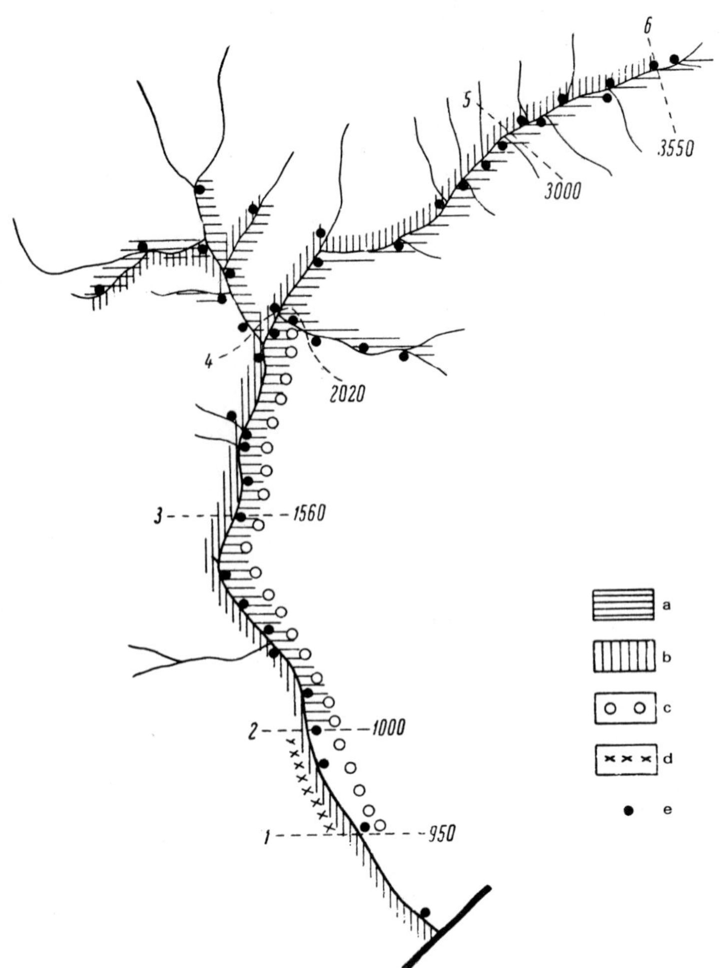

Fig. 58. Mayfly longitudinal distribution profile in Akbura River and its headwaters, the Kichik-Alai River.

1. lower limit for *Ephemerella submontana* and *Ephemerella* sp. 1; 2. upper limit for *Ephemerella* sp. 1 and lower limit for *Iron* spp.; 3. lower limit for *Ameletus alexandrae*; 4. upper limit for *Ephemerella submontana*; 5. upper limit for *Ameletus alexandrae*; 6. upper limit for *Rhithrogena tianschanica*. a, *Iron* spp., b, *Rhithrogena tianschanica*, c, *Ephemerella submontana*, d, *Ephemerella* sp., e, sampling sites. The numbers are absolute heights in m.

stones, rocks, smooth boards in bridges and similar conditions which are so common in torrents.

The younger the larvae of the *Iron* species, the nearer to the bank they occur. The mature nymphs, except for the date of subimago hatching, inhabit deeper waters of 30–60 cm and such fast currents that they become almost inaccessible for the observer.

The imago can be seen very rarely, they keep to the very surface of the torrent, usually near its centre. They fly upstream. Their unusually long filaments are generally stretched with wind.

118

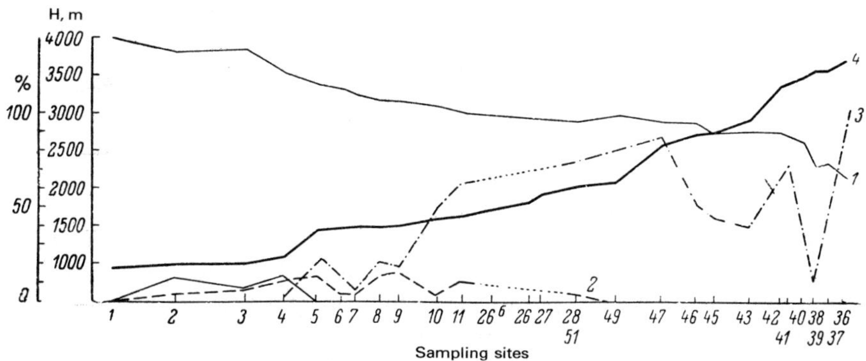

Fig. 59. Mayfly distribution in Akbura River and its headwaters, the Kichik-Alai River. Specimens of each species are shown as percentage of the mayfly nymph total per sampling site.

1. *Ephemerella* sp. 1; 2. *Ephemerella submontana*; 3. *Iron* sp.sp.; 4. elevation H in metres above sea level. Distance between the sampling sites is in a scale corresponding to the real distance on the river.

Year-round observations made in several sites of the Akbura River have shown that nymphs of the *Iron* genus occur at any time but they are much less numerous in winter than in summer. It may be assumed that the *Iron* produces at least two generations per year, in the middle and at the beginning of the lower reach. At the downstream distribution boundary of nymphs of this genus the peak of their abundance is observed in May-July, whereas upstream it is delayed until September and, farther upstream, until the end of September.

The first generation appears to develop during the spring time and the second one starts in late June, which accounts for the fact that the highest

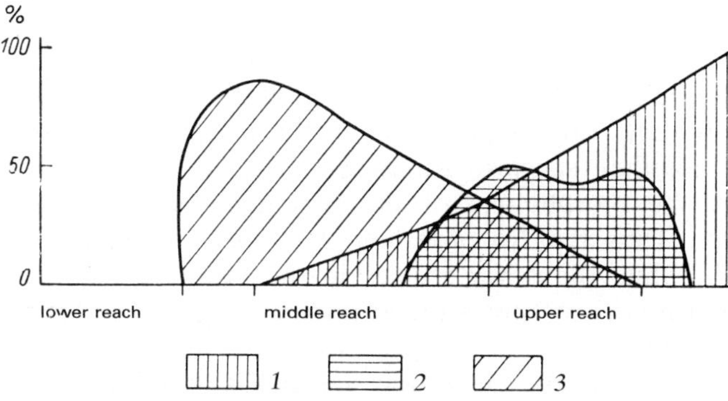

Fig. 60. Distribution of the Iron mayflies in Akbura River. Specimens of each species are shown are expressed as percentage of the total of specimens in all species of the genus.

1. *Iron montanus*; 2. *Ir. rheophilus*; 3. *Ir. nigromaculatus.*

119

abundance of nymphs is observed from April to October. Farther upstream there is probably only one generation per year.[9]

Genus *Rhithrogena*

In the Issyk River nymphs and larvae of the *Rh. tianshanica* occur from the headwaters (more precisely, downstream of its affluent Djarsu) down to the lower reach and even in the initial sections of irrigative canals, provided the canals have the character of a mountain river. At absolute elevations, the distribution range of this species is about from 1100 to 2500 m (Fig. 57). The general character of the torrent sections where nymphs and larvae of this species occur is very typical of habitats of rheobionts: fast, but some-what less rough current in the nymphal habitats with large rounded stones and boulders at the bottom.

It is interesting that in the uppermost Issyk River there are no nymphs either of this or any other species, whereas they are abundant, at the same elevation, in the Djarsu affluent of the Issyk River. The difference between the Djarsu and the upper Issyk River above their confluence is this. The Issyk River bed is formed by rounded boulders of gray or red granite; the current is fast, about 2 m/sec in average but much higher in some jets; 95–100% of the river surface is covered with foam; the water is transparent (because of a low content of suspended particles); water temperature at 13 hours in August is 7.0°. The Djarsu River is shallower, very turbid, with the temperature of 5.0° at the same hour.

If compared with other species, the *Rh. tianshanica* is most abundant in the range 1780 to 2000 m at water temperature of about 6.0–7.5°. Oxygen content in the water is 8.7–9.3 mg/l. At elevations from 1320 to 2500 m in the temperature range of 5.0–11.0° there is lesser amount of the nymphs.

Another species, the *Rhithrogena* sp. (probably *Rh. brodskii*) occurred only in the lowest Issyk River and in the heads of its irrigation canals.

In the Akbura River the *Rhithrogena* genus is widespread over its whole length from the headwaters down to its inflow into the Southern Ferghana canal (Fig. 58) and is represented by numerous nymphs and larvae. In the upper, the middle, and at the beginning of the lower reaches the *Rh. tianshanica* is very common, whereas in the lowest section this species is replaced by another species (*Rhithrogena* sp.1), as evident from some difference in the morphology of the two nymphs and other ecological characteristics. Thus the *Rh. tianshanica* is most abundant in the upper middle reach, while the other species in the lower reach. In the Akbura River *Rh. tianshanica* occupies the altitudinal range from 1560 to 3000 m, i.e. about 1500 m.

At all sampling sites the mean flow velocity is very high: the min. about 1.47 and the max. over 3.0 m/sec. Mature nymphs occur at depths from 30 to 60 cm, but when the time comes for the subimagoes to hatch, the nymphs

---

[9] Data on the seasonal cycle of mayflies from mountain torrents are almost absent in the literature (Lyman, 1955).

move towards the shore where the hatching takes place on the nearshore stones (underwater).

The younger the larvae, the nearer to the shore they occur. The *Rh. tianshanica* nymphs were found exclusively on stones and pebbles, they usually prefer the lower and the lateral surfaces of stones. In no case were they found in moss, on sand or on a silty substrate.

The adults appear to emerge not earlier than June; the emergence period is long and no massive aggregations of the imago were observed. In the insectary, the mature nymphs with black covers on the wings became subimagoes in 3 or 4 days. However, no imagoes could be obtained in the insectary where the nymphs had been contained for some days.[10]

The data obtained on the seasonal dynamics of the fauna in the middle Akbura River suggest that the *Rh. tianshanica* has either one or two generations per year, depending on the absolute height of locality. Well-pronounced is the peak of nymphal abundance which occurs in June and July, but upstream it is delayed until September. This peak is probably due to the growth of the second (summer) generation which follows the winter generation that gives the abundance maximum in April and May. Upstream the abundance of nymphs in spring time is not great, which seems to be due to the lower water temperature; in summer months there is only one generation here, the autumn-winter. Larvae and nymphs of this generation stay for the winter in the river, and the adults emerge, probably, during May and June the next year.

Genera *Ecdyonurus, Cinygma* and *Cinygmula*

Because of the absence of any species differentiation of nymphs of these genera, specifying their ecology would be of little value. We can say only that the *Ecdyonurus* and *Cinygma* nymphs were found in the Issyk River at elevations from 830 to 2230 m at water temperature of 5.0–20.5°. The preliminary study of the nymphs allows to distinguish between three species: *Cinygma* sp.1 and *Ecdyonurus* sp.1 and sp.2. Their distribution throughout the Issyk River can be characterized as follows: *Cinygma* sp.1, at 1100–2330 m, water temperatures 5.0–15.8°; *Ecdyonurus* sp.1, absolute height 1280 m, water temperature 14.2°; *Ecdyonurus* sp.2, at 830–1120 m, temperatures 12.5–20.5° (Fig. 57c).

The *Ecdyonurus* in the Issyk River is a coldwater stenotherm, rather than a rheobiont. In nymphs of this genus, as well in the *Cinygma*, microbiotopes are stones and boulders; however they prefer to hold on the lower surface or the downstream side of the stone. The torrential sections where species of the genera were found have quieter and slower current than the habitats of the *Iron* and the *Rhithrogena* nymphs.

In the Akbura River, the *Ecdyonurus* genus is widespread from the lower reach to the lower middle reach, *i.e.* approximately up to the absolute height of 1500–1700 m. The temperature range for this genus is wider than for the

[10] The imago of this species has been grown recently by Kustareva (1976).

Cinygma, viz. about 6° and up to 17–18° in total. In the same river the *Cinygma* genus is restricted to the upper reach, occurring only sporadically. It seems that here this genus is represented by a species other than in the Issyk River.

Genus *Ephemerella*

Two species, the *Eph. submontana* and the *Ephemerella* sp.1 found in the Issyk and Akbura Rivers, notably differ in their ecology (Figs. 57–59). In both rivers, their longitudinal distribution is nearly the same. In the Issyk River the *Eph. submontana* occupies the middle and part of the lower reaches (absolute heights 880–1560 m, water temperatures 10.5–20.5°), whereas the *Ephemerella* sp.1 occurs only in the lower reach (absolute heights 900–950 m, water temperature 15.8–21.2°).

In the Akbura River the *Ephemerella* sp.1 is a form inhabiting the lower waters, and it is the only section where this species was found. The species is confined to low elevations (about 1000 m) and high water temperatures (15–20°). The population of this form is very small. The upper limit of the *Ephemerella* sp.1 distribution coincides with the lower distribution limit of all the *Iron* species which were found in the Akbura River. Thus, this form of the *Ephemerella* genus, incidentally, having a much more delicate constitution than the *Eph. submontana*, is representative for the lowest Akbura River, *i.e.* where the torrent turns into a mountain river.

The distribution of the *Eph. submontana* in the Akbura River is confined from beneath to the area occupied by the *Ephemerella* sp.1, *i.e.* again to the lower waters (absolute heights of 950–1000 m), but from here the former species spreads upstream to the height of 2020 m above sea level. The upper distribution limit of the *Eph. submontana* in the Akbura River is at the lowest part of the Kichik-Alai River (the Akbura River heads), but this species is absent from the Akbura River right affluent, the Chogom River, lying always higher than 2000 m above sea level. The temperature range for the *Eph. submontana* is 9.5 to 19–20°. Being very abundant in the lower waters and in the lower middle reach, this species justifies its name of a typical submontane form.

All the species of the *Ephemerella* genus are rheobionts to a lesser degree, and they are not such typical dwellers of mountain torrents as the mayflies previously described. The two species of *Ephemerella* live both in the nearshore zone of the torrent and somewhat further off the bank in deeper places. As a rule, the younger larvae hold closer to the bank. As the substrate to live on, the nymphs use pebbles of different size and prefer the lower and lateral surfaces. The gills in nymphs are placed on and pressed closely to the upper surface of their bodies; therefore, the gills are prevented from being injured with vegetation, sharp stones, *etc.* Their legs are long, set widely apart and very robust, the fore femora are broadened and provided with small projecting structures. The strong claws in nymphs serve as a very efficient means of attachment to roughness on the substrate. The *Ephemerella* nymphs are not numerous in winter, but they are present in the

122

river all year round. It is still hard to say how many generations the species of this genus produce every year, but there is a suggestion, which needs verification, that there is only one generation in the Akbura River. All available data indicate that the *Eph. submontana* is a rheobiont and a coldwater stenotherm to a higher degree than the *Ephemerella* sp.1.

Genus *Baetis*

Nymphs of this genus in the Middle Asiatic fauna are not adequately known and now we have no possibility to compare them with the species in which only imagoes have been studied. For this reason, the available ecological data characterize the genus *in toto.*

Distribution over the longitudinal profile in the Issyk and Akbura Rivers is very wide, and nymphs and larvae of different species of this genus can be found from the headwaters to the lower reach because of the presence of ecologically different species.

Larvae and nymphs of this genus prefer such substrates as stones and pebbles, their lower and upper surfaces, but the *Baetis* is also encountered in moss where the young stages, the larvae, sometimes very numerous, dominate. In the Issyk and Akbura Rivers all the species of this genus occupy the nearshore zone in which they dominate in numbers greater than any other mayfly species.

If considered throughout a year, the curve of abundance in larvae and nymphs of the *Baetis* appears three-humped with the peaks in the following periods: the first one falls on March-April, the second on August and the third on October. Until identification of each species by their nymphs is made, it is impossible to discover the number of generations in them.

Genus *Ameletus*

In the Issyk and Akbura Rivers the *A. alexandrae* and the *Ameletus* sp.1 were found from this genus.

The first species was described by us with reference to the imago from the Issyk Lake (the Zailiiskii Alatau; Brodsky, 1930a). Nymphs of this genus were described in detail on the basis of records by the Japanese expedition to the Karakoram and the Hindu Kush in 1955, from the Shiva Lake, the Ishkashim and the Pagman (Uéno, 1966). But, despite the fact that this species has been more fortunate in this respect than any other mayfly species, its ecology is a riddle. In the Issyk River this species is encountered in a wide range of heights and temperatures (Fig. 57): from 2230 to 850 m and from 5.0 to 21.2°, *i.e.* from the river headwaters down to the irrigation canals. The mayfly list for the Issyk River contains still another species of this genus, the *Ameletus* sp.1 with a very restricted distribution area: the headwaters or, more correctly, the affluent, the Djarsu River (2900 m above sea level, water temperature 2.0°).

In the Akbura River the *A. alexandrae* occurs more frequently than in the Issyk River, almost at every study site, but its distribution is more limited than in the Issyk River: from 2710 to 2560 m at water temperatures 3–4 to 9–10°. In an affluent of the Akbura River, the Chogom, this species is restricted in its vertical distribution to the heights about 2600 m, *i.e.* like in the Akbura main channel.

A peculiarity of the ecology of this genus is that its nymphs are of well-defined swimmer type, but they frequently occur not only in lakes, but also in mountain torrents. It is true that they generally live near the shore, but they do not avoid the river sites with rough and fast currents. In the Issyk River this species was found only in pools and in the nearshore zone with very slight current, among stones covered with silt. Many nymphs were encountered both in the large and small Issyk lakes. On the shores of these lakes the massive emergence of adults (males) was observed, which is not typical of true rheobionts.

As for the seasonal distribution, the observations in the Akbura River have shown the nymphs occur, besides the summer months, in February-April and in October and November.

On the basis of the above data, one can give the mayfly distribution pattern in the Akbura's affluent – the Chogom River. This affluent runs at absolute heights from 2000 to 2800 m and therefore lacks species which are typical of the lower and the middle reaches: *Eph. submontana* and, certainly, *Ephemerella* sp.1. The *Iron* dominates quantitatively here; the nymphs of the *Ecdyonurus* genus which are not found above 1500–1700 m in the Akbura are present in smaller numbers; *A. alexandrae* is restricted to the heights of 2300–2600 m, as in the Akbura River.

There is a certain zonality in mayfly distribution in the Akbura River proper and in its upper part, the Kichik-Alai River. The following zones can be distinguished: (i) the lower zone which is characterized by the presence of the *Ephemerella* sp.1 and absence of the *Iron* genus; then (ii) the middle zone where the *Eph. submontana* is present; and (iii) the upper zone with quantitative dominance of the *Iron* mayflies and absence of the *Ephemerella*. This zonality is similar to that found in the Issyk River with one difference, however, that in the Akbura River these zones are more extended and displaced upwards in the absolute heights. The latter seems to be due to the more southern position of the Akbura River compared with the Issyk River and to a more arid climate in the Akbura basin. There is little evidence concerning the vertical distribution of mayflies in torrents of mountain regions adjacent with the Tien Shan and the Alai-Pamirs. There are only random references in descriptions of new species (*e.g.* by Kapur and Kripalani, Uéno, *etc.*), but no generalizations about the mayfly distribution with height are available. Some references one can find in Mani (1968), namely: (i) the *Baetis chandra* larvae encountered under stones in glacial torrents and at the glacier edge are confined to 2750–3660 m above sea level; (ii) the *B. hymalayana* occurs at height of 3200 m; and (iii) the *Iron* sp. in the north-eastern Himalaya at absolute height of 3970 m. But these single data by no means imply that the indicated heights represent the boundaries of the vertical distribution of mayflies. Thus the above author (*loc. cit.*) notes that in his numerous expeditions to the Himalaya the insects were found as high as 6800 m, but he suggests that they may occur still higher. On the Tien Shan the upper vertical distribution boundary of mayflies is mostly determined by the permanent snow line (see Chapter I).

2. *Trichoptera* (caddisflies)

a. Species composition

The caddisflies, the second most significant, after the mayfly order, group of invertebrates in "true" mountain torrents, are present in swift waters in great abundance both in respect to the number of species and especially to the number of individuals.[11]

If one compares the mayfly order with caddisflies, one is able to see also some difference in their ecology and, therefore, in abundance. But we shall discuss this later; here we only point out that the mayflies are capable to inhabit and live in torrents with a periodically moving substrate, because they are very mobile themselves, whereas the caddisfly larvae, and, especially the pupae, are fixed in most species in their cases and need a stable substrate. In torrents the latter condition is met with not everywhere. That is why the caddisflies rank second against the mayflies in terms of their significance in the fauna of swift waters. Frequently the abundance ratio of these two groups allows identification of the watercourse type.

Passing over to the discussion of the caddisfly species composition in mountain torrents of the Tien Shan, we emphasize again what we have said about the mayflies that the preimaginal stages in caddisflies from torrents of this region (and in general in the global caddisfly fauna) are still poorly understood and that many caddisfly species lists based on analyses of the larvae and pupae collections mostly give only generic names. "The number of species with the known larvae of the *Rhyacophila* genus is much less than that of the world-wide caddisfly fauna" (Lepneva, 1964, p. 141). Incidentally, the *Rhyacophylidae* family, so representative of fast waters, amounts to over 360 species, from which only a small proportion is known with larvae.

In the USSR, the adult winged caddisflies, including those from fast watercourses, in particular, from the Middle Asian streams, were an object of study by Martynov (1927a, b, 1930a, and others). The preimaginal stage of this order has been studied for many years by Lepneva (1964, 1966). Unfortunately, the caddisfly fauna of the Tienshanian streams at the moment of guide compilations was rather inadequately known with respect to the preimaginal stages and has been reported only partially in the "Fauna of the USSR". For example, from the *Agapetus* genus such species as the *A. kirgisorum* Mart., *A. tridens* McL., *Agapetus* sp. were found in Tienshanian streams, whereas the guide describes larvae of only two, but different, species (*A. fuscipes* and *A. comatus*, both from western Russia). Among the

---

[11] With respect to this order the Hymalayan glacial torrents occupy a more modest position in the "community" of streams (Dubey and Kaul, 1971).

mountain regions of the USSR, the Caucasus is better represented than Middle Asia.

One can get some idea of the available knowledge of caddisfly larvae of Middle Asia from the few statistical data and the list of species presented in Lepneva's above-mentioned monograph. For Middle Asia only 24 species are reported, among them 10 species from mountain streams, brooks and rivers, which makes up 0.4% of the total number of caddisfly species described in the "Fauna of the USSR" with reference to the larvae. The total list of species (larvae) from swift waters of Middle Asia is so small that it is useful to give it here:

Species	Locality
Annulipalpia	
Rhyacophila sp. "larva praebranchiata" Lepn.,	Chirchik R., Talasskii Alatau
Rh. obscura Mart.,	Middle Asian Mts., Iran, Pakistan
Himalopsyche gigantea Mart.,	Middle Asian Mts.
H. "larva *hoplura*" Lepn.,	Middle Asian Mts.
Agraylea pallidula McL.,	Middle Asia, Caucasus, Europe, Iran
Integripalpia	
Apatania[12] *copiosa* Mart.,	Middle Asian Mts.
Micropterna nycterobia McL.,	Middle Asia
Dinarthrum pugnax McL.,	Middle Asia
D. reductum Mart.,	Middle Asia
Oligoplectrodes potanini Mart.,	Northern Kazakhstan, Siberia, Mongolia (in middle Asia, a variety of this species)

Some contribution to the knowledge of caddisfly fauna from the Tienshanian torrents has been made by Sibirtzeva (1958) who examined samples of caddisfly larvae and pupae from the Chirchik River.

The question arises: how can we identify caddisfly larvae and pupae from swift waters in Middle Asia if their fauna is so poorly known? In this, some assistance may come from comparison of habitats of the known adult forms and from the similarity to the described larvae; but there is still no more or less complete species list which would allow, as in the case of *Blepharoceridae* and some mayflies, a detailed ecological analysis of caddisfly larvae and pupae (except for a few forms).[13]

To give an idea of the species composition of caddisflies from swift waters of Middle Asia before 1966, we cite the general description by S. G. Lepneva (1966, p. 115, 117), an expert in the preimaginal phase of this order:

---

[12] Like Vallengren, A. V. Martynov has divided the *Apatania* Kolenati genus into the *Apatania* s. str. Wall. and the *Apatelia* Wall. Schmidt (1953) returns to the initial state of the *Apatania* genus (after Lepneva, 1966, p. 110).

[13] We acknowledge the great assistance from O. L. Kachalova (Riga) in determining caddisflies from the Akbura River.

"The springs of the Hissarskii Range and other regions of Middle Asia are inhabited by larvae of *Rhyacophila obscura*,[14] *Glossosoma dentatum, Agapetus* sp., *Dolophilodes* sp., *Hydropsyche* sp., *Apatania copiosa, Pseudostenophylax secretus, Limnophilus asiaticus, Dinarthrum pugnax*, while in more elevated zones, randomly, there occur the *Himalopsyche gigantea* (= *Rhyacophila gigantea*) (Martynov, 1927b; Lepneva, 1945)".

"The Middle Asia: high-mountain and mountain brooks, streams, and small rivers with alimentation from springs, snowfields and, partly, glaciers. One example of a large and powerful snowy stream in the upper zone is the Mazordarja River (Hodga-Obi-Garm region) with water temperature in August about 8.0–10.5°; here the *Himalopsyche gigantea* and *Himalopsyche* sp. "larva *hoplura*" are typical. This genus is widespread in the mountains of Central and Southern Asia, in particular on the Hymalaya; in the USSR it is known up to 3200–3800 m (Lepneva, 1945, 1956, 1960).

There are small rivers with spring alimentation, water temperature 14 to 18° in summer (the Kondarinka and Kolondue Rivers on the Hissarskii Range and rivers in the vicinity of Tashkent city).

Species common to such streams are: *Annulipalpia – Rhyacophila obscura, Glossosoma dentatum*, species of the *Agapetus* (*A. kirgisorum, A. bidens, A. tridens*), *Hydroptila* sp., *Stactobia olgae, Dolophilodes ornata, Tinodes turanica, Hydropsyche stimulans, Hydropsyche* sp.; *Integripalpia – Apatania copiosa, A. zonella, Pseudostenophylax secretus, Astratus alaicus, Dinarthrum pugnax* (Martynov, 1927a; Lepneva, 1951; Sibirtzeva, 1953)".

One can easily see from the citations that the list of caddisfly species covers different types of watercourse, from springs to powerful "snowy" streams, and it is difficult to identify forms representative for the mountain torrent – the type we are concerned with.

At present, it is possible to compare caddisfly faunas from four mountain torrents on the Tien Shan: one from its northern part (the Issyk River), another from the southern region (the Akbura River), still another from the western region (the Chirchik River) and a fourth from the south-western Tien Shan (the Zeravshan River). However, this comparison is rather restricted in view of the fact that the fauna includes a lot of forms not yet identified to a species, which is characteristic of the field material consisting mostly of preimaginal stages (Table 29). The Table also gives the caddisflies from the Chu River and adjoining waterbodies, however some of these caddisflies can not be considered as true torrential dwellers.

One can see from Table 29 that hymarobionts should include certain species from the genera *Rhyacophila, Himalopsyche, Glossosoma, Agapetus, Dinarthrum* and some others. But this is only a preliminary appraisal of the caddisfly species. Certainly, with more detailed study of the preimaginal composition of torrential caddisflies, other hymarobionts may be discovered. The ecology of the forms listed in Table 29 is discussed in the next section of the chapter.

---

[14] The *Rhyacophila obscura* Mart. is a common form in mountain torrents of Middle Asian mountain regions, Iran and Pakistan.

Table 29. List of caddisflies from the Tienshanian rivers.

| Name of species | Chirchik River | | | Zeravshan River | | | | | | |
| | | | | | affluents | | | | | |
	Mountain section	Submontane section	Lowland section	Zeravshan R.	Magian-darya	Kshtut-darya	Issyk R.	Akbura R.	Chu R.	Turgen R. and Chilik R.
Rhyacophila extensa Mart.	–	–	–	–	–	–	+	+	–	–
Rh. obscura Mart.	+	–	–	–	–	–	–	+	–	–
Rhyacophils sp. "larva praebranchiata" Lepn.	–	–	–	–	–	–	–	+	–	+
Rhyacophila sp.	+	+	–	–	–	–	+	–	–	–
Himalopsyche gigantea (Mart.)	+	–	–	–	–	+	+	+	–	+
Himalopsyche sp. "larva hoplura" Lepn.	+	–	–	–	+	+	–	+	–	+
Glossosoma dentatum McL	+	+	–	–	–	–	–	+	–	
Gl. unguiculatum Mart.	–	–	–	–	–	–	–	–	–	+
Glossosoma sp.	+	+	–	–	–	+	–	–	+	–
Agapetus kirgisorum Mart.	+	+	–	+	+	+	+	+	–	–
Ag. tridens McL.	–	–	–	–	–	–	+	+	–	–
Agapetus sp.	–	–	–	–	–	–	+	–	+	–
Hydroptila insignis Mart.	–	–	+	–	–	–	–	–	–	–
Hydroptila sp.	–	–	+	–	+	–	+	+	+	–
Agraylea pallidula McL.	–	–	+	–	–	–	–	–	+	–
Stactobia sp.	+	–	–	–	–	–	–	–	–	+
Dolophilodes ornata Ulm.	–	–	–	–	–	–	+	+	–	–
Dol. trinodes (?)	–	–	–	–	–	–	+	–	–	–
Dolophilodes sp.	+	–	–	–	–	–	–	–	–	–
Hydropsyche stimulans McL.	+	+	–	–	–	–	–	+	–	–
Hyd. ornatula McL.	–	–	–	–	–	–	–	–	–	+
Hyd. gracilis Mart.	–	–	–	–	–	–	–	+	–	–
Hyd. guttata Pict.	–	–	–	–	–	–	–	–	–	+
Hyd. pellucidula Curt.	–	–	–	–	–	–	–	–	–	+
Hydropsyche sp. sp.	+	+	+	+	+	+	+	+	+	–
Arctopsyche sp.	–	–	–	–	–	–	–	+	–	–
Psilopterna pevzovi Mart.?	+	–	–	–	–	–	–	–	–	–
Apatania zonella Zett.	–	–	–	–	–	–	–	–	–	+
Apatelia copiosa McL.	+	+	–	–	–	+	+	+	–	–
Ap. arctica Bch.	–	–	–	–	–	–	+	–	–	–
Brachycentrus maracandicus Mart.?	+	+	–	+	+	+	–	–	–	–
Br. montanus Klap.?	–	–	–	–	–	–	+	–	–	–
Br. subnubilis Curt.	–	–	–	–	–	–	–	–	+	–
Brachycentrus sp.	–	–	–	–	–	–	–	–	–	+
Dinarthrum reductum Mart.	+	+	+	–	+	+	+	+	–	+
Din. pugnax McL.	+	+	–	+	+	+	+	+	–	+
Din. chaldirense Mart.	–	–	–	–	–	–	–	–	–	+
Olig. potanini Mart.	–	–	–	–	–	–	–	–	–	+
Oligoplectrodes potanini var. *excisa* Mart.	–	–	–	–	–	–	–	–	–	+

Note: Chirchik R. after Sibirtzeva, 1958; Zeravshan R. after Sibirtzeva *et al.*, 1961; Issyk R. after original identification by Martynov; Akbura R. after Omorov, 1975 in L. O. Kachalova's identification; Chu R. after Ovchinnikov, 1936, identification by S. G. Lepneva; Turgen R. and Chilik R. after Kurmangalieva, 1976.

b. Notes on adaptations in caddisfly larvae and pupae

Extensive and detailed evidence on caddisfly preimaginal stages can be found in the above cited Lepneva's work. The section on caddisfly larvae and pupae ("Introduction" in Lepneva, 1966) contains much valuable data on their adaptations to different environmental conditions and, specifically, to those encountered in swift waters. These data were gathered from the literature or represent original observations; so, with a clear conscience, we refer to her works all readers who are interested in the way of life and adaptations in caddisflies to swift-water habitats.

Here we shall cite Lepneva's work (1966) and give a few data from Scott (1958), Hynes (1970) and Pennak (1953) in order to compare the adaptations in different groups of the lithorheophilous fauna from mountain torrents.

It is known that larval caddisflies have become adapted to the current factor and widely inhabited all types of running waters from springs to rough mountain torrents. Some species have developed high specialization in the process of adaptations to the highly dynamic environment of torrents, including the powerful rough mountain brooks and small rivers.

As will be seen later, the adaptation in caddisfly larvae is of different character to that in mayfly nymphs. It is largely due to the difference in the mobility of larvae and nymphs of the mayflies and the preimaginal caddisflies. The former possess good mobility and are capable of leaving hard substrate to inhabit stones, pebbles and rock fragments dislodged by current, while caddisflies are devoid of such property. That is why in the caddisfly larvae we do not encounter the so-called false suckers, so typical in some mayfly nymphs, and the caddisflies do not have true suckers like the *Blepharoceridae* larvae. Other adaptations in caddisflies are mainly firm fixation to the substrate, the "ballast" in their cases and formation of silken threads, meshes and elaborated interlacements.

In torrents the most stable, and not infrequently the only type of substrate on which the caddisflies can live, is stones, so it is the stony environment that is so characteristic of larval caddisflies from swift waters.

Larvae in some higher *Rhyacophila* have a peculiar flattened body, whose ventral surface they use for creeping by holding on the roughness of stones and periphyton with claws on the widely sprawled legs and with hooks on the posterior prolegs: the basal hook together with the bended claw forms something like tongs to grip firmly on the object they touch. Such claw-like hooks develop on the hind projects not only in larval *Rhyacophila*, but also in the related species of the *Himalopsyche* genus (Fig. 61).[15]

"The entire gill system in larval *Rhyacophila* is developed on the pleuron section of the abdomen and the thorax (on the dorsal and abdominal surfaces the gills are absent): the elongated clustered gills in these insects are spread outward, trimming the lateral surfaces of the thorax and abdomen. Larval *Glossosomatidae* are algivorous animals which live like the

[15] Larvae of these genera also build meshes of silk.

129

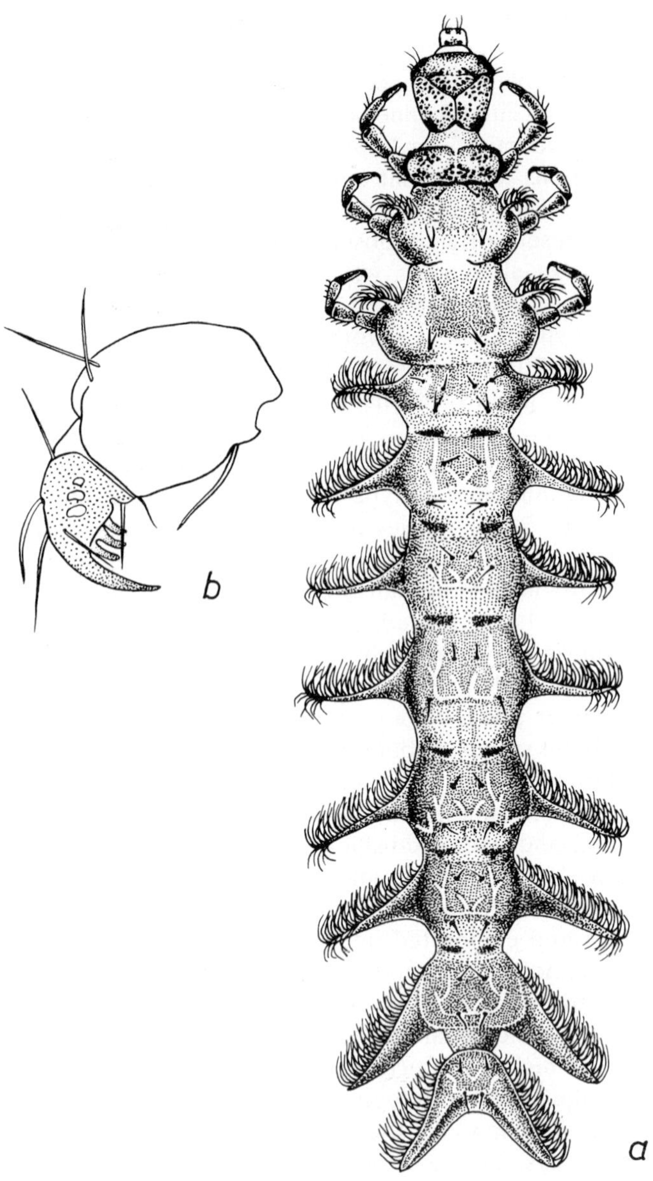

Fig. 61. Caddis-worm *Himalopsyche gigantea* MARTYNOV. a. dorsal view (the rear with hooks is curved ventrally); b. the hook on the right posterior proleg. (a, original; b, after Lepneva, 1964).

Rhyacophilidae larvae on stones in swift waters but are differently adapted to this life; just like the larval *Rhyacophila* which attach themselves to the stone with their ventral surface, the larval *Glossosomatidae* fix themselves to the stone with their cases having a flat lower surface and oval streamlined upper surface. Being between-stream-stone dwellers, the larvae of the genera *Philopotamidae*, *Psychomyiidae*, and *Polycentropodidae* with their

130

lissom bodies devoid of tracheal gills, like the larval *Hydropsychidae*, keep to the structures they build: in *Psychomiidae* these are long and narrow passages or trenches made in the coatings on the stony surfaces, in the *Philopotamidae* these are short tubes slightly broadened at the base, in the *Polycentropodidae* and *Hydropsychidae* variously designed catching nets with a shelter in the form of a special cell or funnel" (Lepneva, 1964, p. 110, 111).

Figures 62 and 63 show caddisfly cases which attract the observer's attention in the fauna sampling because they stand almost vertically resisting the force of the fast current, with one end of the case firmly attached to the stone. Some cases are made of vegetational detritus (Fig. 62a) and seem to belong to the larval *Oligoplectrodes*. The other sandy cases (Fig. 62b and 63) are probably of the *Dinarthrum* larvae. But such identification is very approximate because larvae of different species build similar cases.

Before pupation, the larvae from swift waters attach their cases of fine gravel and sand to a "solid" (*i.e.* immobile) substrate, shortening them and sealing at the ends with small stones and sand grains, as well as perforated membrane. Such cases are made even by caddisflies which do not have cases in the larval stage, *e.g. Rhyacophilidae, Polycentropodidae* and *Philopotamidae* which make pupal shelters using small stones and silk.

A special problem for aquatic stream-dwelling insects is that of hatching which is solved in the following way. On completion of its development, the pupa, using the galeal teeth, destroys the membrane or case wall and leaves the shelter. In some species, the pupae, using their legs, swim rapidly to the bank where the hatching occurs. "In most genera that inhabit swift streams the pupae swim rapidly to the surface where the adult emerges from the pupal integument in a matter of a few seconds" (Pennak, 1953, p. 581). The

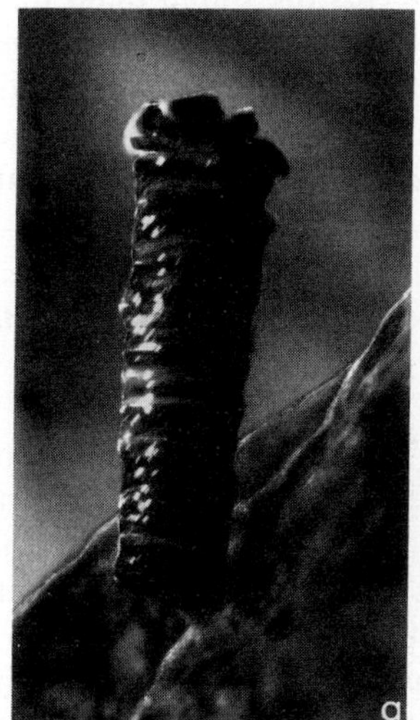
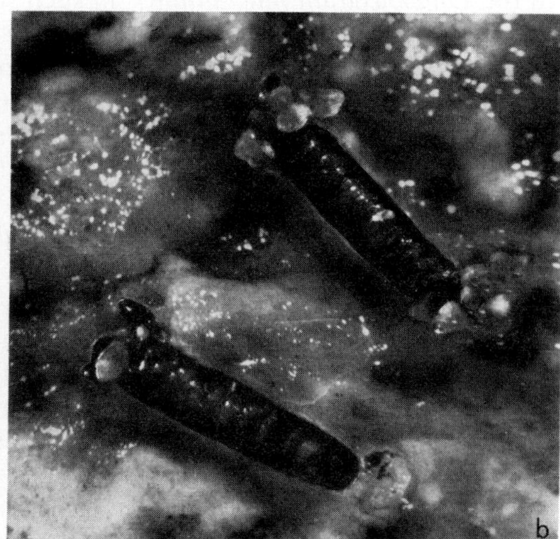

Fig. 62. Cases of caddis-worms (Photo by V. Bukin).
a. *Oligoplectrodes* sp.; b. *Dinarthrum* sp.

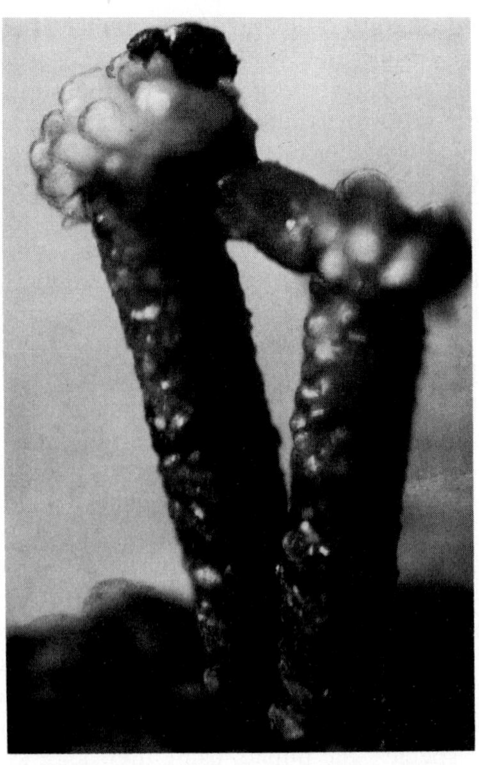

Fig. 63. Cases of caddis-worms (?*Dinarthrum* sp.) (Photo by V. Bukin).

most complex problem arises where the pupa is itself fixed to the substrate
(*e.g.* in the *Oligoplectrum maculatum*). In such species the pupal head in the
case is directed against the current; just in front of the case there is a space
of "dead water" enabling the imago to hatch in the water. For oviposition
many females of caddisflies creep into water and deposit their eggs under
stones. Some caddisfly imagoes, *e.g. Hydropsyche*, are capable of swimming
and diving.

c. Distribution of caddisfly larvae and pupae along torrent and some other ecological data

We have described those genera, some species of which should be considered as torrential dwellers (see Table 29). But the rivers presented in the
Table pertain to different types and vary in respect to the ecological
significance to the lithorheophilous fauna. Thus the Chirchik River in the
western Tien Shan, the Chatkalskii Range, is a large river whose power of
discharge is far superior to the Issyk River; it drains into the Syrdarya River
and has a number of power plants, which regulate its flow. The Chirchik
River has a well-developed mountain portion, the submontane and lowland
sections. In the mountain river we observe fast currents, stony substrate and

132

water temperature between 4.9–15.5° in June and August. Therefore, the caddisfly forms which will be encountered in the mountain section of the river are typical of torrents.

The Zeravshan River at the border between the Tien Shan and the Western Pamirs represents a very large river running between the Zeravshanskii and Turkestanskii Ranges. It is a typical intermontane river having large extension and ecological conditions varying with the distance from the headwaters. The caddisflies known from the Zeravshan River include the forms encountered in the river affluents – the Kshtutdarya and Magiandarya Rivers – which have fast current and stony substrate but relatively high water temperature from 7.5 to 19.0° at sites of caddisfly sampling. Most representative among the torrential dwellers are caddisflies found in the affluents of the river. The caddisfly treatment from the Chu River and its accessory waterbodies is not completed and, therefore, it is hard to get an idea of this order in these waters.

The Issyk and Akbura Rivers and the mountain section of the Chirchik River should be considered as the most typical torrents among those included in Table 29 and their caddisflies as definite hymarobionts. However, even in the Issyk River not all of the species were found in "true" torrential conditions, *e.g. Rhyacophila* sp., *Hydropsyche* sp., *Hydrophila* sp. and *Agapetus* sp. have been recovered only from the lowermost section of the river and from the irrigative canals. *Agapetus* sp. was especially abundant in the canals with stony bottom. The caddisflies occurred in such vast numbers that, forming a continuous layer of their cases on stones, they changed the loose substrate into a "solid" one, *i.e.* they themselves created favourable conditions to produce dense population at some places of the canals. In this one can see a manifestation of the biocenotic factor.

Thus the analysis of the caddisfly list (Table 29), taking into account their ecological conditions, makes possible a general picture of the species composition of typical torrential dwellers, *i.e.* hymarobionts, and of their ecological peculiarities.

First of all, one should single out the species of the *Hymalopsyche* genus, *viz.* the *H. gigantea* and the *Himalopsyche* sp. "larva *hoplura*". These are widespread species in Central and Southern Asia, the Himalaya, in the USSR on the Kirghizskii Alatau, the Dzhungarskii Alatau, Zailiiskii Alatau and the Talasskii Alatau and on the Ferghanskii Range up to absolute heights of 3200–3800 m. Lepneva (1966, p. 106, 107) writes: "Among the well-defined stenotherms in the caddisfly order there are larvae of the Middle Asiatic genus *Himalopsyche* (*H. gigantea, Himalopsyche* sp. "larva *hoplura*") which inhabit snowy and spring waters at temperature of 5–9° in summer. ...The larvae of both species were found in similar thermal and altitudinal conditions in streams on the Hissarskii Range: in the Obi-Hilf (4.6–8.5°, 17–21 VII 1941) and in the Mai-Hur (1.5–6.8°, 25–27 V); the diurnal temperature variation in the Mazordarya stream, one of the places inhabited by *Himalopsyche* larvae, was (according to the nearshore observations at a depth of 20–40 cm, elevation of 2300 m, the Hodja-Obi-Garm health resort, August 1941) as follows: 0800 15 VIII – 8°, 1200 – 12.1°,

1600 – 14°, 2000 – 12.2°, 0800 16 VIII – 8°. Maximum water temperature in habitats of the *H. gigantea* larvae in the Mazordarya was 14.2° (12 VIII 1954) and farther downstream, at temperature of 18.5°, they did not occur".

For the *Himalopsyche* genus one species, the *H. gigantea*,[16] was found in the Issyk River. This larva (see Fig. 61) is very large, 32 mm long (and even up to 34 mm in the Chirchik River), greenish in colour. It builds the case only at the moment of pupation from coarse sand and small stones. The larva lives on stones, in shelters under stones and rocks, more rarely on pebbles, on their lower and upper surfaces. It was never encountered in moss. At higher temperatures, the *H. gigantea*, as a stenotherm, was not found in the Issyk River, but it was found in water with temperature 5.5 to 8.0° at elevations 1740 to 2230 m, *i.e.* up to the very headwaters of the river (Fig. 64). The *H. gigantea* can hardly be called a rheobiont in the strict sense of the word; it should be more properly described as a rheophil, but not as a krenobiont, as also pointed out by Lepneva (1964, p. 117) when she objected to Shadin. We found this species in cold brooks and, more rarely, in springs near the permanent snow line. In spite of a certain ecological plasticity of the species larvae, their abundance in the most typical mic-robiotopes of the mountain torrent makes us refer the *H. gigantea* to the true torrential dwellers, *i.e.* describe them as hymarobionts.

The species of the *Rhyacophila* genus (*Rh. obscura*, *Rh. extensa* and others undescribed), which were found in the rivers Chirchik, Zeravshan (Sibirtzeva, 1958, 1961), Kondara (Lepneva, 1951), and the Issyk and which are generally widespread in the Middle Asian mountains, should be clas-sified as representative of the mountain rivers and torrents (of course, only the above named forms; the genus includes many species with a different ecology). The *Rh. extensa* larva has not yet been described, but some idea can be inferred from the drawing of the *Rh. obscura*: it is a common form in the high-mountain rivers and torrents of Middle Asia (Lepneva, 1966, Fig. 297). The *Rh. extensa* lives on stones and rocks in very fast currents, rarely occurring in the nearshore zone and very rarely in moss. In the Issyk River this species occurs at heights 950 to 1805 m (see Fig. 64a), its temperature range is 7.5 to 17.5°C.

Larva of the variety *Oligoplectrodes potanini* var. *extensa* has not been described and should probably not be classified as hymarobiont: it lives only in the lower Issyk River (see Fig. 64a), mainly in the nearshore zone in relatively slow current.

Among the *Dolophilodes* only one larva has been described (a species from Teletskoe Lake). In the Chatkal River the larvae of this genus, not identified to a species, were encountered at flow velocity up to 2.5 m/sec and for the Kondara River they were found in springs. Therefore, some species of this genus can be hardly classified as typical of torrents because of the scanty and uncertain evidence concerning their taxonomic position and ecology, including the aquatic phases.

---

[16] The larvae from this river were first identified by Martynov.

134

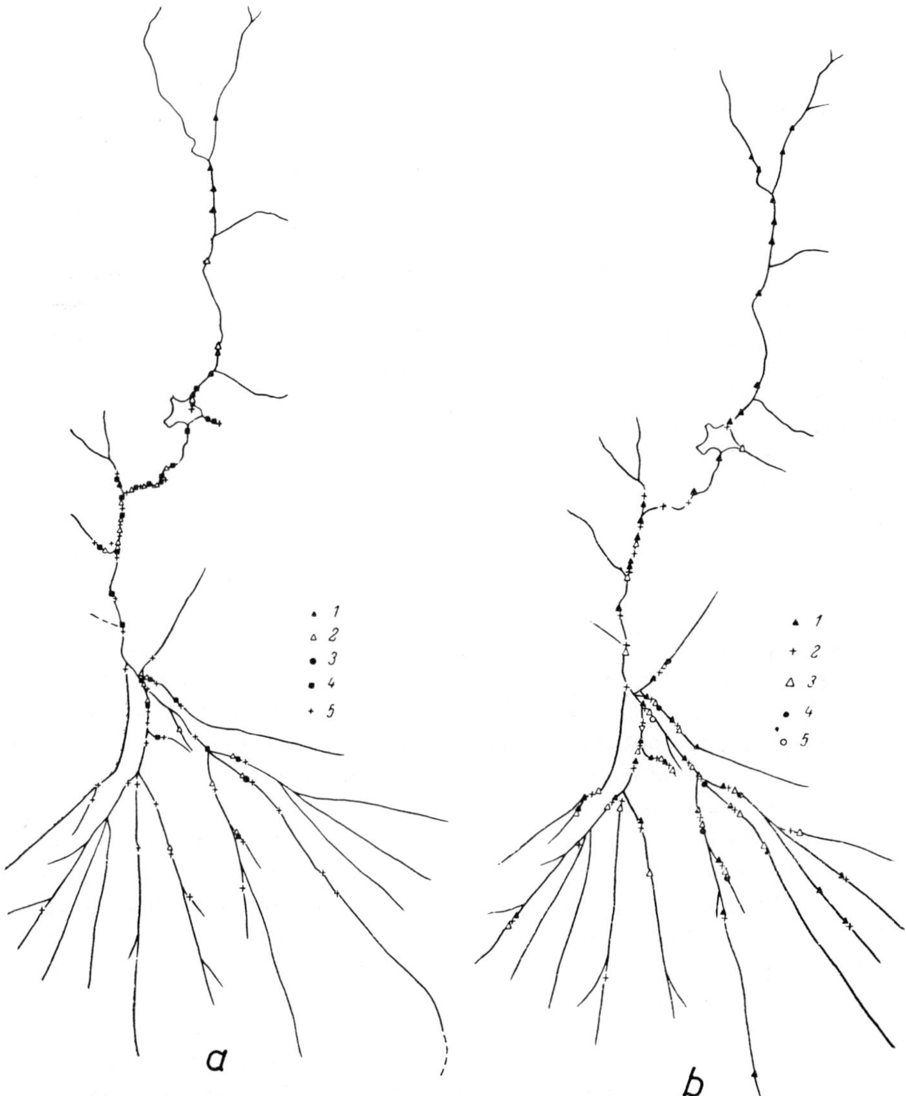

Fig. 64. Caddisfly longitudinal distribution profile in Issyk river.

a. 1, *Himalopsyche gigantea*, 2, *Rhyacophila extensa*, 3, *Oligoplectrodes potanini*, 4, *Dolophilodes* sp., 5, *Dinarthrum* sp.;
b. 1, *Apatelia* sp., 2, *Agapetus* sp., 3, *Brachycentrus montanus*, 4, *Hydropsyche* sp., 5, *Hydroptila* sp.

From the *Dinarthrum* genus, two species, the *D. reductum* and *D. pugnax*, were found in the rivers Issyk, Chirchik, Zeravshan and Akbura. The *D. pugnax* occurs on the Pamirs and in the Kondara River. As Martynov notes, all species of the *Hypodinarthrum* subgenus and almost all species of the *Dinarthrum* genus are generally representative of only Turkestan,

135

namely, of its mountain regions. On the Caucasus there occurs only one subgenus *Paradinarthrum* (Martynov, 1930b). According to Lepneva (1964), both species are common in swift rivers and brooks of Middle Asian mountains. The larval cases are either conical, slightly bended, from fine sand and detritus (*D. reductum*), or straight and tubular, from sand, as shown in Figs. 62b, 63 (the other species: Sibirtzeva, 1958). During samplings from the Kondara River, the *D. pugnax* larvae were recovered from springs at low water temperature. According to Sibirtzeva's and our own observations these two species should be classified as krenobionts rather than hymarobionts.

Into this ecological group of lotic spring dwellers we should also place the *Brachycentrus montanus* and the *B. maracandicus*(?), the latter being found in the Chirchik and Zeravshan Rivers. The larval cases in these two species are tetrahedral, from vegetational remnants and detritus.

The species of the *Agapetus* genus from the Issyk River were also recovered from other mountain regions of Middle Asia; however, among the very diverse ecological conditions they prefer waters with relatively slow current (Fig. 64b). They occur in the Kondara River and springs. Yet we have few grounds to classify these forms as typical torrential dwellers.

Widespread is the *Apatelia copiosa* (= *Apatania copiosa*; Lepneva, 1966), starting from the upper Issyk River, the affluents of the Chirchik River (the rivers Pskem, Ugam, Chatkal), the Akbura River and watercourses on the Hissarskii Range. According to Sibirtzeva, larvae and pupae of this species live on stones at relatively slow current (0.2–0.5 m/sec). Data for the Issyk River show that this species should be considered as representative of mountain brooks rather than of torrents. The larval cases are smooth, conical, from medium-size sand grains. According to Lepneva (1966), typical habitats of this species will be swift cold mountain brooks, the upper reaches of large mountain rivers, rapids near waterfalls, springs and spring brooks. Its distribution area in the Soviet Union is mountainous Middle Asia. Another species of this genus, the *A. arctica*, was found on the Pamirs (Martynov, 1930a). As Martynov believes, this genus is not under any circumstances an arctic one. The larvae of the *Apataniini* tribe occur in small brooks and springs and more rarely in fast rivers. An exception was the *A. arctica* encountered also in standing tundra waters. But the primordial area of its occurrence, as Martynov believes, was the mountainous localities of Middle Asia (Turkestan, the Altai, the Hangai).

It is hard to say anything definite about two members of the genera *Hydropsyche* and *Hydroptila* until we gain some understanding of their species status, but judging from their occurrence in the Issyk River (see Fig. 64b), these are hardly typical dwellers of mountain torrents. Their solitary larvae could be recovered at very few study sites, mainly in places with slight current. Somewhat more frequent was the *Hydropsyche* which was rather abundant in the irrigative canals.

It should be recalled that in the discussion of different caddisfly species from the Issyk and other rivers, their habitats were not always mentioned, but all these forms refer to the lithorheophylous fauna, *i.e.* they live on stones, pebbles, rocks, *etc.*, rather than in vegetation or on other substrata.

To summarize, very few caddisfly forms which have been listed can be characterized as typical dwellers of torrents, *i.e.* as hymarobionts. From all the forms recovered from the Issyk and Akbura Rivers, where their distribution is much alike, only the following species can be described as hymarobionts: two species of the *Dinarthrum* genus, *Himalopsyche gigantea*, *Dolophilodes ornata* and *Rhyacophila extensa*. It should be emphasized again that the caddisflies are a fairly plastic group, and ecologically they are stenobionts to a lesser degree than some mayfly species or some *Diptera*. Vertically, caddisflies of the Tienshanian torrents and brooks occur as high as 4000 m, but this depends on the position of the permanent snow line on the Tien Shan, since in the north-western Himalaya some caddisflies were found at almost 5000 m above sea level (*Pseudohalesus kashmiris* at 4050 m and *Pseudostenophylax micraulax* at 4750 m; Mani, 1968).

The development cycles in caddisflies from the Akbura River have been traced in all-year-round observations (Omorov, 1975) and substantiated the general belief that at small heights many species of this order have one or two generations during the year, whereas at larger heights they have only one generation (such generalization was made also for North American caddisfly fauna; Pennak, 1953).

3. *Plecoptera* (stoneflies)[17]

a. Species composition

The stonefly order plays a less significant role in the torrential fauna than the caddisflies, *Chironomidae* or even mayflies. Judging from the available ecological data, the microbiotopes occupied by the stoneflies in mountain torrents are not typical for true hymarobionts, *e.g. Blepharoceridae* and some mayflies and caddisflies. Stoneflies occur in places with relatively slow current (near the bank, in pools), whereas in rapids they usually live not above but beneath the stone. And still, the stoneflies are a constant component of the lithorheophilous fauna in torrents and deserve certain attention in this essay.

As pointed out above, the invertebrate fauna in Middle Asian watercourses has been studied incompletely and irregularly. But for the species composition of stoneflies, we have fairly comprehensive evidence thanks to Zhiltzova's investigations. In a relatively short time (from 1969 to 1977), "as a result of treatment of extensive information... the species composition of stoneflies of Middle Asia has been elucidated by us rather exhaustively" (Zhiltzova, 1977), while only in 1969 the same author pointed out that "the Middle Asiatic stoneflies are known very inadequately... the stonefly fauna is hardly subject to investigation". Zhiltzova has published a number of

[17] We express our thanks to L. A. Zhiltzova who has critically revised this section of the essay.

works describing 43 species, with 28 new species among them. All this is a result of personal samplings by Zhiltzova who penetrated, mostly alone, inaccessible mountain regions of the Tien Shan and Pamirs where she collected a very large body of data on the stonefly adults and larvae. She travelled largely on foot or on horseback and very seldom by automobile. We fully comprehend all the difficulties connected with such voyages and know from our own experience how hard is the sampling of lithorheophilous fauna in swift mountain torrents and brooks with their icy water. Such selfless activity has been rewarded: at present the Middle Asiatic stoneflies are known quite completely, and description of the species composition of this order is possible thanks to the work of this author (1969, 1970, 1971a, 1972a, 1974, 1976, 1977; Zhiltzova and Zwick, 1971; Grizai and Zhiltzova, 1973).

One can only regret that the author has been largely devoted to the adult forms: of 41 species only 6 have been described and depicted in the larval stage (*Isocapnia aptera, Eucapnopsis stigmatica transversa, Mesoperlina pecirkai, Xanthoperla curta, Rhabdiopteryx tianshanica* and *Capnia pedestris*[18]). *Mesoperlina capnoptera* was found as larvae of different ages, both nymphs and adults, males and females.

We pointed out earlier (Brodsky, 1935) that adults of this species from the Issyk River were somewhat different from the female described by Maclahlan from the Akbura River (the Osh city). These are less intensely coloured; there were very pale individuals and larger than the key specimen. *Isoperla grammatica* was represented only by one, adult specimen. The *Nemoura avicularis* and *Amphinemura standfussi* were found both as larvae and as winged specimens, male and female. The *Leuctra* sp., *Capnia* sp. and *Isogenus* sp. were encountered only as very young larvae, and therefore even identification of the genus is in question.

In Zhiltzova's identification, the following stoneflies were recovered from the Akbura River: *Amphinemura crenata* Kop., *Nemoura* sp., *Phasganophora* sp., *Mesoperlina ochracea* Klap., *M. pecirkai* Klap. (Fig. 65), *Chloroperla curta* McL., *Filchneria mongolica* Klap., *Capnia prolongata* Zhiltzova.

More recently, Kustareva has sampled the lithorheophilous fauna from rivers and brooks in the Issyk-kul Valley. The stoneflies, according to Zhiltzova, include the following species: *Mesoperlina muricata* Klap., *Amphinemura kustarevae* Zhiltzova, *A. mirabilis turkestanica* Zhiltzova, *Mesoperlina pecirkai* Kalp, *Xanthoperla curta* McL.

Although biogeographical data on stonefly larvae from Middle Asia are rather scanty at present, we can use Zhiltzova's data as she was studying in Middle Asia only the fauna of mountain rivers, brooks and, partly, springs. Table 30 lists stoneflies for the Tien Shan and the Pamirs (in the broad sense of the word, *i.e.* both the Western and the Eastern Pamirs). This table does not include only two species described from the Kopetdagh, whose stonefly

---

[18] This species was described earlier.

138

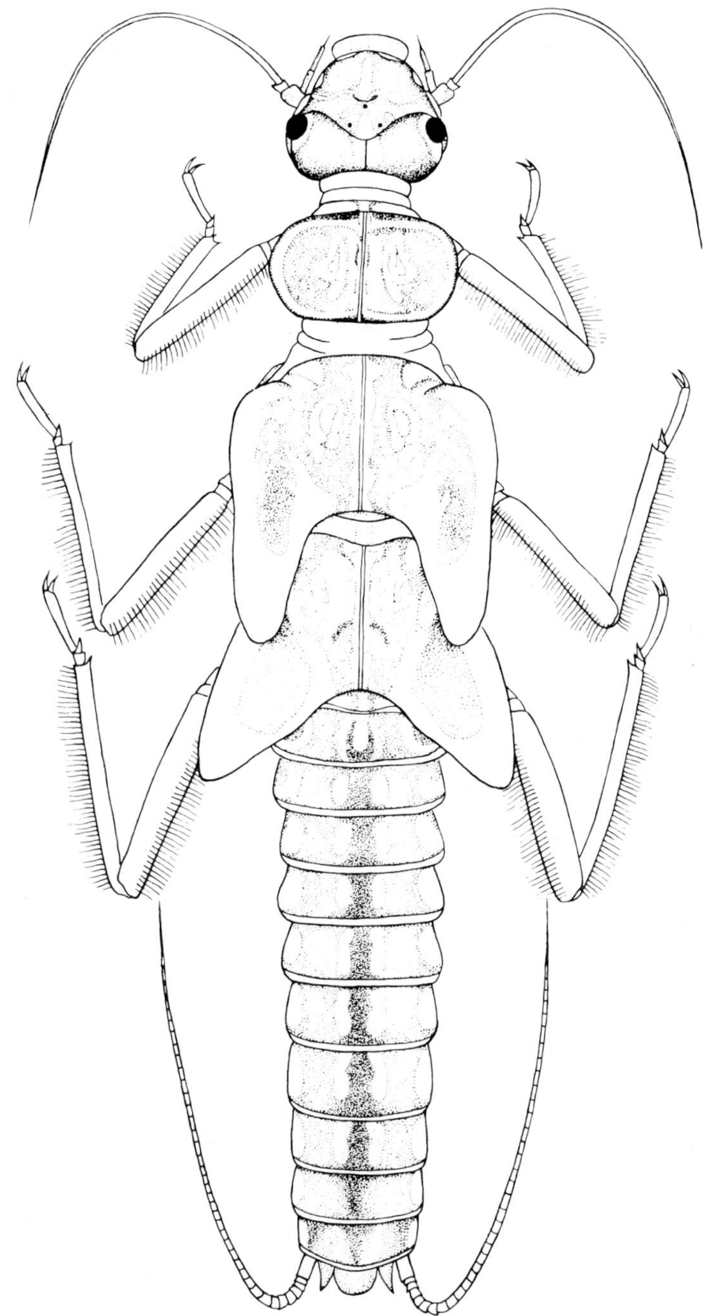

Fig. 65. Stonefly nymph *Mesoperlina pecirkai* KLAPALEK, dorsally. (After Zhiltzova, 1970).

Table 30. List of stoneflies from the Tien Shan and the Pamirs (in broad sense of the name) (After Zhiltzova, 1963; 1970, 1971a, 1971b, 1972a, 1972b, 1974, 1976, 1977; Zhiltzova and Zwick, 1971; Grizai and Zhiltzova, 1973).

Name of species	Locality
Family *Taeniopterygidae*	
Rhabdiopteryx tianshanica ZHILTZ.	Western Tien Shan (Chatkalskii Range)
R. brodskii ZHILTZ.	Northern Tien Shan (Zailüskii Alatau, Issyk R.)
Kyphopteryx pamirica ZHILTZ.	Gorno-Badakshan
Family *Nemouridae*	
Nemoura cinerea RETZ.	Vicinity of Samarkand
Nemoura (Nemoura) hamulata ZHILTZ.	Western Tien Shan
Nemoura (Nemoura) alabeli ZHILTZ.	Western Tien Shan (Talasskii Alatau)
Nemoura (Nemoura) lepnevae ZHILTZ.	Northern, Western, and Central Tien Shan (Kirghizskii Range, Zailiiskii Alatau, Turkestanskii Range, Alaiskii Range). Western Pamirs
Nemoura (Nemoura) ornata McL.	Western Tien Shan, Alaiskii and Kuraminskii Ranges, Hissaro-Darwaz, Gorno-Badakshan
Nemoura (Nemoura) alaica ZHILTZ.	Alaiskii Range (Akbura R.)
Nemoura (Amphinemura) tragula KIMM.	Tien Shan, Western Pamirs (Turkestanskii, Zeravshanskii, Hissarskii, and Alaiskii Ranges)
Nemoura (Amphinemura)maracandica McL.	Western Tien Shan (vicinity of Tashkent and Samarkand)
Nemoura (Amphinemura) gritsayae ZHILTZ.	Western Pamirs (Hissarskii Range)
Nemoura (Amphinemura) crenate KOP.	Northern and Central Tien Shan (Kirghizskii Range, Zailiiskii Alatau, Terskei Alatau, Alaiskii Range), Hissarskii Range
Nemoura (Amphinemura) mirabilis turkestanica ZHILTZ.	All mountain regions of Middle Asia, excluding the Kopetdagh
Nemoura (Amphinemura) zimmermanni JOOST.	Western Tien Shan (Talasskii, Pskemskii, and Chatkalskii Ranges)
Nemoura (Amphinemura) kustarevae ZHILTZ.	Tien Shan, the Issyk Lake basin
Nemoura (Protonemura?) vaillanti (NAVAS)	Western Pamirs (Hissarskii Range), Nanshan, Karakoram, Hindu Kush
Nemoura (Protonemura?) tianshanica ZHILTZ.	Western, and Central Tien Shan (Talasskii Range, Zailiiskii Alatau), Turkestanskii Range, Gorno-Badakshan
Family *Leuctridae*	"No species of this family are present in the faunas of the Tien Shan and Pamirs" (Zhiltzova, 1972a: 1741). For Middle Asia is given only one species from the Kopetdagh (*Leuctra kopetdaghi* ZH. – Loc. cit.)
Family *Capniidae*	
Capnia longicaudata, ZHILTZ.	Western, and Central Tien Shan (Talasskii Alatau, Kirghizskii Range, Terskei Alatau)
Capnia hamifera ZHILTZ.	Western Tien Shan (Talasskii Alatau)
Capnia prolongata ZHILTZ.	Western and Central Tien Shan (Talasskii, Kungei, and Terskei Alatau)

140

Table 30. (continued)

Name of species	Locality
Capnia bimaculata ZHILTZ.	Central Tien Shan, Gorno-Badakshan, Eastern Pamirs
Capnia turkestanica KIMM.	Western Tien Shan (Chatkalskii and Talasskii Ranges), Djungarskii Alatau
Capnia jankowskajae ZHILTZ.	Tien Shan (Terskei Alatau)
Capnia ansobiensis ZHILTZ.	Hissarskii Range
Capnia singularis ZHILTZ.	Hissarskii Range
Capnia badakhshanica ZHILTZ.	Gorno–Badakshan
Capnia bicuspidata ZHILTZ.	Gorno–Badakshan
Capnia shugnanica ZHILTZ.	Gorno–Badakshan
Capnia pedestris KIMM.	Eastern Pamirs, Himalaya
Isocapnia aptera ZHILTZ.	Western Tien Shan (Talasskii Alatau)
Eucapnopsis stigmatica transversa AUB.	Western, and Central Tieh Shan, Gorno–Badakshan
Family *Perlodidae*	
Filchneria mongolica KLAP.	Western, Northern, and Central Tien Shan, Hissaro–Darwas, Northern Mongolia
Filchneria mesasiatica ZHILTZ and ZWICK	Northern and Central Tien Shan (Talasskii, Kirghizskii, Zailiiskii, Kungei, and Terskei Alatau)
Scobeleva (*Dictyopteryx*) *olgae* McL.	Vicinity of Samarkand
Perlodes (*Scobeleva*) *cachemirica* AUB.	Gorno–Badakshan
Mesoperlina muricata KOP.	Northern and Central Tien Sahn, Djungarskii Alatau
Mesoperlina ochracea KLAP.	Tien Shan (mostly the Western, and Central), Hissaro–Darwaz, Kuen Lun, Northern Iran
Mesoperlina capnoptera McL.	Western and Southern Tien Shan, northern piedmonts of Ferhgana Valley
Mesoperlina pecirkai KLAP.	Tien Shan (chiefly the Western and Central), Ranges: Karatau, Ugamskii, Pskemskii, Zailiiskii, Kungei, Zeravshanskii, Hissarskii
Mesoperlina martynovi ZHILTZ.	Western Pamirs (Turkestanskii and Hissarskii Ranges)
Family *Perlidae*	
Perla cadaverosa McL.	Turkestanskii and Zeravshanskii Ranges
Phasganophora spp.[1]	
Family *Chloroperlidae*	
Xanthoperla kishanganga (AUB.)	Western Pamirs, Darwas, Badakshan
Xanthoperla curta (McL.)	Northern, Central, and Southern Tien Shan (Ranges: Alaiskii, Zeravshanskii, Talasskii, Zailiiskii, Chatkalskii, Kirghizskii, Turkestanskii), Hissarskii Range
Xanthoperla gissarica ZHILTZ. et ZWICK	Zeravshanskii, Turkestanskii, and Hissarskii Ranges

[1] The *Phasgonophora pedata* KOP. (from Issyk-kul Lake basin) and other species, not reexamined by Zhiltzova, are not included in this table.

and Blepharocerid faunas are entirely unknown but which, as inferred from
the other groups, differ from the Tienshanian faunas.

Middle Asiatic stonefly fauna can be characterized as follows (Zhiltzova,
1969, 1972, 1973, 1976): (i) high species endemism (about 80% endemics!);
(ii) a large proportion of the high-altitude Asiatic species known from the
Hindu-Kush, Karakoram and the Himalaya, which indicates a faunistic
relation between these regions and the Tien Shan and, especially, the
Pamirs; (iii) almost complete absence of any common species, the more so
the "european" ones.

These conclusions match those made by us with respect to the relatively
well-known groups (*Blepharoceridae* and the mayfly family *Heptageniidae*).

Unfortunately, the stonefly fauna from the mountain regions adjacent to
Middle Asia is inadequately known, and particularly few data are available
on the larvae of this order. Only a few examples can be given. In particular,
on the base of treatment of samples taken in expeditions to the Nepal,
Hindu Kush and Karakoram, descriptions were published of some stonefly
larvae (Kawai, 1963, 1966).

The stoneflies from the Issyk River have been identified by Martynov, but
the stonefly fauna of the Tien Shan is still poorly known, therefore, the
available identifications (Brodsky, 1935, 1976) are not reliable enough to be
given here. More recently, Zhiltzova (1972b) has described a new species
from this river, the *Rhabdiopteryx brodskii.*

b. *Notes on adaptation and biology of stonefly nymphs*

Unfortunately, there is little one can say about the adaptations in stonefly
larvae. Although a point of view exists, that "lithorheophilous stonefly
larvae are adapted to life in much the same way as mayfly larvae, because
they have a flattened body, widely sprawled legs, rows of contact setae and a
flattened head" (Shadin and Gerd, 1961, p. 177, 178); however, the stonefly
and mayfly nymphs are rather different in their adaptations. The special
adaptions for life on stones exposed to fast current, which are so well-
defined in such mayfly genera as the *Iron* and *Rhithrogena,* are absent in
stonefly nymphs. Gills in stoneflies are designed quite differently (in the
form of small tufts on the thorax, between the leg attachments and at the
abdominal end or, not infrequently, entirely reduced) and never form the
false suckers. The body flattening is less defined: stoneflies do not have such
a broad head shield as in some mayflies. Among the characteristic adaptations in stonefly nymphs are only their stout, widely sprawled legs with
robust claws and the firm pressing of the body to the stone. But some
predatory stonefly nymphs are capable of moving about fast, without
"creeping" but by holding on with the claws to the roughness of stone. The
absence of well-defined morphological adaptations in stoneflies can be
conceived if one takes into account those microbiotopes which are typical of
the larvae of this order in mountain torrents (described above). Therefore,
the adaptations which have developed in stoneflies to the torrential conditions must be considered as ecological and ethological rather than morphological ones.

As for the biological peculiarities of stoneflies, mention should be made of the following.[19]

The problem of imago hatching from nymphs is solved very simply, like the terrestrial insects. Just before this act, mature nymphs creep out of the water onto the nearshore stones, grass, brushes: the imago emerges in the air environment. On the shore, on stones, tree trunks, bushes, grass stems and leaves one can frequently find the cast skins.

The dates (season) of imago hatching vary with species from early spring to late autumn or even winter, on snow. The flight occurs in different seasons, but from Zhiltzova's observations in stoneflies from the Tien Shan and Pamirs the flight is observed mainly in spring or early summer. For example, the flight of species of the *Capnia* genus was recorded in early spring, but in high mountains it was observed in May and early June or later, in summer (*C. bimaculata, C. pedestris*). Species of the *Eucapnopsis* genus fly from April to June, whereas in brook headwaters the flight occurs later; some species of the *Mesoperlina* genus fly in spring (May, June) and some others from May to August (*M. pecirkai* and others). Species of the *Nemoura* genus, occurring in the moderate and lower altitudinal zones, fly out at the end of winter and in early spring (*e.g. N. maracandica*); but on the high mountains this happens during May and June. Just about this time the species of the *Filchneria* and *Xanthoperla* fly out.

What is to be stressed when describing the flight period in stonefly imagoes is a time lag between the flight at high altitudes and that at lower heights (*i.e.* in the moderate and lower altitudinal zones), a phenomenon characteristic also of other groups, *viz. Ephemeroptera* and *Trichoptera*.

Females lay their eggs into the water either while flying above the water surface or touching it with their abdomens, or even on submerged stones, in which case the females creep into the water. Stonefly larvae spend from one to three years in the water. In this instance, the number of molts varies with the sex, water temperature and species affiliation from 12 to 36, mostly 12–16.

Larvae of different genera and species, as well as of different ages, are very diversified in their feeding. Usually they have "biting" mouth parts, many adapted to predatoriness, whereas many others are vegetarians which feed on disintegrated plants, moss or living vegetation. In torrents most vegetarians are algivorous, and feed on the periphyton. The large-sized species are carnivorous and feed on the larvae of mayflies and other aquatic insects.

The ecological data on stoneflies from the Tien Shan and the Pamirs collected by Zhiltzova make it possible to make some generalizations about this order.

The altitudinal distribution of stoneflies shows that they are mainly mountain dwellers. Thus species of the *Capnia* genus are chiefly confined to the heights from 1000 to 4000 m: *C. bimaculata* from 2800 to 4500 m, *C. hamifera* above 2000 m, *C. pedestris* at 4500 m, and very few species were

[19] For a good concise account of the biology of this order *in toto*, see Hynes, 1975

143

encountered at lower heights (*C. turkestanica* up to 2000 m). In contrast to the *Capnia*, the species of the *Mesoperlina* genus were found at relatively low heights, and for some species there are available indications that they do not reach great altitudes, but occur only between 300–2000 m (*M. muricata*, *M. ochracea*, *M. capnoptera*); only *M. pecirkai* was not found in the submontane region, ranging in distribution from 800 to 3000 m.

Species of the *Nemoura* genus were found within the range of 800–3000 m and only one species (*N. maracandica*) was encountered in the lower altitudinal zone. A similar vertical distribution is characteristic of species of the genera *Filchneria, Xanthoperla* and *Kyphopteryx*.

The above data show that stoneflies on the Tien Shan and the Pamirs are representative for the upper and, partly, the middle altitudinal zones, whereas only few species were encountered at low heights.

When one attempts to relate different types of torrents to species, one is faced with great difficulties. It seems that either the data on larval ecology for the Tien Shan and the Pamirs are scanty or, which is less likely, the larvae are very flexible as to the preference of certain types of waters; it is therefore difficult for us to name species which are representative of mountain torrents. Some indications can be found only for some species of the *Capnia* genus, *e.g.* "headwaters of mountain rivers and at the foot of glaciers" about *C. longicaudata*; "headwaters of mountain rivers" about *C. hamifera*; and "the upper reaches of rivers" about *C. prolongata* (according to Zhiltzova).

Perhaps, a certain difference can be revealed for species of the *Mesoperlina* genus described as occurring "in different watercourses and large rivers" (*M. muricata*); "mostly in large rivers" (*M. ochracea*); "warm brook and large rivers" (*M. capnoptera*). From these still scanty data we can conclude that the *Capnia* genus is more typical of torrents than the *Mesoperlina* genus. However this conclusion should, of course, be taken with caution.

It is hard to make any conclusions about the preferential waters for the species of genera *Nemoura, Filchneria* and *Xanthoperla* which occur "in various mountain rivers, brooks and springs"; "in various mountain types of waterbodies"; and "lotic mountain waters of diverse nature". *Xanthoperla gissarica* is said to prefer spring brooks. For the time being, one can claim that we deal with rheobionts, but to identify hymarobionts is a hard thing to do.

The microbiotopes occupied by stoneflies were described earlier, and we can only cite some data to confirm that these microbiotopes are by no means preferred by hymarobionts. Thus for the *Capnia* genus it was noted that their larvae were found beneath the stones and in quiet water near the bank (*C. bimaculata, C. singularis*). For *Mesoperlina pecirkai*: "most of the larvae are found in places where current is not strong, largely near the bank, more rarely in swift waters, but under stones". The same about the *Xanthoperla* genus: under stones, mainly in places unexposed to fast current (*X. curta*); and under stones, usually near the bank (*X. gissarica*).

It is worth making one more comment on the stonefly ecology associated with the "organic drift". The resistance to the downstream removal is accomplished not only, or rather, not so much due to the upstream flight of adults (as is typical of the *Ephemeroptera*), but observations are available on larval upstream migration (Bishop and Hynes, 1969b).

Below are given observations of stoneflies in the rivers Issyk and Akbura.

c. Distribution of stonefly larvae and nymphs along the torrent and some other ecological data

In the Issyk River the *Mesoperlina capnoptera* nymphs were found in small numbers at few sampling sites. The substrate preferred by the nymphs were pebbles and stones. On moss this species was never found. The *Nemouridae* preferred moss fringing the torrent banks, mainly in places of the strongest current. *Mesoperlina* and *Nemouridae* did not occur far from the bank in the deeper waters and were absent from adjacent brooks and springs as they preferred only mountain torrents of greater strength.

In the Issyk River the *Mesoperlina* is not found beyond 1980 m above sea level, where the water temperature is about 6°. Adults of *Mesoperlina* and *Nemouridae* were encountered beginning from early June; the flight of the *M. capnoptera* begins at this time and continues until mid August. It was a mass emergence and adults of both sexes were encountered in vast numbers. They were scurrying about on nearshore vegetation, crawling on the lower surface of leaves and on stones. When scared, they fly away slowly for a short distance. The flight of another species of the same genus (*M. ochracea*) occurs in the Akbura River much earlier than in the Issyk River, since April or May, which seems to be due to higher temperatures of the air in the lower Akbura River.

In the Akbura River the most common stoneflies are nymphs and adults of the *Mesoperlina* and *Amphinemura* genera. The stoneflies were encountered mainly in the middle and the lower reaches.

Since in all these rivers the stoneflies were found as solitary individuals and only sporadically, it is difficult to present a pattern of their vertical distribution. In the Issyk River the stoneflies were encountered up to 2900 m above sea level and a little farther up in the Akbura River. But in other rivers and torrents (see Table 30) the altitudinal distribution of the Middle Asiatic stoneflies is confined to the high-mountain zone – up to 4000 m above sea level.

For the mountain regions adjacent to Middle Asia the following examples of altitudinal occurrence of stoneflies may be given. *Nemoura vaillanti* was found in the Nanshan within 2000–4000 m and on the Pamirs within 1680–3000 m. On the Himalaya, where stoneflies are better known than mayflies, the former were found at elevations from 4000 to 5500 m (*Capnia pedestris*) and on the Rongbuk Glacier in the Mt. Everest area at elevation of 5030 m (*Rhabdiopteryx lunata* (Mani, 1968).

According to Dubey and Kaul (1971), in glacial torrents on the northwestern Himalaya the stoneflies occur mostly within the upper reaches, in the alpine zone.

4. Diptera

4.1 Family *Blepharoceridae*

a. Species composition

The *Blepharoceridae* are a very peculiar and highly specialized family of two-winged flies, which has a world-wide distribution. The preimaginal phases in this family are everywhere confined to swift waters. In spite of the great attention given by systematizers and ecologists to the *Blepharoceridae* because of adaptive peculiarities of the larvae and pupae, the taxonomy of *Blepharoceridae*, with respect to all their stages, is very poorly developed. The largest difficulties arise in identification of the species represented in samples only with larvae and immature pupae (if mature, then adults could be extracted). As a result, a number of species, described only by their preimaginal stages, are known under conventional designation (with numbers or letters). Unfortunately, the list of *Blepharoceridae* from Middle Asia, a subject of our special attention, also contains unrefined species, *i.e.* those represented only with larvae and immature pupae. The forms so far found in torrents, brooks and rivers of Middle Asia are listed in Table 31. It should be borne in mind that some of these have been shifted from one genus to another; in this instance, the initial species and, occasionally, the new taxonomic position of the forms are specified.

Table 31. List of species and forms of *Blepharoceridae* from Middle Asian torrents (After Brodsky, 1930c, 1936, 1954, 1972b).

		Locality of recovery	
Name of species or form	Old name (synonym)	Issyk R.	Akbura R.
Blepharocera asiatica Brodsky	*B. fasciata asiatica* Brod.	+	+
Tianschanella monstruosa Brod.	No synonym	+	+
Asioreas tianschanica (Brod.)	*Philorus tianschanicus* f. *typica* Brod.		+
A. nivia (Brod.)	*Ph. tianschanicus nivia* Brod.		+
* *A. turkestancia* Brod.	No synonym	−	−
* *Philorus asiaticus* Brod.	No synonym	−	−
Larvae:			
Blepharocera sp. A.		+	+
* *Blepharocera* sp. C			
* *Blepharocera* sp. C$_1$			
* *Asioreas* sp. B			
* *Asioreas* sp. B$_1$			
* Genus? (*Philorus?*) sp. G			

* The larvae were found in other rivers on the Tien Shan and the Pamirs.

146

The generic status of the species was discussed in our paper on the finding of a Blepharocerid member in Mongolia (Brodsky, 1972a). Here we shall recall shortly the main grounds for placing species in certain genera. The generic status of species from the genera *Blepharocera* and *Tianshanella* has remained unchanged. The genus *Asioreas* is stated by us (Brodsky, 1972a) for species which were earlier placed in the genus *Philorus* (Brodsky, 1930c, 1936, 1954). In 1958 the species with the transverse vein *m-cu* were assigned by Alexander (1958), instead of their previous genus *Philorus,* to the new genus *Dioptopsis* separated from the *Philorus* by Enderlein (1936; see also Mannheims, 1954). However, in our opinion and that of Zwick (1968, 1970), the *Dioptopsis* genus must include only the respective species from Europe, whereas the Middle Asiatic species, previously placed in the *Philorus* genus by us and in the *Dioptopsis* by Alexander, must be isolated in a new genus, the *Asioreas.* Diagnosis of this genus was given in Brodsky's work cited above (1972a).

Thus, at present there are six precisely defined species which refer to four genera, and six forms designated by means of the *nomenclatura aperta* as relating to three genera. The above list takes into account not only our samples (from 60 streams on the Tien Shan), but also samples collected by Omorov and myself from streams in the northern Alai Mountains.; by Sibirtzeva from the Chirchik River basin (rivers Pskem, Ugam, Koksu, Chatkal) and from the Zeravshan; by Gurvitsch from the Sary-Djaz basin and the Pamirs; by Yankovskaia from the Tien Shan and the Pamirs, and by Zhiltzova from the Western Pamirs.[20]

Among the Middle Asiatic *Blepharoceridae* all the six species identified are endemics to Middle Asia, two of the four genera being also endemics and the other two of wide distribution.

Distribution of adults, larvae and pupae of all *Blepharoceridae* so far found are given in the author's works (Brodsky, 1930c, 1936, 1972b). In this essay we present only the general views of larvae and pupae, and one adult; all from the Tienshanian torrents (Figs. 66–69).

Of great interest is the geographical distribution of the *Philorus asiaticus,* whose larvae and pupae were recovered from typical mountain torrents and only once from a brook. Its distribution is restricted to the Western Tien Shan and the Western Pamirs. Somewhat unexpected was the recovery of one pupa of the new species in the Central Tien Shan. Despite numerous samples of *Blepharoceridae* from this region, the species was found only in one instance. In contrast to the other *Blepharoceridae,* the range of distribution of the new species is very limited. The genera *Blepharocera, Tianshanella* and *Asioreas* are found throughout the Tien Shan and the Alai-Pamirs, while the *Philorus asiaticus* is known only from four rivers; one may assume that *Ph. asiaticus* which is so sporadically distributed in the Tien Shan and the Pamirs seems to be a derivate of a species group more widely distributed on the Himalaya. We regret the fact that there are no data on the Blepharocerid fauna for the Hindu Kush and the Karakoram.

---

[20] The author is very thankful to them all.

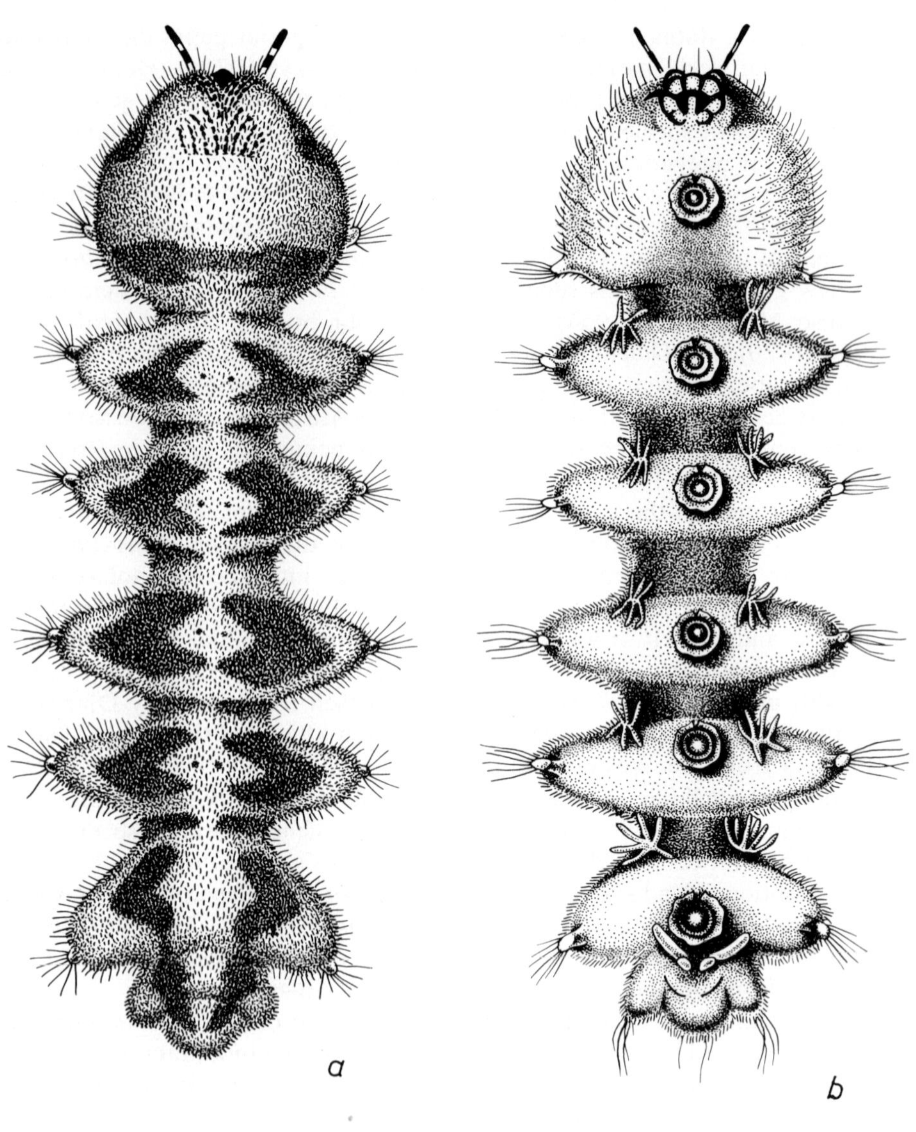

Fig. 66. Net-winged midge (*Diptera*) *Blepharocera asiatica* BRODSKY.
a. larva, dorsal view; b. larva, ventrally; c. pupa, dorsal view; d. female imago, lateral view.
(original).

148

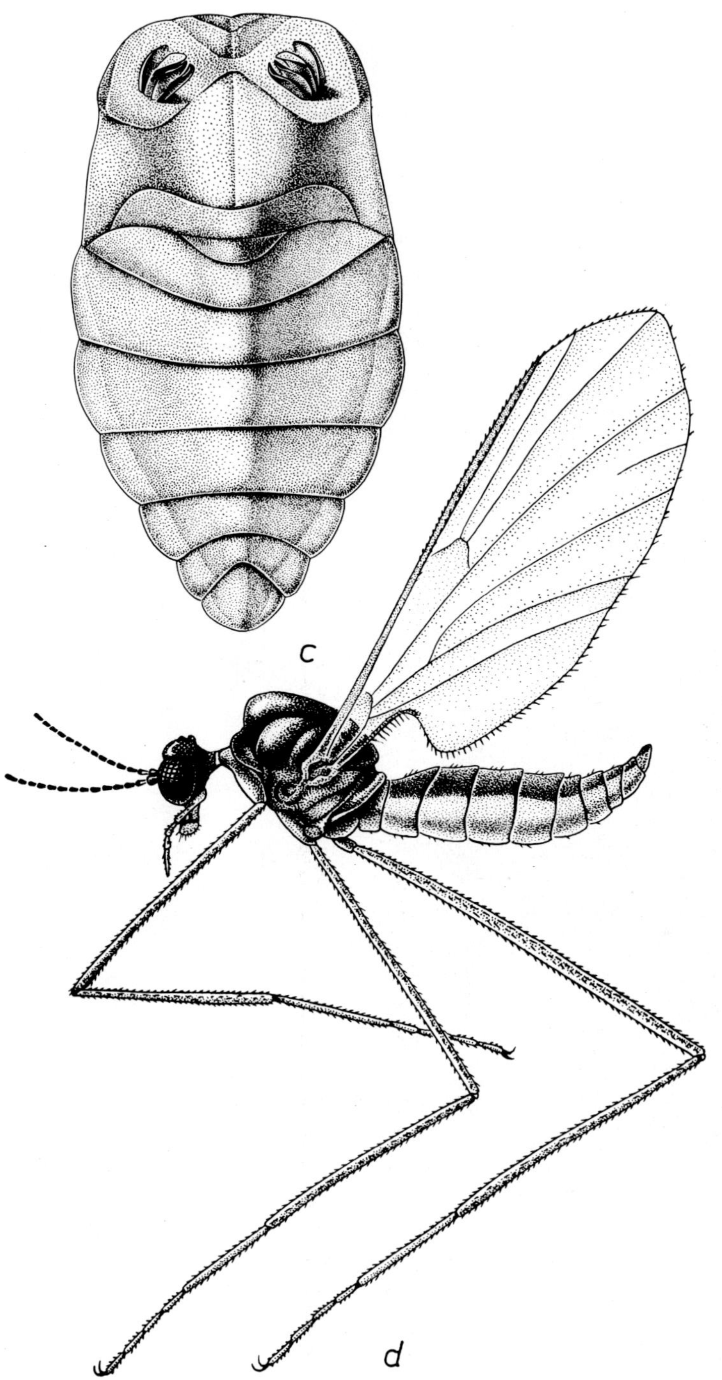

Fig. 66. (continued)

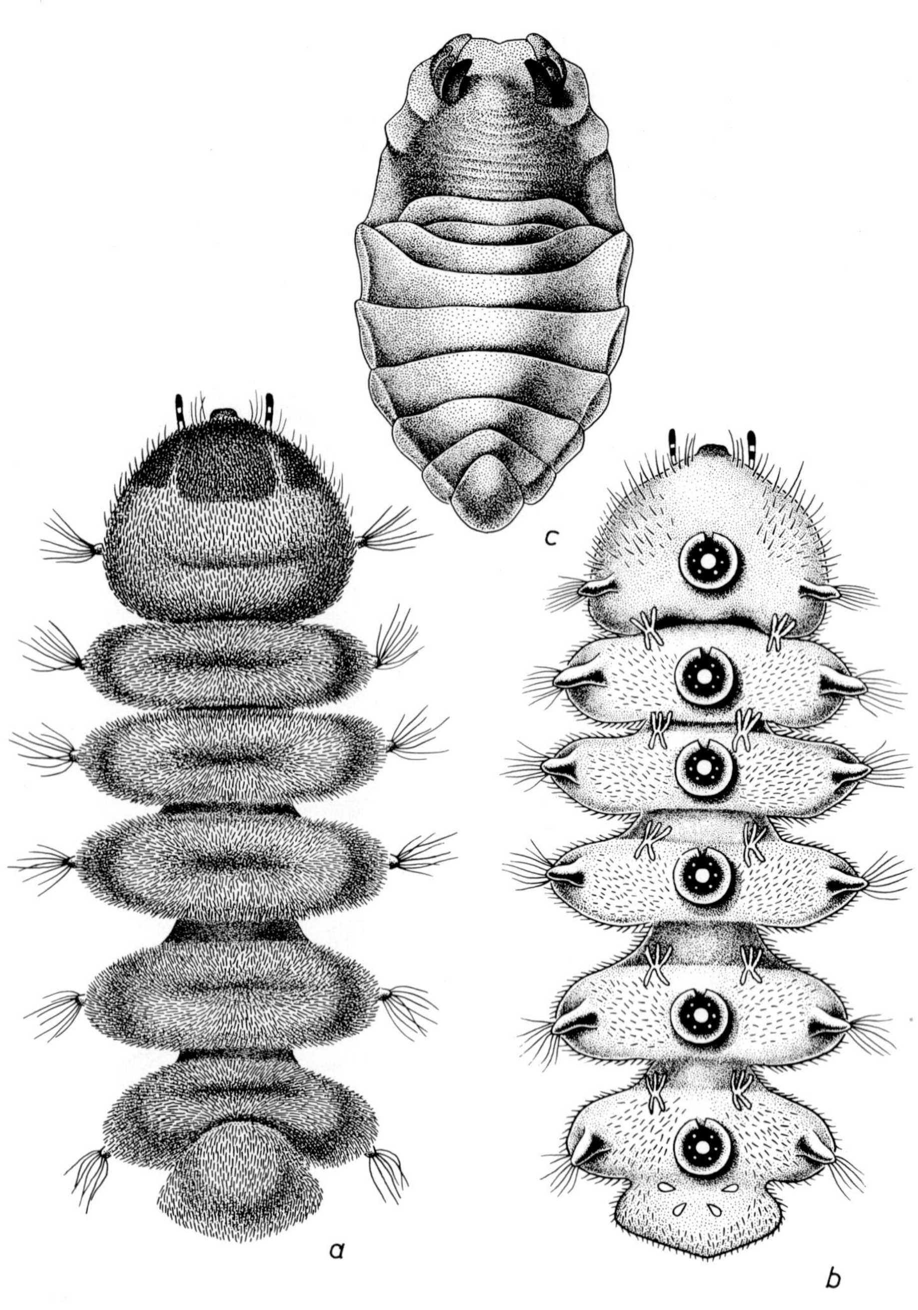

Fig. 67. Net-winged midge (*Diptera*) *Tianschanella monstruosa* BRODSKY.
a. larva, dorsal view; b. larva, ventral view; c. pupa, dorsal view. (original.)

150

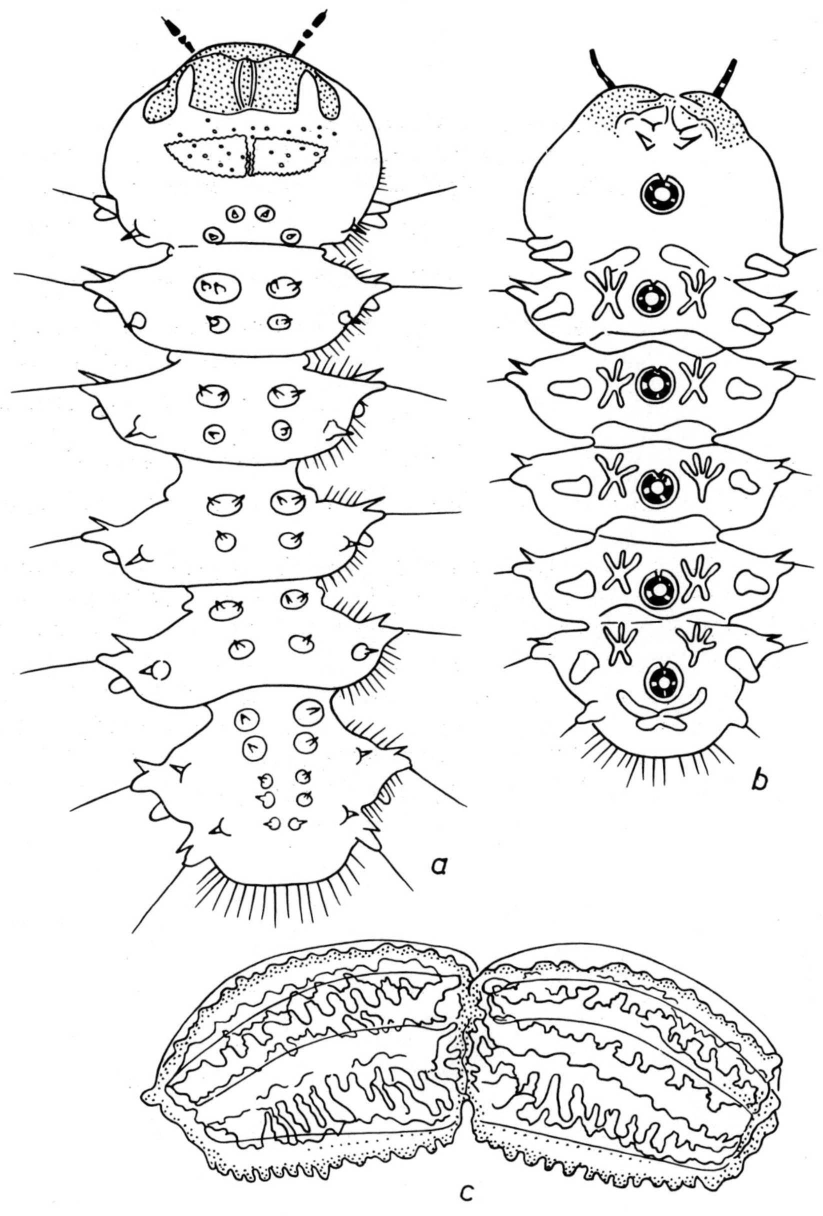

Fig. 68. Net-winged midge (*Diptera*) *Philorus asiaticus* BRODSKY. (After Brodsky, 1972b.)
a. larva, dorsal view; b. larva, ventral view; c. the tracheal gills of pupa.

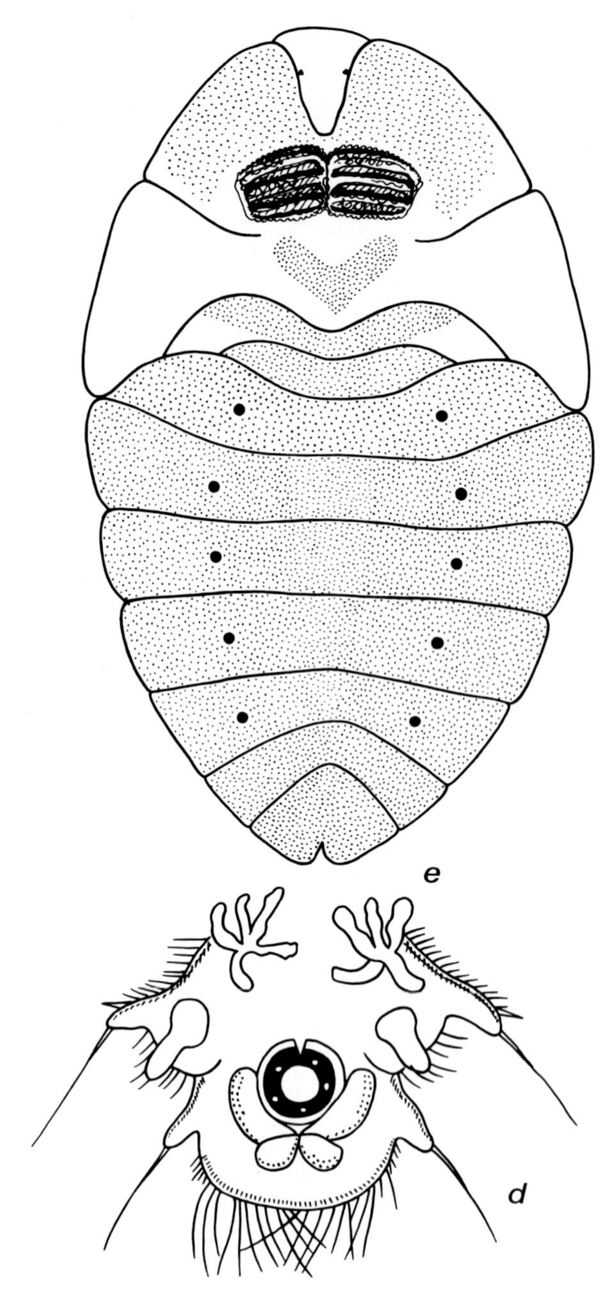

Fig. 68 (continued). d. posterior end of larva; e. pupa, dorsal view.

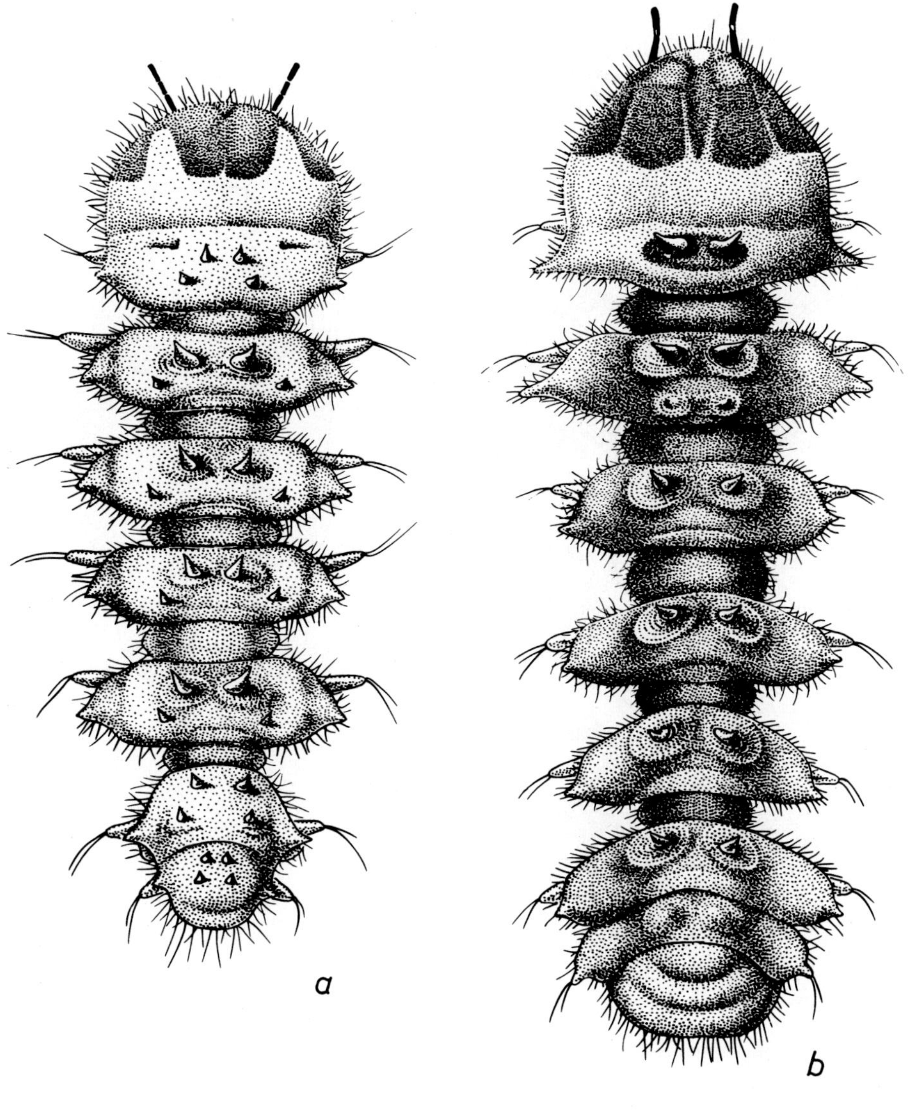

Fig. 69. Net-winged midge (*Diptera*) *Asioreas tianschanica* BRODSKY, a. larva, dorsal view; b. larva, another example of spiny armour.

153

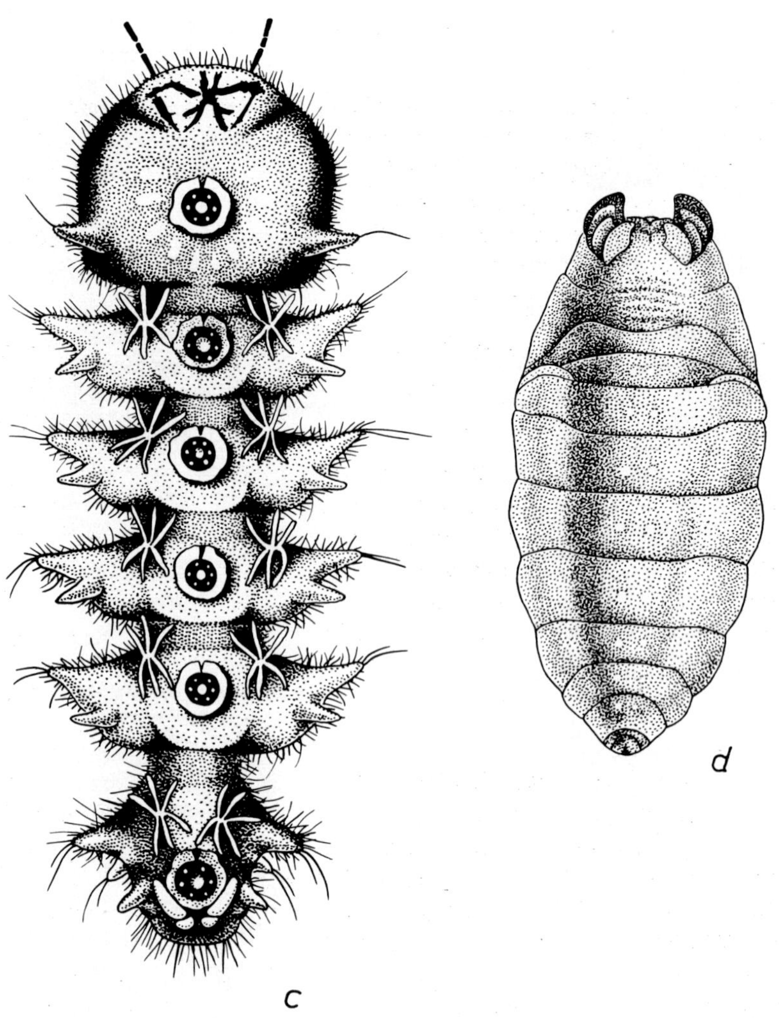

Fig. 69 (continued). c. larva, ventrally; d. pupa, dorsally. (original).

It is of importance to compare the Middle Asiatic Blepharocerid fauna with that of adjacent mountain regions. Using a summarizing work by Alexander (1955) and publications of Agharkar (1914) and Tonnoir (1930, 1931, 1932), as well as a more recent work of Kaul (1971), all devoted to the *Blepharoceridae* from the north-western Himalaya, we have compiled a list of the *Blepharoceridae* from India, Pakistan and Shri Lanka (Ceylon):

Species or form	Locality
Adults:	
Apisotomyia trilineata Brunetti	Northern India
A. *santokhi* Kaul	North-western Himalaya
Blepharocera indica Brun.	Northern India

154

B. tertia Kaul	North-western Himalaya
B. automnalis Kaul	North-western Himalaya
Philorus assamensis (Tonn.)[21]	Assam
Ph. horai (Tonn.)[22]	Northern India
Ph. novem Kaul	North-western Himalaya
Ph. thorus Kaul	North-western Himalaya
Ph. dubeyi Kaul	North-western Himalaya
Asioreas bionis (Agharkar)[23]	Kashmir
Horaia longipes Tonnoir	Northern India
H. montana Tonn.	Northern India
Hammatorrhina bella Loew.	Southern India, Sri Lanka

Larvae and, occasionally, pupae:

Blepharocera, 2 species	Kashmir
Euliponeura, 2 species	Kashmir
Blepharocera, 5 species	Pendjab
Horaia, 3 species	Pendjab
Euliponeura, 2 species	Pendjab
Apistomyia, 4 species	Pendjab
Genus? 2 species	Pendjab
Blepharocera, 4 species	North-eastern India
Horaia, 4 species	North-eastern India
Apistomyia, 2 species	North-eastern India
Genus? 1 species	Southern India

Judging by the appearance of the larvae and pupae designated by Agharkar and Tonnoir with letters as unidentified to a species, one may suggest that some of the forms described from India may be identical to the Middle Asiatic ones so far unidentified to a species. However, the comparison of these forms is not yet completed and we have to satisfy ourselves with a suggestion.

The Tibetan fauna is unknown. In Mongolia we found a species known from the Altai (*Asioreas altaicus* Brod.; Brodsky 1972a). Besides this species, three others were found in the Altai: *Bibiocephala maxima* Brod., *B. decorilarva* Brod. and *Asiobia acanthonympha* Brod. (the latter is a new genus, probably, a synonym for the genus *Neohapalothrix* Kitakami; Brodsky, 1954).

The comparison of the Middle Asiatic fauna of *Blepharoceridae* with the forms from the Far East, the Altai and Mongolia shows that the distribution of the Middle Asiatic species is restricted to the Tien Shan and the Alai-Pamirs. The fauna of the Hindu Kush, the Karakoram and the Chinese Tien Shan is poorly known, but available descriptions of the Blepharocerid members from the Himalaya suggest endemism of the fauna in this region.

---

[21] Described as *Euliponeura assamensis*.
[22] Described as *Euliponeura horai*.
[23] Described as *Philorus bionis*.

b. Distribution of the Middle Asiatic Blepharoceridae

In the mountain regions of Middle Asia the *Blepharoceridae* have a very specific distribution. Two types of distribution areas are recognizable. One corresponds to a wide distribution covering nearly all torrents and rivers of the Northern, Central and Southern Tien Shan and the Pamirs (*Blepharocera asiatica, Tianschanella monstruosa, Asioreas tianschanica* and *A. nivia*), the other is a restricted, local type (*Philorus asiaticus* and *Asioreas turkestanica*).

The question may arise whether the data are sufficient to draw conclusions about the limited distribution of these species. With respect to *Philorous asiaticus* the answer will be positive, since this species is described not only with reference to adults, but also to preimaginal stages; the larval samples of the Tienshanian *Blepharoceridae* are numerous. The sampling covered almost all mountain regions, but the species was found only in few sites. (Several hundred of these samples have been re-examined by us, but the species was not found in the other sites.) As for *Asioreas turkestanica*, the recovery and description of its preimaginal stages might make its distribution wider.

c. Notes on adaptations in Blepharoceridae

Of all aquatic insects only *Blepharoceridae* have true hydraulic suckers. The larval body is shortened in such a way that it centers around the suckers: the head is joined to the thorax and the first abdominal segment to form the first false segment with one sucker, then follow the four abdominal segments, each with a sucker, and the posterior false segment also with a sucker, consisting of six abdominal segments soldered to the remainder of the abdomen. Thus, the larva has altogether six suckers. The number of suckers is remarkably constant and representative of all the species of the *Blepharoceridae* family. Each part of the larval body bears one or two pairs of lateral appendages ("parapodia", in some authors) which are directed ventro-laterally at every side of the sucker.

By its origin, the sucker (Fig. 70)[24] was derived, probably, from the typical pseudopodium (proleg) of the dipterans. This is a shallow funnel-shaped chitinous formation, the soft rim of which holds onto the supports with hooks at their end. At the front every sucker has a V-shaped notch. The central part of the sucker is a slightly dome-shaped piston, which is capable of moving up and down against the chitinous circular part of the sucker (Huboult, 1927; Hora, 1930). When the soft free edge of the sucker touches the substrate, the piston moves down, the water is pushed out through the frontal V-shaped notch which works as a valve and then, with lifting the piston, a negative pressure results. If the piston has been lifted, the hooks, directed inwards the sucker, engage roughness on the substrate.

In addition to the described devices, there are in the sucker six tubular

---

[24] The picture of the sucker (lateral view only) is given, after Komarek, in "Freshwater Life of the Soviet Union" by Shadin (vol. VI, p. 153, Fig. 17).

156

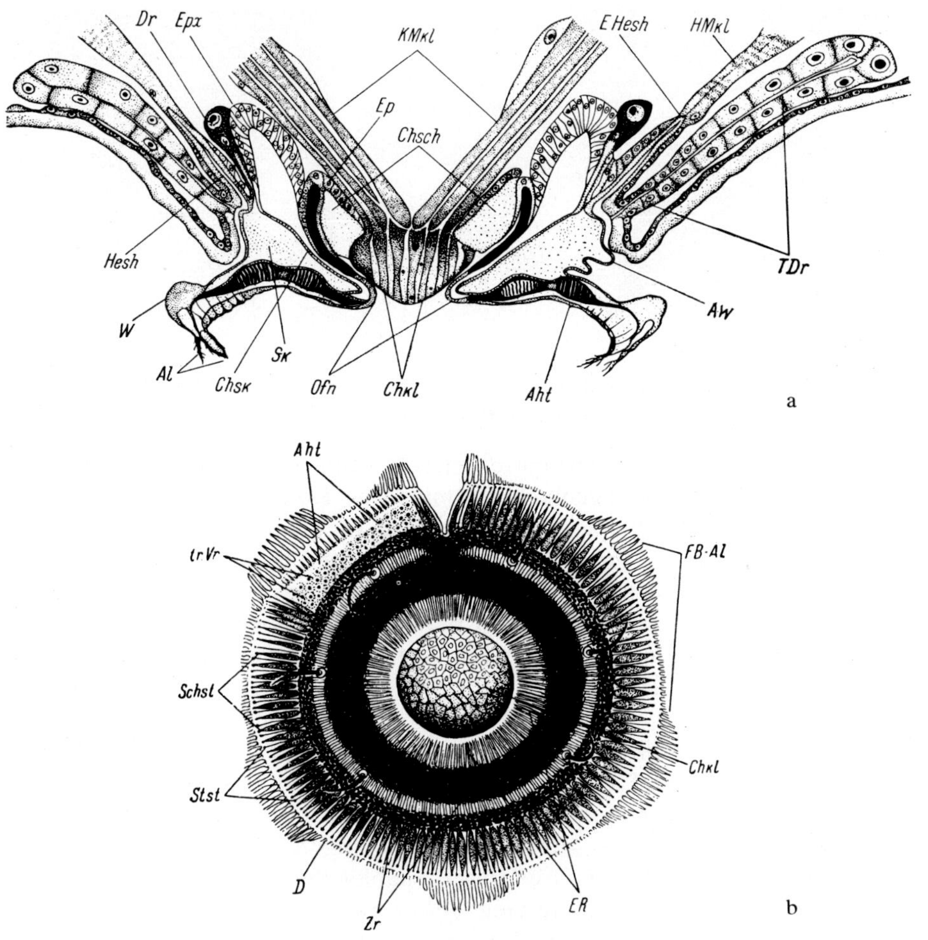

Fig. 70. Abdominal sucker in the *Blepharoceridae* larva. (After Komarek, 1914).
a. median section of the sucker across the body's axis: Aht, outer ring; Hesh, epidermis cut-off; EHesh, end of epidermis cut-off; Aw, outer wall of sucker; W, thickened end; Al, outer appendages; Chsk, chitinous bugs; Ofn, opening of sucker; Chkl, chitinous piston; Chsch, chitinous thickening (rings) on piston; Ep. epithelium of piston; Epx, epithelial cellular thickening; Dr, two-cellular gland; TDr, tubular gland; Sk, secret of gland; HMkl, muscles; KMkl muscle of piston.

b. piston rings, ventral view (central opening is closed with piston): ER, first light-coloured ring; Zr, second dark-coloured ring; FB–Al, hair tufts; D, spines; Chkl, chitinous piston; Aht, outer cover of disk; tr and Vr, funnel-shaped hollows; Schst, swordlike beams; Stst, stylet-like beams.

glands, a secretion of which emerges on the inner surface of sucker and is accumulated between the outer rim and the inner part of sucker; this secretion is sticky and assists the larval fixation to the substrate. But the secretion and hooks, of course, are only subsidiary devices and the major role is that of the piston, whereas vacuum maintaining is possible owing to both the "paling" layer and the chitinous "stringers" and to the specially adjusted outer covers of the sucker, providing for the hermeticity of the

157

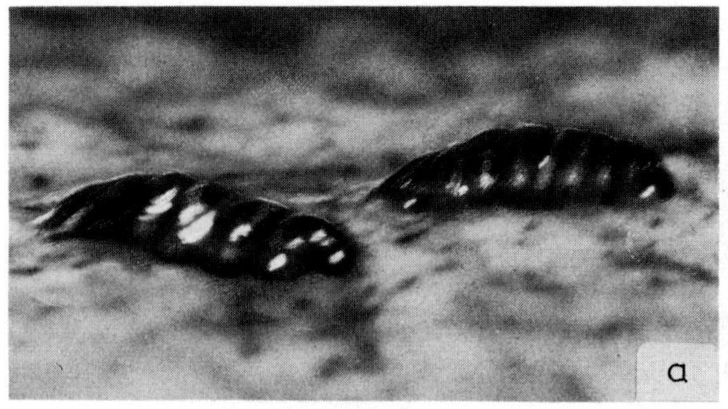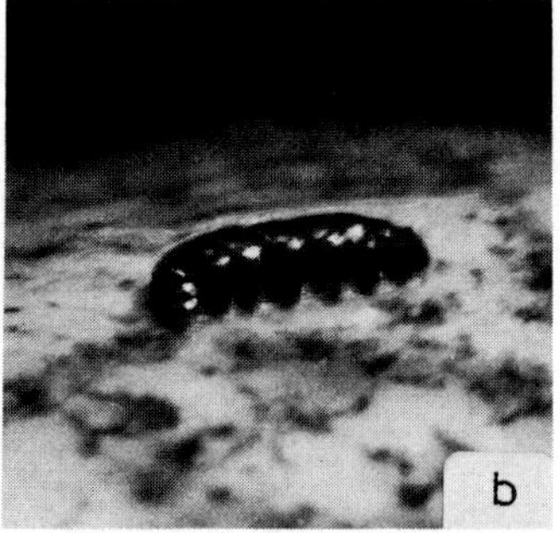

Fig. 71. Blepharocerid larvae in their natural poses on granite boulder. Photo by V. Bukin.
a. *Blepharocera fasciata*; b. *Tianschanella monstruosa*.

inner cavity of the sucker (Komarek, 1914a). Specially adapted is also the muscular mechanism, exhaustively examined by Komarek.

The locomotion of larvae with suckers is of great interest. The firmly attached larva releases three fore-suckers, lifts the upper part of its body, bends itself aside and attaches the released suckers. Then the next three suckers are released and the second part of body moves as above. So, by releasing and attaching its suckers in turn, the larva creeps slowly and sideways (Komarek, 1914a).

It seems, however, that the Blepharocerid larvae have a variety of movement. For example, Hora (1930) writes that they maintain their length across the current and Dorier and Vaillant (1955) that their heads are always downstream. The larvae can turn by releasing suckers 1–4, attaching 1 and 2 to the substrate laterally and moving 5 and 6. This series of movement, repeated two or three times, enables them to turn through 180°. Or they can walk forward moving either two suckers at a time in the sequence 1 and 4, 2 and 5, 3 and 6, or one sucker at a time, starting at the rear. Huboult describes this last gait as similar to that of the caterpillar.

Of interest is the use of the lateral appendages. When a sucker of one body segment is detached from the substrate, the appendages move forward and engage the substrate, stabilizing the larva, but if the larva is moving across the current, then the hind (*i.e.* situated downstream) lateral appendages act as bearings. Therefore, the Blepharocerid larvae, being exposed to strong current, are capable of moving in any direction and at a notable speed.[25]

Our observations of the larval movement in *Blepharoceridae* from the Tienshanian torrents confirm the idea that the larvae are capable of moving with different motions and in various directions to the current, but in our opinion, they prefer very slow movement, if any (Fig. 71). In this respect, larvae of *Deuterophlebia* are much more active.

Absence of true suckers in torrential insects, excluding the

[25] The larvae of the first age have on the lateral appendages a bundle of small hooks which they lose after the second age (Brodsky, 1936, Fig. 11).

158

Blepharoceridae, is somewhat unexpected. Hora explains it by the fact that *Blepharoceridae* are absent from streams of India (the Himalaya), because the rock surfaces here are roughened by deposition of travertine or growths of algae and moss. Thus, in his opinion, the area of sucker application is limited, while the other adaptations to life at fast current are more universal. But the various groups of aquatic insects have evolved, of course, in different ways and developed diversified adaptations of varying complexity and perfection. Of particular interest are the adaptations in the blackfly larvae which cross exposed stone surfaces in fast current by webby "trails" and webby "mats". But in comparison with different adaptations inherent in torrential insects, the larval suckers in *Blepharoceridae* seem to be the most relatively perfect device for providing independence in their movements on exposed surfaces of rocks, stones, *etc.*

There is not much to say about the pupae of *Blepharoceridae* except that they have rigid dorsal covers affected by currents, a streamlined shape and six attaching fields for tight stabilizing on the stone. Of interest is that the tracheal gills in the pupae work equally well both in the water and in the air; they are not therefore threatened by exposure to the air environment because of the water level fluctuation in the torrent. This phenomenon is frequent in and typical to the snowy and glacial watercourses.

The hatching of winged adults in *Blepharoceridae* is connected, of course, with the difficulties common to other aquatic insects, whose pupae firmly attach themselves to stones. This process was described as early as 1845 by Comstok and observed many times by other workers and by ourselves in mountain torrents on the Tien Shan. For example, in the Issyk River we found the Blepharocerid pupae immediately under the water surface or almost above it. The pupae were constantly watered by the spray and waves when water jets hit the rocks and stones. However, some pupae were found at depths down to 40–50 cm. If the pupae are near the water surface, the hatching proceeds without any difficulties. The pupal covers are torn in two directions. The major rupture passes along the median line of the thorax and the second one detaches the thorax from the abdominal covers of the pupa perpendicular to the first rupture. Through this orifice the adult squeezes itself, with its mesothorax forward and the head bent ventrally. The imago gradually disengages its body from the pupal covers, while the fast current forces it vertically. Up to the last moment, *i.e.* until complete disengagement, the imago grips by the exuvia with its claws. If the imago hatches in deeper waters, its emergence is facilitated by an air bubble which lifts the imago to the water surface. The body of the adult thus remains dry. Of course, the fast current catches the imago and takes it downstream, but this is compensated during oviposition by adult insects when the females fly upwards the stream.

d. Notes on the ecology of Blepharoceridae

The ecology of the *Blepharoceridae* family has been studied extensively (Tonnoir, 1923a,b, 1924; Hubault, 1927; Hora, 1930; Brodsky, 1930c, 1935, 1936, 1972b, and others). All the investigators characterize the larvae of this family as typical dwellers of swift torrents and brooks, which inhabit the roughest waters in foamy cascades and small waterfalls and live on stones, rocks, etc. Most authors do not consider the possibility of the preimaginal phases of *Blepharoceridae* existing in the lowland rivers and in the torrent's sites with slow current. Of course, there exists a certain difference in ecological conditions in which various orders and species of the family live, but in general all the *Blepharoceridae* are restricted to rough torrents which are abundant in cascades and have high content of dissolved oxygen. On the Tien Shan, the Altai, the Himalaya and adjacent mountain regions, the *Blepharoceridae* are cold-water stenotherms, while in the tropics they inhabit the water with relatively high temperature, but again necessarily with rough current. The question now arises why the Blepharocerid larvae choose such unsuitable places as cascades, rough torrents, etc? Indeed, in order to maintain themselves on the stones exposed to a current of 3 m/sec and more, they need some special and very complicated adaptations. We attempt to answer the question at the end of this section, but here we take the opportunity of referring again to the great significance of the biocenotic factor (particularly, absence of predators) in the habitat selection by larvae of *Blepharoceridae*, which is frequently underestimated in the study of the torrential fauna.

In torrents of the Tien Shan, the Blepharocerid larvae and pupae, including even the form inhabiting the middle and the lower reaches (*Blepharocera asiatica*), were encountered exclusively on stones and rocks. None was found in moss, pools or calm water near the bank. They were mostly on the upper and, partly, on the lateral surfaces of stones and rocks. They were absent from the lower surface and preferred large stones and boulders or, more frequently, huge fixed rocks. Thus, the *Blepharoceridae*, when choosing microbiotopes, avoid the places of slower current (on the downstream side of the stone or in whirlpools). The groups of different age are somewhat differentiated in their choice of habitat. Younger larvae may be found both nearer the bank and on smaller stones. The same was observed by Hora (1930) in the Himalayan *Blepharoceridae*. It seems that a certain relation exists between the larval age and the flow velocity.

Mature larvae (IV stage), pupae and very young larvae (I and II stages) sometimes occurred altogether on the same rock on which adults probably laid the eggs. In some instances, when the eggs were deposited on the stone which, because of the lower water level happened to be in slow current, the young larvae travelled across the stones to rocks exposed to faster current.

Hence, despite some differences in ecological conditions preferred by different age groups of the larvae, it may be stated that the preimaginal phase of *Blepharoceridae* are true hymarobionts, a kind of typical, or "landscape-confined" group in rough mountain torrents. Analysis of their distribution in the torrent shows clearly that when either creeping (larvae) or staying attached (pupae) on the upper and lateral sides of the stones, they are affected by fairly strong water pressure. They use their adaptive devices continuously, not occasionally as do some lithorheophylous animals which choose biotopes with slower current.

The life cycle of *Blepharoceridae* is not completely known. According to Kellogg (1903), some *Blepharoceridae* from North America have two generations: (i) the first one in June, from eggs which were deposited at the end of the last summer and spent the winter as larvae; (ii) the second from eggs deposited at the beginning of summer by wintered females. Komarek (1941b) and Hubault (1927) suggested only one generation in *Blepharoceridae* from the Caucasus, Armenia and the Alps. According to Hubault, they have a very simple cycle: there is one prolonged generation, like in many *Arthropoda* from cold swift waters. Females deposit their eggs during all warmer months of the year, particularly at the beginning of summer. The larvae hatch a month later, spend the autumn, winter and spring in the torrent and then pupate.

Our observations are in favour of both one or two generation cycles. The matter seems to be that there is a general regularity in the development of the lithorheophylous torrential fauna, *viz.* in the lower waters many forms, even from different groups, have two generations per year, but upwards the stream, in the upper middle reach and, the more so, at the headwaters they have only one generation.

For instance, the emergence of pupae of *Blepharocera asiatica* in the Issyk River was confined to a certain date. They were found only in mid June; then, until late August they were absent and only larvae were present. From the end of August, pupae again occurred in large numbers, persisting in the river during the whole winter. Adults appeared only after 9 July. In the Akbura River adults were found even in April. From this one may suggest that the *Blepharocera asiatica*, a dweller of the middle and lower reaches of the torrent, has two generations. The presence of pupae in the Issyk River after 11 July and adults after 9 July is an indication of imago hatching and appearance of the generation whose larvae have wintered. The occurrence of pupae from the end of August suggests that they are from adults which have wintered and deposited their eggs early in spring.

Adults of *Blepharoceridae* usually occur not far from larval habitats near the torrents. They can be found along the stream shores in vegetation, on leaves of the poplars and bushes, but they chiefly prefer the places inaccessible for the investigator: above the water surface at the midstream, above cascades and waterfalls or on stones constantly watered by spray.

Males and females of *Blepharocera asiatica* have been many times encountered in the Issyk and Akbura Rivers. But the "dancing" of adults over the water can be observed only at certain times of the day. This usually occurs in the evening (at five or six o'clock, but occasionally earlier, at two or three o'clock) when small swarms of adults arranged "dances" just over the water surface, flying so close to the water that they are certainly sprinkled with spray. During the rest of the day the adults crowd in the recesses of rocks, under the bridge abutments and in other heavily watered shelters or in the vicinity of cascades. In nearshore vegetation they were only in small numbers. In tranquil state, the *Blepharoceridae* adults have a very characteristic posture, in which they are completely different from the *Tipulidae* adults. When sitting down, the *Blepharoceridae* keep their body high on their long legs; the rear is raised a little. In contrast to the *Limonia*, their body is never parallel to the plane on which they sit.

Except for *Blepharocera asiatica*, adults of other *Blepharoceridae* can be rarely observed as they inhabit inaccessible and specific places. As is known, adults of most *Blepharoceridae* species have been described from the specimens extracted from mature pupae.

Blepharoceridae females feed on blood of small insects (*Chironomidae* and others of similar size). In females their mandibles and hypopharynx are well-developed, large and serrated, but in males these are smaller and smooth (males probably feed on the nectar).

Blepharoceridae deposit their eggs on stones in the torrent near the very water surface.

Let us briefly consider their reasons of preference for extreme conditions in a torrent rather than for the swift waters in general. The fast-flowing watercourses are generally considered as the primary ones to many groups of aquatic insects, for example, to the *Chironomidae* family (Brundin, 1966), and it is no wonder that the Blepharoceridae prefer such waterbodies. Here they find much food (diatomic growths on stones), abundance of dissolved oxygen, rather stable water temperature (many streams are free from ice in winter) *etc.*, but the fact that the preimaginal phase of this dipteran family chooses, among all possible microbiotopes, most extreme conditions with fast turbulent current needs to be explained in some way or other.

Blepharoceridae in Tasmania have serious enemies: of the 40 pupae collected in the National Park 10 had holes which had caused their death; two of the few pupae collected in the Burma locality were empty. Somewhat later, in this locality three out of five pupae were found empty. There were two species in these cases: *Edwardsina fluviatilis* and *E. feruginea*, which inhabit moderately fast waters. On stones in the places where these species live a numerous population of caddisworms and stonefly larvae was found (Tonnoir, 1924). In Tonnoir's opinion, the frequent findings of the demolished pupae could explain the rarity of grown forms in some species. For example, in a torrent from the crater lake Cradle Mountain (Tasmania) he could find no adults or larvae over a period of three weeks. Similar observations were made by Needham and Lloyd (1916) with respect to *Blepharoceridae* in North America.

Our observations in the Issyk and Akbura Rivers and in other streams on the Tien Shan suggest the following regularity. The more numerous the mayflies, stoneflies and caddisflies on the stones in a torrent, the less numerous the larvae and pupae of *Blepharoceridae*. These microbiotopes and torrent sections are exactly those where current is relatively slow and the water temperature is a little higher than is typical of the torrent in its middle and lower reaches. But where these orders of insects were in few, if any, numbers, here was the realm of *Blepharoceridae* which alone represented the whole lithorheophylous fauna of invertebrates. One result of the ousting of *Blepharoceridae* into conditions unusual for them was the frequent occurrence of larvae and pupae, particularly from the *Blepharocera asiatica* and the *Tianschanella monstruosa* species, covered with diatomic and green-blue algae and all kind of growths (see, *e.g.* Brodsky, 1936, Fig. 20) and, literally, dress-coated by fine mineral particles, more often by loess-like grains. In such places many dead pupae were found dressed in fungal coatings.

From all this one may conclude that the extremes of biotopes occupied by preimaginal *Blepharoceridae* are a way of protection against their enemies (carnivorous caddisworms and stonefly larvae) and, perhaps, against the need to compete for habitat and food, *i.e.* in this instance, the main role was played by the biocenotic factor.

Such ousting of the aquatic *Blepharoceridae* from more suitable (less extreme) places has resulted in formation of perfect adaptations, as a result of which the Blepharocerid larvae became capable of living on stones exposed to strong current which can wash away any other aquatic animals.

When explaining the absence of *Blepharoceridae* from some watercourses which one would think to be appropriate biotopes for them, not only abiotic factors (roughness of stones)[26] must be considered, but also the composition of the lithorheophylous fauna should be analyzed, *i.e.* it is essential that "neighbours" of *Blepharoceridae* be studied taking into account their quantitative relation to *Blepharoceridae*.

Their larvae are food not only for carnivorous nymphs of aquatic insects, but also for fishes. Thus the analyses of fish from the Akbura River, *viz.* the marinka (*Schizothorax intermedius*) and the acclimated Amu-dar trout (*Salmo trutta oxianus*) showed large numbers of Blepharocerid larvae in their guts. The larvae were found also in the guts of *Diptyhus dybowskii* from the Chirchik headwaters (Sibirtzeva, 1961). The significant role of *Blepharoceridae* in the forage of the Amu-dar trout from the Amudarya headwaters is described by Shaposhnikova (1950).

4.2. Family *Deuterophlebiidae* (mountain midges)

a. Species composition

This is a very peculiar family in the *Diptera*, whose aquatic phases can be with certainty assigned to typical dwellers of swift waters with relatively low temperature. Species of this family should be described as true hymarobionts, or members of the lithorheophylous fauna of mountain torrents. Previously considered as monotypical, the one genus *Deuterophlebia* of the family is represented now by several species, however not adequately examined or comparatively analyzed. At present, the species of this genus, after their discovery in Kashmir (Edwards, 1922) and later in the Altai (Pulikovskaya, 1924) and Middle Asia (Brodsky A. and Brodsky K., 1926;

---

[26] Another reason has been suggested for the absence of *Blepharoceridae* from some (specific) places of torrents: no suitable "shelter" for adults (small caves and recesses in rocks, nearshore woody vegetation, bushes, *etc.*). This suggestion was made from the study of the Alhni River in the north-western Himalaya where it runs in broad moraine amphitheatres (Dubey and Kaul, 1971). But this represents a particular case of dominance of abiotic conditions.

Brodsky, 1930b), were found in North America (Pennak, 1945; Kennedy, 1958, 1960), in Canada, north Korea, in Japan and, more recently, in the Himalaya (Lahaul region) at elevations 3500 to 4000 m in glacial torrents with temperature of water 4.5° (Santokh and Singh, 1961).

We found the species described as *Deuterophlebia mirabilis* Edw. in the Issyk and Akbura Rivers and in many typical mountain torrents on the Tien Shan: from the collections by Gurvitsch, Yankovskaia and Sibirtzeva in the Western Tien Shan and the Western Pamirs. However the species, as we pointed out above, was not found in the Eastern Pamirs. With this exception, it may be said that the mountain midge is found throughout the mountain regions of Middle Asia.

To conclude the section, we refer to Mani (1968), who writes that at high elevations in the Himalaya the same families of the *Diptera* are common as in the Tien Shan and the Alai-Pamirs: *Simuliidae, Chironomidae, Tipulidae, Bibionidae, Culicidae, Deuterophlebiidae* and *Blepharoceridae, etc.* The *Deuterophlebia* was encountered at heights ranging from 3000 to 4000 m above sea level.

b. Notes on adaptations in Deuterophlebia

The larvae and pupae of *Deuterophlebia*, as previously pointed out, are found only in mountain torrents, but, in contrast to the Blepharocerid larvae, they prefer places with very fast, but more regular (*i.e.* less turbulent) current. Observations on habitats of the preimaginal *Deuterophlebia* showed: "The mountain torrents which gave a shelter to larvae and pupae of *Deuterophlebia* are not very rough and have no large boulders with strong churning of water in their vicinity. In this, the larvae and pupae of *Deuterophlebia* differ from those of *Blepharoceridae*, since the latter usually prefer rough foamy torrents and situate themselves towards the water churning. Their way of attachment to the substrate seems to be different from that in the Blepharocerid larvae and their resistance to washing out is less strong. We noticed that the *Deuterophlebia* larvae occur on horizontal stone surfaces where diatomic slimes are common" (Brodsky A. and Brodsky K., 1926, p. 77).

Of interest is the larvae's characteristic means of movement over stone surface: the larvae regularly swing their heads like a pendulum and slowly, beginning with the first pair, move their lateral appendages forward in turn. Such mode of movement is so typical that, despite the small size (from 1 to 5 mm) the larvae are fairly visible on the stone surface. The *Deuterophlebia* larvae do not have the suckers which are so typical for the Blepharocerid larvae and – perhaps as a result – their body is not fused like in *Blepharoceridae*. The head, three sections of the thorax and eight abdominal segments bear lateral appendages (prolegs), one pair on every segment (Fig. 72). By protruding or removing the appendages, the larvae cling tightly to the substrate and are capable of fairly fast movement at flow velocity over 2.5–3.0 m/sec.

The normal position of the body in a moving larva is lateral, *i.e.* the larva

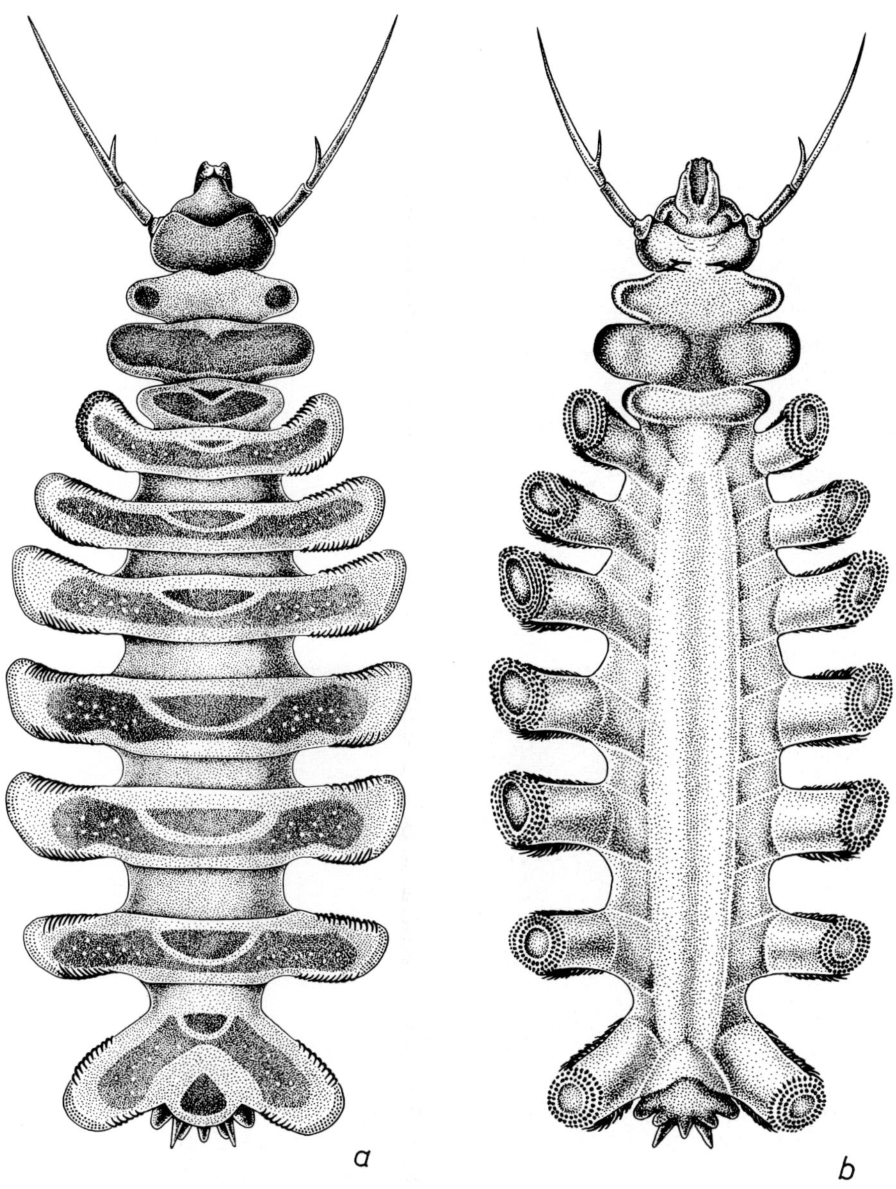

Fig. 72. Mountain midge (*Diptera*) *Deuterophlebia mirabilis* EDW. a. larva, dorsal view; b. larva, ventral view. (a and b, original).

alternately frees its fore or hind part of the body and, before attaching itself, rotates the free end of the body in a semicircle. The pupae firmly attach themselves to the substrate (stone, rock) (Fig. 74) with their six oval attaching fields situated on the abdominal segments 3–5. The pupa attaches itself with a sticky secretion so tightly that any attempt to take it off the stone with pincers leads most commonly to injury of the pupa (it must be

165

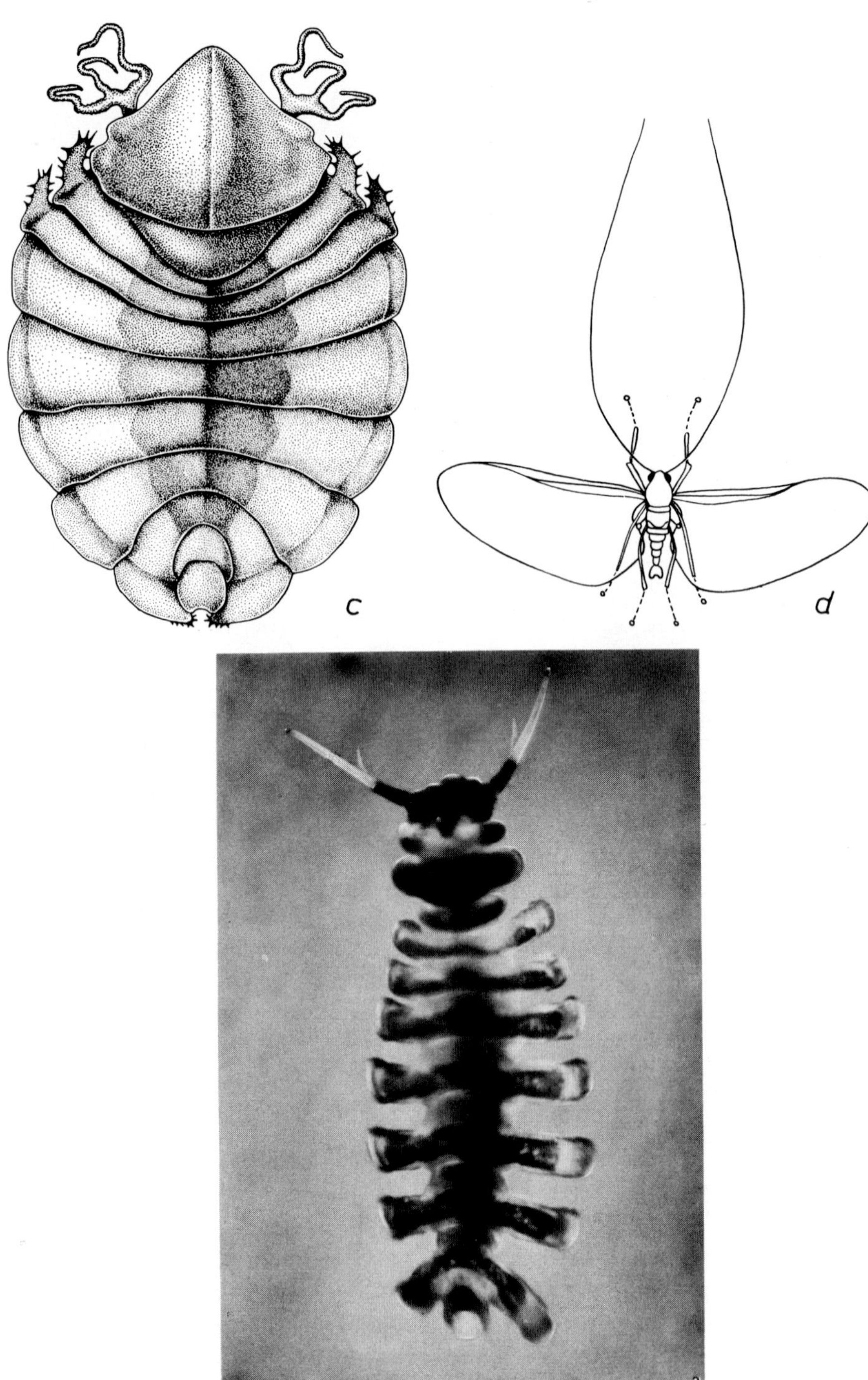

Fig. 72 (continued). c. pupa, dorsal view; d. male adult, dorsal view; e. larva, dorsal view. (c and e, originals; d, from Edwards, 1922).

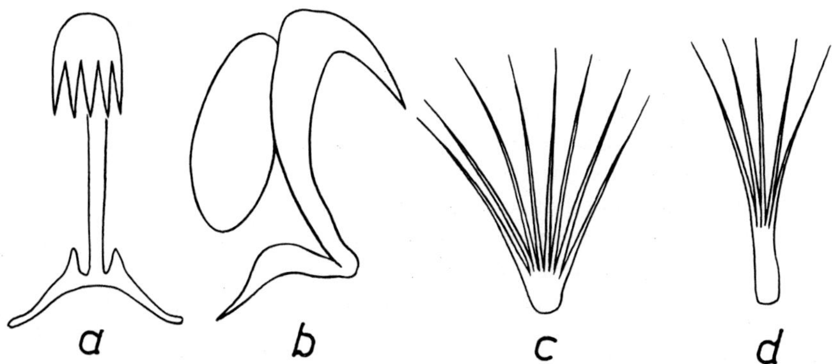

Fig. 73. Hooks and setae on lateral appendages in larva of *Deuterophlebia coloradensis* PEN. (After Pennak, 1945).

a. frontal view of claw; b. lateral view of claw; c and d. brancheal setae.

detached with a razor or sharp penknife). The *Deuterophlebia* pupa is very flat, more so than the Blepharocerid pupae (except the *Philorus*) and the upper surface of the hard pupal cover is ideally streamlined.

Nobody was able to observe the adult hatching, but this problem is probably as difficult for *Deuterophlebia* as for other members of the torrential lithorheophylous fauna with attached pupae. It seems that the hatching is somewhat encouraged by a sharp change of the water level in the stream, because of which the pupae previously located at depths of 30–40 cm and deeper appear near the water surface or are only wetted by the spray. Thus, the water level drops significantly, for example, towards the autumn.

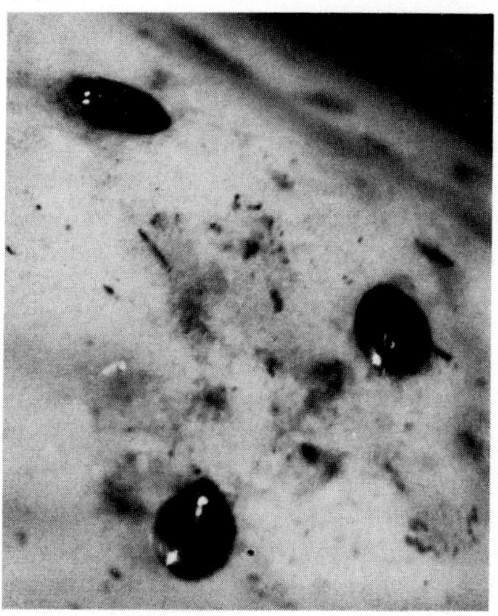

Fig. 74. Pupa of the mountain midge on granite boulder (Photo by V. Bukin).

c. Some data on the ecology of Deuterophlebia larvae and pupae

The substrate selected by the larvae is exclusively stony: pebbles, stones, rocks; they have never been found in moss or on sandy or silty deposits. The depths vary from several to 30–40 cm. The larval sensitivity to oxygen was described above: they are encountered only in waters of high oxygen content. The flow velocities are up to 3–4 m/sec and higher; the minimum velocities at which single larvae were found were about 1 or even 0.71 m/sec (one instance). The thermal range is fairly wide, from 5 to 11°, but most numerous the larval and pupal populations are at water temperature of 5.0–5.8°, much less numerous at 10°. In rare cases the larvae were found at the temperature of 15.0 and even 17.0° (one instance). For the Issyk River the following relation was found between the larval and pupal population and the water temperature (°C) from simultaneous samplings:

Water temperature	Numbers of larvae and pupae
5.0–5.8	8.6
7.0–7.5	4.0
8.0	6.3
10.6–11.0	1.0

The seasonal cycle is not traced, but the larvae and pupae could be recovered from June till October. This does not mean, of course, that the larvae and pupae are absent from torrents in winter. The number of generations is most probably one per year. Adults are found very rarely (single instances).

d. Vertical distribution of larvae and pupae of Blepharoceridae and Deuterophlebia along the torrent

The vertical distribution of *Blepharoceridae* and *Deuterophlebia* has been traced in the Issyk and Akbura Rivers in more detail than in any other stream. Comparing the data on these two families, one can see a difference in their vertical distribution patterns, which depends on the river lengths and, mostly, on the different altitudinal position of the headwaters.

Even a very rough comparison of the species and various river sections shows that the species of *Blepharoceridae* are not distributed over the entire length of the rivers, but each is restricted to a certain portion of them (Figs. 75 and 76). In the Akbura River at about 1900–2000 m above sea level, only two species occur – *Blepharocera asiatica* and *Tianschanella monstruosa*. Farther upstream one species is found, *Asioreas nivia*, which is confined to elevations from 2600 m to the headwaters (about 4000 m above sea level). Sporadically the *B. asiatica* is also met with at elevations 2600–2800 m. At higher elevations the species of the genera *Blepharocera*, *Tianschanella* and *Deuterophlebia* disappear and the river is inhabited only by the species *A. nivia* of the *Asioreas* genus.

In the Issyk River which runs, as pointed out earlier, at lower elevations and farther north than the Akbura River, one species of *Blepharoceridae*

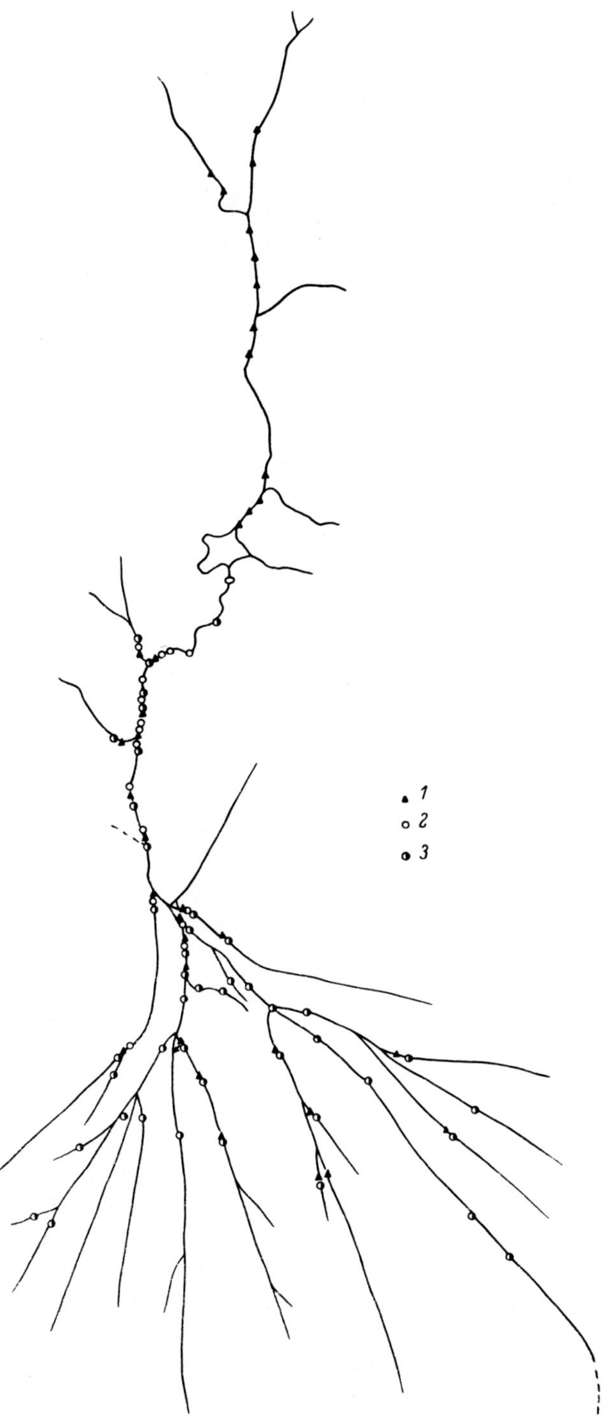

Fig. 75. Longitudinal distribution profile of the net-winged and mountain midges in Issyk River.
1. *Deuterophlebia mirabilis*; 2. *Tianschanella monstruosa*; 3. *Blepharocera asiatica.*

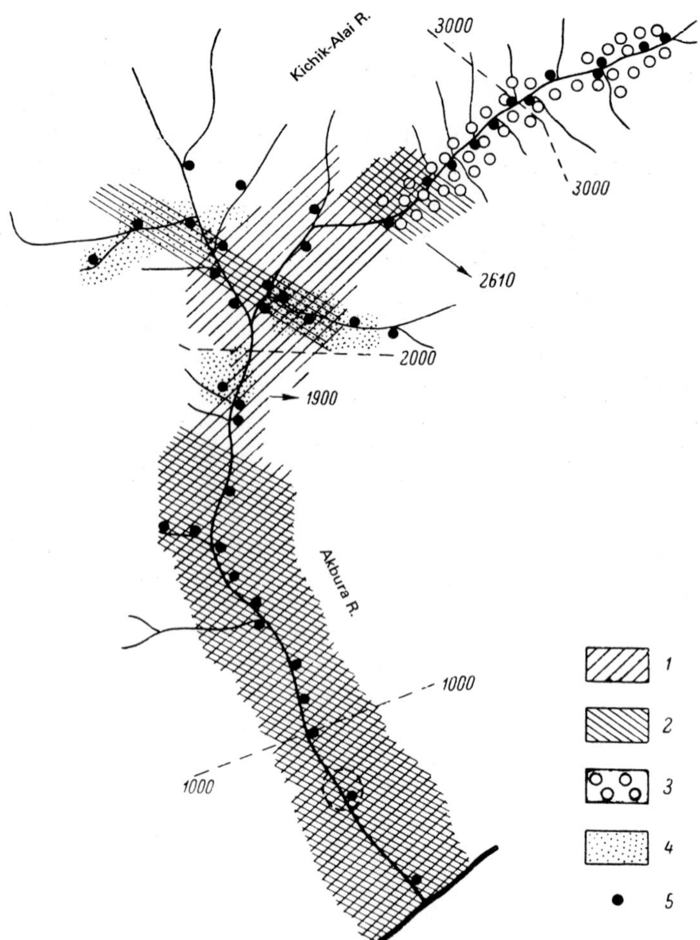

Fig. 76. Longitudinal distribution profile of the net-winged and the mountain midges in Akbura River and its headwaters, the Kichik-Alai. (After Brodsky and Omorov, 1972).

1. *Tianschanella monstruosa*; 2. *Blepharocera asiatica*; 3. *Asioreas nivia*; 4. *Deuterophlebia*; 5. sampling sites. The numbers are absolute elevations in metres.

which is known in the Akbura River is absent, *viz. A. nivia.* Two species of this genus (*A. nivia* and *A. tianschanica*) were found on the Zailiiskii Alatau, but at higher absolute elevations, up to 3800–4000 m. Thus *A. nivia* was repeatedly found in cascades below the snow-fields and glaciers and in the headwaters of glacial torrents at absolute height about 4000 m (at the Djure Pass in the Zailiiskii Alatau none of the numerous larvae and pupae were found dead or injured as it was at lower elevations). In the Issyk River the *Tianschanella monstruosa* and *Blepharocera asiatica* occupy a part of the irrigation system and the lower and middle reaches; therefore, it is difficult to differentiate between their vertical distribution patterns from the longitudinal profile of torrent as is possible in the Akbura River. The

170

Table 32. Percentage ratios of immature stages (larvae plus pupae) of Blepharoceridae and Deuterophlebiidae from Akbura river.

Elevation in m	*Blepharocera asiatica*	*Tianschanella monstruosa*	*Asioreas nivia*	*Deuterophlebia mirabilis*
<1000	70.5	29.5	—	—
1000–1500	77.5	22.5	—	—
1500–2000	88.5	8.8	—	2.7
2000–2500	80.0	8.4	—	11.6
2500–3000	1.1	14.8	1.5	82.6
3000–3500	—	—	100.0	—
>3500	—	—	100.0	—

vertical distribution pattern of *Blepharoceridae* and *Deuterophlebia* for the Akbura River is given in Fig. 76.

The patterns made on the basis of a mere presence or absence of species give a simplified picture of the species distribution. A quantitative criterion is needed to evaluate the abundance of individuals. We have used the percentage ratio of the amount of individuals of every species to the total amount of individuals of all species of both *Blepharoceridae* and *Deuterophlebia* (Table 32).

From a plot of these data versus the longitudinal Akbura profile (Fig. 77) one can get a very representative pattern. In the lower reach two species, *B. asiatica* and *T. monstruosa*, occur; the population of the first species increases up to an elevation of 2200 m, then drops down to a negligible value. The second species is present in relatively small numbers of larvae and pupae, but it goes up beyond 1000 m along the river, *i.e.* nearly up to 3000 m and higher. Starting from quite small numbers at the lower reach,

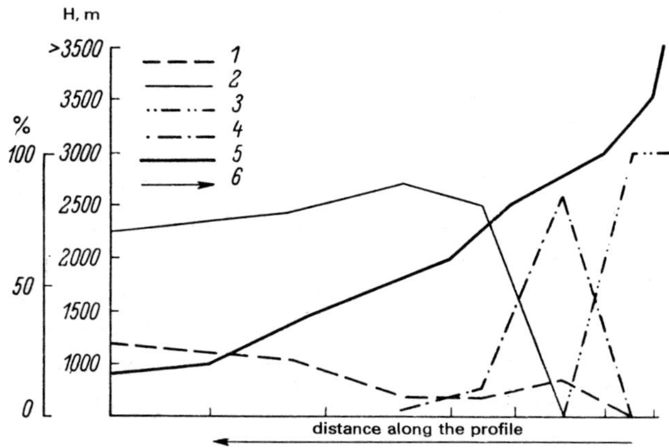

Fig. 77. Percentage abundance of each species of *Deuterophlebiidae* and *Blepharoceridae* along longitudinal profile of Akbura River and its headwater, the Kichik-Alai. (After Brodsky and Omorov, 1972).
1–4. see Fig. 76; 5. longitudinal profile; 6. direction of flow; 7. absolute elevation in metres.

the larvae and pupae of *Deuterophlebia* increase sharply between 2500–3000 m as the *Blepharocera* population decreases. Lastly, the upper river, especially its headwaters, is occupied by the only species, *Asioreas nivia*, whereas the other two species of *Blepharoceridae* have disappeared.

For comparison of distribution boundaries of *Blepharoceridae* and *Deuterophlebiidae* in the Issyk and Akbura Rivers, we present Table 33 where absolute elevations and water temperature are given for zones inhabited by different species. It must be borne in mind, however, that the extreme limits could be broadened if data on other torrents of the Tien Shan will be used, but we must always take into account the quantitative aspect of distribution.

Table 33. Altitudinal and thermal limits of distribution of the net-winged and the mountain midges.

Name of species	Issyk River		Akbura River	
	absolute height in m	water temperature in °C	absolute height in m	water temperature in °C
Deuterophlebia mirabilis	1050–2230	5.0–17.5	1500–3500	9.0–11.0
Blepharocera asiatica	900–1439	11.0–21.2	900–2750	9.0–19.0
Tianschanella monstruosa	1050–1554	11.0–17.5	900–3000	9.0–19.0
Asioreas nivia	—	—	2750–3000	0.5–6.2

Apart from the description of different river sections with respect to the presence or absence of certain forms of the dipteran families under study, these data, although preliminary, make it possible to give a generalized ecological account of the forms proper which can be used in the study of other mountain rivers and torrents of Middle Asia.

For example, it is clear that *B. asiatica* is a dweller of the lower reach and the lower middle reach, but *A. nivia* is a form typical of the upper reach or even of headwaters, if the river is either of glacial or snowy and glacial origin.

As is seen from the Table, different forms of *Blepharoceridae* have different tolerance and the most flexible form is the one from the moderate mountain zone or even the submontane zone, the *Blepharocera asiatica*. Indeed, this species appears to be most tolerant of calcium content in the water. In some brooks and river with high calcium content which can be easily recognized even from the calcerous scale on the stones, the larvae and pupae of this species were also covered by the scale. No other forms of this family or *Deuterophlebiidae* were found under similar conditions.

As for the genera of *Blepharoceridae* then the most typical of mountain torrents are the genus *Asioreas* described earlier and the *Philorus* not found in the Issyk and Akbura Rivers, but found in the Western Tien Shan and the Western Pamirs. The respective ecological data indicate that the latter genus avoids slow current and relatively high water temperature, and for this

172

reason, together with the *Asioreas* genus, it should be regarded as a particularly characteristic hymarobiont.[27]

After the short account of the distribution of the two families, it is necessary to pass over to a physiographical description of individual river sections which are occupied by different species of the families; the more so because the distribution has so far been considered by generalized and purely conventional sections of the river, subdivided on the elevation basis rather than by their natural physiographical conditions. The sections differed by 500 m in elevation, which may be more or less appropriate to the lower and middle reaches, but in the upper reach this interval may cover very different zones, both ecologically and physiographically.

It is clear that disappearance or advention of species and variations in their proportions do not depend directly on the elevation above sea level, but on the water temperature, current, substrate and, particularly, on its mobility and stability, on nearshore vegetation, *etc.* To exemplify this idea, one can point out the absence of Blepharocerid larvae and pupae from rivers or river sections which would seem to satisfy entirely the ecological requirements (including the biocenotic ones) of these organisms. Absence of this family and *Deuterophlebiidae* from some affluents and river sections is accounted for by flash mudflows and sharp fluctuations of the water level. Quite another matter is the other groups of inhabitants of mountain torrents and rivers, for example, mayflies, stoneflies, side-swimmers and aquatic beetles which do not attach themselves to stones like pupae of *Blepharoceridae*, or which are more mobile than the larvae of *Blepharoceridae* and *Deuterophlebiidae*. Not only winged stages, but also larvae and nymphs of mayflies are very mobile, easily transported by current and, therefore, are capable of inhabiting rapidly of a river section just flashed by mudflow. This is impossible for *Blepharoceridae* and *Deuterophlebiidae* and not all rivers and their affluents are inhabited by larvae and pupae of these families. The development cycle of one generation of *Blepharoceridae* and *Deuterophlebiidae* lasts in a river about a year. And all this time the stones where the larvae and especially the pupae abide must be immobile.

What zonality is to be attributed to the Akbura River and its headwaters, the Kichik-Alai River, from the distribution of the *Blepharoceridae* and *Deuterophlebiidae*? Comparing Figs. 76 and 77, one can distinguish between a zone of the lower reach with a relative abundance of *T. monstruosa* and *B. asiatica* and a zone of the middle reach which is characterized by a sharp increase of larval and pupal population of the second species. The latter zone should be divided into two subzones. Lastly, a third zone is the upper reach with abundant *A. nivia* (Table 34).

The general characterization of these zones and subzones is based on the averaged data for August. One should remember that the zonal boundaries

---

[27] The above account of the individual species of *Blepharoceridae* may be rather perplexing, since we repeatedly assign the entire family *Blepharoceridae* to hymarobionts, but the question is now of differentiation within the group of hymarobionts, and that is why the genera are named as hymarobionts in a "higher degree" than the other genera of the family.

Table 34. Vertical zonality in Akbura River with respect to *Blepharoceridae* and *Deuterophlebiidae*.

		Absolute height in metres	Mean water temperature in °C	Mean velocity of flow in m/sec	Substrate	Scenery vegetation	Characterization of *Blepharoceridae* and *Deuterophlebiidae*
Upper reach		4000–2700	6.2	2.5	Rough fragments	Bare moraine, then alpine herbaceous vegetation below it	Only *Asioreas nivia*; at the lower limit of the zone occur *Blepharocera asiatica* (solitary), *Tianschanella monstruosa* (more) and *Deuterophlebia mirabilis* (very abundant)
Middle reach	Upper section	2700–1900	9.9	3.2	Badly-rounded pebbles, in some places rocks and fragments	Juniper zone	Two species: *Blepharocera asiatica* and *Tianschanella monstruosa*, the former more abundant; numerous *Deuterophlebia mirabilis* becomes rarer towards the end of the section
	Lower section	1900–1100	11.6	2.5	Rounded pebbles	Lower limit of the Juniper zone at 1700 m; below it hardwood trees and bushes	Two species: *Blepharocera asiatica* and *Tianschanella monstruosa*, the former more abundant; at the beginning of the section *D. mirabilis* becomes less abundant
Lower reach		1100–900	19.0	1.47	Rounded pebbles, silted in some places	Hardwood trees and cultural landscape	Only two species: *Blepharocera asiatica* and *Tianschanella monstruosa*, the former more abundant

do not represent distinct, sharply-defined lines and that the vertical subdivision of a torrent by the distribution of species of any group means only a principal subdivision which varies even in the same river with the hydrological conditions of the year (different water discharge) and season.

4.3. Family *Simuliidae* (blackflies)[28]

a. Species composition

Like other groups of torrential dwellers, the blackflies can be described in much the same way: systematics of the lithorheophylous fauna of this group from the alpine zone is far from adequate, which strongly impedes the study of their ecology. "The fauna of the mountain and high-mountain areas of Europe, as well as of the south Caucasus and Middle Asia is very poorly known..." (Rubzov, 1956, p. 106).[29] Further he notes the immeasurably poorer knowledge of the fauna of mountain lotic waterbodies compared with lowland ones and that very little evidence is available on the genus *Montisimulium* – an ecologically high-altitude group. "Very few valleys have so far been examined up to the elevation of 3000 m. In nearly every gorge between 1000 and 1500 m new forms and species are discovered in the northern Caucasus and the Trans-Caucasus, in Kirghizia and Tadjikistan. Only two or three samplings were made at about 4000 m in torrents flowing down from the snow-fields. In every sample two to four new and quite original species were found... among them such peculiar ones as *Montisimulium alpinum* (Rubz.), *M. keiseri* Rubz., *M. luppovae* Rubz., *M. planipuparium* (Runz.) and others. Their way of life and occurrence are almost entirely unknown" (op. cit., p. 118, 119). About the Kirghizian fauna we read in Konurbaev and Tadjibaev's work (1970, p. 119) that it is "extremely poorly known", but this work contains a list of 40 blackfly species from 7 genera which inhabit a number of Kirghizian rivers (Kara-darya, Yassy, Kugart, Zarkent, Kara-su, Kara-ungur, Akbura, Abshirsai, Chilisai, Isfairamsai and others).

A few years later we again learn from specialists in the blackfly fauna that the fauna of high-elevated areas on numerous ranges in Middle Asia and Kazakhstan is almost untouched, and that every new sample from the Pamirs, the Tien Shan and the Gorno-Badakhshan Region at elevations of 3000–4000 m contains new blackfly species. Especially frequent are new species from the endemic Middle Asiatic genera *Montisimulium, Sulcicnephia, Foretodagmia,* and the pan-palaearctic *Metacnephia* and *Eusimulium.* It is this high-mountain fauna, although poorly covered, which contains the largest number of endemics representative of the individual mountain systems (Rubzov and Shakirzyanova, 1976). Konurbaev also stresses the unusual abundance of endemics in Middle Asian mountain regions. Most of these species are well adapted to the conditions of rough

[28] The author is very thankful to I. A. Rubzov who has read the manuscript and given his considerations on this part of the chapter.

[29] Since 1956 the data on the mountain fauna have considerably increased (see Rubzov, 1976).

mountain torrents and differ in their morphology from the species known from mountain valleys. These are the genera *Montisimulium, Obuchovia, Odagmia* and *Tetisimulium* (Konurbaev, 1976b).

The scanty knowledge of the mountain and high-mountain blackfly faunas is especially evident against the background of submontane and lowland blackfly data (Rubzov, 1976). Rubzov has kindly provided us with a list of blackfly species typical of this dipteran fauna of torrents, mountain rivers and brooks of Middle Asia (to identify the torrential fauna proper is still impossible). Using these data and his work of 1976 and a joint work of Shakirzyanova and himself (1976), it is possible to give the following list. As Rubzov notes, the genus *Montisimulium* (detached from the *Eusimulium*; Rubzov, 1976) is especially representative for these waters:

Montisimulium montium Rubz., *M. ocreastylum* Rubz., *M. shadini* Rubz. (the Pamirs), *M. stackelbergi Rubz.* (Kondara River, the Western Pamirs), *M. duodecimcornutum* Rubz. (Mazor-darya River, the Pamirs), *M. kirjanovae* Rubz., *M. decimfiliatum* Rubz. (*s. lato*), *M. quatturodecimfilum* Rubz. (*s. lato*), *M. djabaglense* Rubz., *M. lepnevae* Rubz. (Kondara River, the Western Pamirs), *M. inflatum* Rubz., *M. octofiliatum* Rubz. (*s. lato*), *M. longifiliatum* Rubz., *M. decafilis* Rubz., *Obuchovia albella* (Rubz.), *Ob. versicolor* (Rubz.), *Obuchovia* spp., *Wilhelmia veltistshevi* (?) (Rubz.), *Metacnephia multifilis* Rubz. (*s. lato*), *M. tetraginata* Rubz. (*s. lato*), *M. pedipupalis* Rubz., *M. kirjanovae* (Rubz.), *M. persica* Rubz., *M. sedecimfistulata* Rubz., *M. pamirensis* Petr., *Simulium flavidum* Rubz. (a collective species: from the alpine to the submontane zone), *S. multistriatum* Rubz., *S. subornatoides* Rubz., *Tetisimulium alajense* (Rubz.) (some forms of a collective species), *Prosimulium phytofagum* Rubz., *Eusimulium kazahstanicum* Rubz., *E. brachyantherum* Rubz., *Eusimulium* spp., *Odagmia* spp., *Sulcicnephia stegostyla* (Rubz.), *S. ovtshinnikowi* (Rubz.), *S. argulacea* (Rubz.), *Cnetha keizeri* Rubz., *C. kugartsuense* Rubz., *C. planipuparium* Rubz., *C. luppovi* Rubz., *Chelocnetha montschadskii* Rubz., *Ch. longpipes* Rubz., *Ch. crassiusculum* Rubz.

This list gives only some idea of the blackfly fauna in fast watercourses on the Tien Shan and the Pamirs and is, of course, only a first attempt to understand the rich, diverse and principally endemic fauna of these regions.

The systematizers know that the family *Simuliidae* presents a special difficulty. Like some other insects with aquatic preimaginal phase, the blackflies, outwardly very similar in their morphology, are however remarkable for their significant biological flexibility, diverse development cycles and different number of generations per year. Hibernation and development halt may occur at the egg stage or in the larval one (sometimes in one and the same species). As a result, the larvae, pupae and adults may have significant seasonal differences, by some characters more remarkable than the ecological and local differences (Rubzov, 1974). But the seasonal, local, interpopulational and the individual variabilities are relatively large and the largest variability affects the characters which are functionally related to the highly variable environmental factors (Rubzov, 1970). Such variability is found in mayflies, *Blepharoceridae* and caddisflies (for example, see Lepneva, 1964). The variability, particularly in larvae and pupae, results in almost every watercourse having its own "forms" of these insects. What are they? – Usually ecotypes, sometimes subspecies and species. Variability of the preimaginal phase in the Middle Asiatic blackflies was a subject of special study (Konurbaev, 1973, 1976a) which has revealed a dependence of some structures on the ecological conditions. But this work has just commenced, and at present the *Simuliidae* systematics is subject to changes and intensive development: the genera change their coverage, the collective species are subdivided into species in their own right, *etc.*

These comments are made here only to show that the identification of blackfly species which typify the mountain torrent of the Tien Shan is still a matter of the future.

For the Issyk River unfortunately, we do not have a complete list of blackfly species, but they were a subject of special investigation in the Akbura River and some other regions of Kirghizia (Logacheva, 1959; Konurbaev, 1965, 1970; Konurbaev *et al.*, 1972), from which one can find appropriate data both on the species composition and distribution of the blackflies.

The material on which the paper of Konurbaev *et al.* (1972) is based was samples and observations in 1966–1968 and, partly, in 1969 and 1970. The treatment of these data and identification of blackfly species was performed by Konurbaev. Systematic sampling was carried out at 12 sites in the lower and middle reaches of the Akbura River within 1400 m above sea level and in its affluents, and at 64 sites at higher elevations towards the river headwaters (3670 m). The list of blackfly species identified from the Akbura River is given in Table 35.

As is seen from the Table, the Akbura blackfly fauna is not rich in species.

Table 35. List of blackfly species from Akbura river and its affluents (after Konurbaev *et al.*, 1972).

Name of species	Main channel	Affluents
Wilhelmia mediterranea PURI*	–	+
Tetisimulium alajensis RUBZ.	+	+
T. desertorum RUBZ.*	+	–
T. kerisorum RUBZ.*	–	+
Phoretodagmia ex. gr. *ornata* MG.	+	+
Ph. ephemerophila RUBZ.	+	+
Sulcicnephia ovtshinnikovi RUBZ.	+	+
S. jankovskiae RUBZ.	+	+
S. undecimata RUBZ.	–	+
Metacnephia kirjanovae RUBZ.	+	+
Metacnephia sp.	–	+
Tetisimulium sp.	–	+

Note: *are those of uncertain identification, as Rubzov informed us.

The overwhelming majority of species found in the Akbura River, except for *W. mediterranea* Puri and *Odagnia ex. gr. ornata*, are dwellers of the mountain and high-mountain lotic waters. They are seldom, if at all, encountered in the lowland rivers.

b. Notes on adaptations in blackflies

The morphological structure of the organs of attachment and locomotion and their functioning are fairly well-known in blackfly larvae, and we use here the data summarized by Hynes (1970) and Rubzov (1956), as well as our own observations of blackflies in the Tienshanian torrents.

On their front "leg", or "proleg", and at their hindquarter the blackfly larvae bear a circlet of fine "prehensile" hooks (Fig. 78). These structures are similar to those found at the lateral appendages in *Deuterophlebia*. In addition, the blackfly larvae have two well-developed web glands which, in combination with the hooks, form a rather reliable system to resist and move in fast currents over smooth rocks, stones or tree trunks.

What is the mechanism of attachment and locomotion in blackfly larvae? The larva spins on the substrate a kind of tangled mat from the secretion of the web glands, which it then attaches itself to by hooks on the highly modified anterior adjunct of the "leg". This adjunct is on the ventral side of the first thoracic segment, and it projects forwards under the head. Its distal portion can be retracted by special muscles or erected by the blood pressure when the larva is moving. The adjunct bears circlets of outwardly directed hooks. When the "proleg" touches the web mat, the hooks are erected outwards and engage in the web threads. The posterior organ is similar to that on the anterior adjunct ("leg") but is much more powerful and bears from 500 to 6000 (in rheophils) hooks; it has also a set of retractor muscles and prehensile hooks which are capable of making a series of outward circular movements.

At rest the larva is normally attached by its posterior organ to the web mat and hangs free in the current (in slack waters it sticks up vertically); if moving, it bends over so that its head and the openings of the web glands touch the substrate, where it spins another mat and attaches the anterior adjunct; then it moves the posterior organ to the newly made mat, releases its anterior adjunct and repeats the process. In this way it progresses slowly in a manner of caterpillar of the geometrid moth. If the larva is disturbed or the current becomes very strong, it bends over and attaches its fore proleg to the mat. In this way it doubles its attachment to the substrate and brings its body into the boundary layer. Attaching a web thread to the substrate, the larva moves by it like a spider and it can, in the same manner, climb back up the thread, by gathering it between the front proleg and the head which bears a number of pectinate spines engaging the web.

Therefore the blackfly larvae are well adapted to life in rapid running waters where they often occur.

The blackfly pupae are housed in cases of the very different construction from the web threads (a secretion which sets in the water); although

178

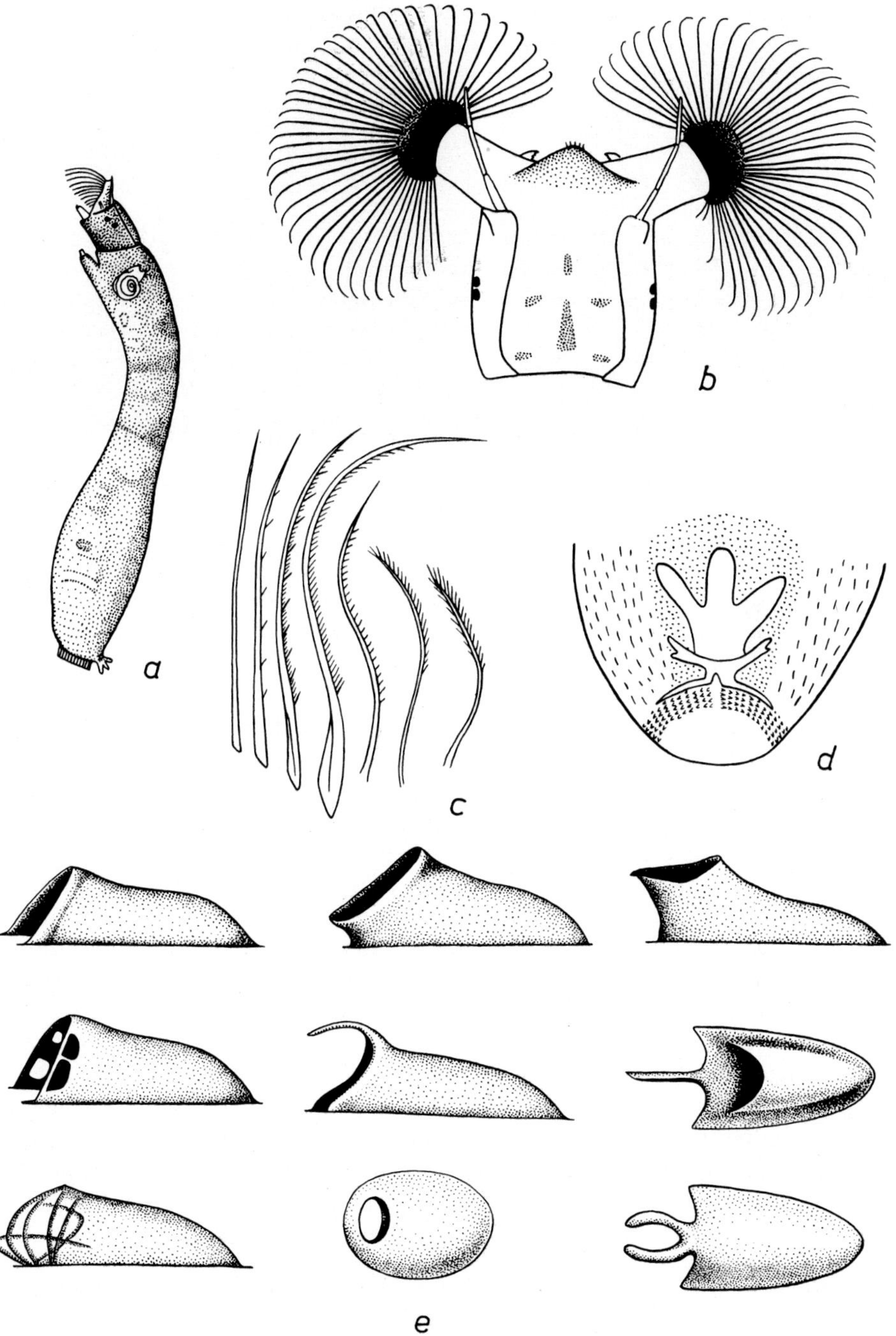

Fig. 78. Larva of blackfly *Simulium alajense* RUBZ and various cases. (After Konurbaev, 1965.)
a. larva laterally; b. larval head with "large fan"; c. setae of the fan; d. posterior end of larval body, the anal gills and rings of the attaching organ are seen; e. types of cases made by blackfly pupae, lateral view.

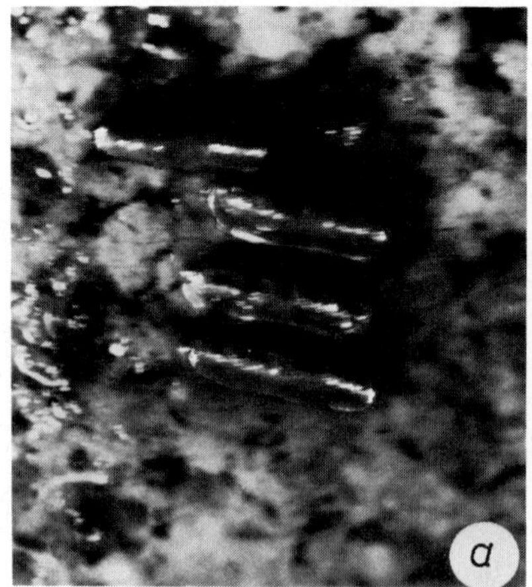
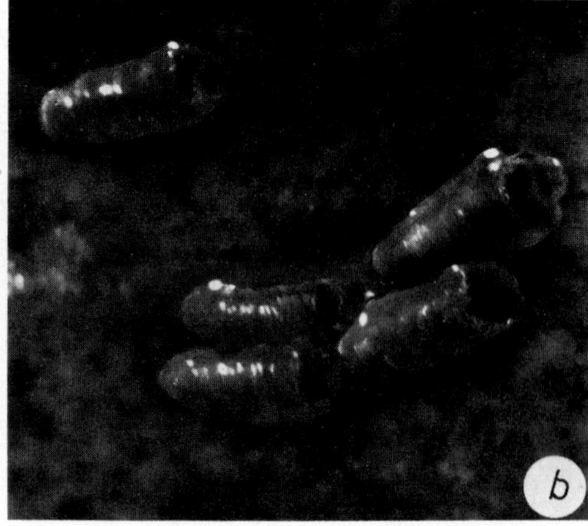

Fig. 79. Blackfly (*Simuliidae*) pupae on stones. The flow direction is from left to right. (Photo by V. Bukin).

varying in their form (the shape of a slipper, *etc.*; Fig. 79), the cases have a streamlined upper surface and their posterior end directed upstream. The pupae are firmly attached to the substrate and are capable of withstanding strong water pressure.

Hatching of the winged phase proceeds in the following manner. In the pupal coverings a T-shaped rupture is formed. Its transversal line detaches the head capsule and the longitudinal one separates the covers of the midback starting from the front side. The insect quickly escapes into the water and, thanks to its waterproof covers, comes up to the surface; only a few seconds after emergence from the case, it flies away.

After the description of adaptations in the preimaginal stages it is worthwhile mentioning an interesting phenomenon peculiar to some blackfly species. We have in mind the phoresia, or cohabitation of larvae and pupae of these species with the larvae and nymphs of mayflies (Rubzov, 1972). Older larvae of three species (*Phoretodagmia ephemerophila* Rubz., *Ph. obikumbensis* Rubz. and *Ph. alajensis* Rubz.) climb from stones onto the dorsal surface of mayfly nymphs of the genera *Iron* and *Rhithrogena*, where they pupate. Rubzov believes that the attachment of blackfly larvae to mayflies helps them to avoid destruction from movement of stones in rapid torrents and improves their feeding conditions. The larvae of the phoretic species have modified organs of case spinning and food catching, as well as the structure of their head adjoints and the posterior attaching organ.

c. Vertical distribution of blackfly larvae and pupae along the torrent and other ecological data

Our observations show that, in contrast to the Orthocladiin larvae, the blackfly larvae prefer a substrate free from vegetation – stones, wood, but they also use branches hanging over the water, stems of terrestrial plants, *etc.* Rubzov (1956, p. 86) notes what "... the largest aggregations of larvae

180

occur at maximum velocities of the water flow". Their preferential substrate is stones or, if the stones are mobile, the vegetation which hangs over the torrent. However, Hynes (1970) writes that the blackfly larvae prefer fastflowing watercourses with laminar current and are absent from the turbulent currents since these tend to remove food particles from larval fans. We do not see any discrepancy in this, because they seem to refer to different species and neither Rubzov nor Hynes have in mind definite blackfly species.

Thus our observations show that the Issyk blackflies occur in sections of the torrent with very swift and rough water, sometimes in such places where no other torrential dwellers live, except the larvae of this dipteran family and the *Chironomidae* (*Orthocladiinae*). That the blackfly larvae prefer the stone, but not vegetation, is confirmed by the comparative counting of blackfly larvae from ten sites of the Issyk River where there were both stones and moss. It turned out that on stones there were 2340 larvae and in moss 108, which gives, respectively, 95.0 and 5.0%.

In some parts of torrents and brooks the blackfly larvae occur in such vast numbers that it can be said that they create a continuous cover in suitable microbiotopes. The following are the observational data for the Issyk River.[30] In a site of the middle reach where the river runs out of the Issyk Lake, 300 individuals were found on the upper surface of a stone of $70\,\mathrm{cm}^2$, or 4.3 larvae per $1\,\mathrm{cm}^2$; on another stone in the same site were 150 animals per $18\,\mathrm{cm}^2$, or 8.7 larvae per $1\,\mathrm{cm}^2$. A little upstream but in the same reach, almost every stone was covered with larvae from 13 to 20 individuals per $1\,\mathrm{cm}^2$, in average, 16.5 larvae per $1\,\mathrm{cm}^2$, which gives from 165,000 to 200,000 individuals per $1\,\mathrm{m}^2$.

The abundance of larvae in specific sites of the mountain torrent may be explained as follows. The large numbers of larvae were found in the immediate vicinity of the lake from which the stream runs out. Since there are no "filters" through which the water would pass, the plankton washed out of the lake occurs in vast numbers past the lake (it disappears 0.5 km downstream). Besides, in the conditions of numerous cascades (large slope of the moraine over which the stream is running), the blackflies have no enemies, *viz.* carnivorous larvae of caddisflies and stoneflies, whose absence or presence is an important factor in the distribution of lithorheophylous fauna (as early as 1922 Miall considered the caddisfly larvae as the blackfly's most dangerous enemy).

Thus, the extreme conditions which make the lithorheophylous fauna poor ensure, with the absence of enemies and competition, rich blackfly larval fauna, *i.e.* they provide shelter, or refuge, for these animals, where their population density can reach a maximally possible value.

The data on development cycles in blackflies are summarized in Rubzov's monograph (1956) and in Hynes (1970). Here we only point out that the

---

[30] We draw attention to the fact that the observational data characterize individual microbiotopes, and these are, therefore, not the same values as the averaged data for the entire river sections with which, unfortunately, one has to deal when determining biomass of different groups in the lithorheophylous fauna.

number of generations is larger in blackflies than in other hymarobionts: there are species with two or three generations per year, and some species are multicyclic. However here again the dependence exists: the number of generations decreases with increasing elevation.

Thus the blackflies of the genera *Metacnephia* and *Sulcicnephia* from the upper Akbura River are monocyclic. They pupate and fly out since after mid June and August. However the species from the middle and lower reaches are multicyclic and give at least two generations per year.

In the Issyk River, the blackfly larvae make their appearance only in July, but as single individuals; in August they increase their quantity and in late August the pupae become as numerous as the larvae (50%) and even more (100%). According to Rubzov (1956), the blackflies hibernate at the larval or pupal stage, but Konurbaev and Tadjibaev (1970) point out that the larvae hatch from and pupate all year round, *i.e.* also in winter, even at water temperature 1.5 to 4.0°C.

The blackfly distribution along the Akbura River was studied by Konurbaev and Tadjibaev (1972). Analysis of the vertical occurrence of blackfly species and inventory of their quantities per 10 cm^2 or as percentage in each sample (where the larvae and pupae were numerous) were made, as noted above, in 1966–1968 and, partly, in 1969 and 1970. Regular sampling was performed over the entire length of the river up to its headwaters (3670 m above sea level).

The species abundance ratio varies with the vertical belts. Some difference in the species composition is observed between the faunas in the main channel and in the affluents. Thus in the lower zone (900–1100 m above sea level) upstream from Osh city, *Tetisimulium alajensis* dominates (84.3%) in all the sampled sites of the main channel. The accompanying species are *T. desertorum* (5.7%), *Odagmia ex. gr. ornata* (9.4%) and *Phoretodagmia ephemerophila* (0.6%). The presence of the latter in the zone seems to be occasional, perhaps, as a result of its washing out or migration of mayflies of the *Iron* genus, to which the blackfly larvae attach themselves. The samples from this zone contained only two larvae of *Ph. ephemerophila*.

What attracts one's attention is the fact that in the main river downstream from Osh city the blackflies are entirely absent.

Repeated observations made in different seasons of the year have shown that in this zone of the main channel the blackflies do not breed. This may be due to the sewages and other wastes, *etc.* which enter into the Akbura River near the residential area; the water becomes muddy and dirty downstream of the city. Perhaps the blackflies avoid this river site.

Quite a different picture is observed in the main, more or less large irrigative canals and ditches flowing from the Akbura main channel. Water turbidity here increases a little, the water temperature sharply varies both during the day and with seasons from 1 to 32°. In these canals *Wilhelmia mediterranea* prevails (75%) and develops all year round (it hibernates in the larval stage), being in some places very numerous (up to 400 larvae and pupae in June per 10 cm^2). The accompanying species are *T. alajensis* and *O. ornata.*

182

In the lowland and submontane regions, on the flood plain of the Akbura River, *W. mediterranea* is an ill-natured bloodsucker which attacks farm animals and man, but in the main channel it occurs very rarely, only where the current becomes slower near the shores overgrown with vegetation.

The species composition of blackflies in the middle reach zone (1400–2000 m above sea level), both in the main channel and the affluents, is somewhat different from that in the lower reach. The dominants here are *T. alajensis* (67.4% in the main river and 33.5% in affluents) and *Sulcicnephia ovtshinnikovi* (18.2 and 51.0%, respectively). Among the accompanying species in this zone are *Odagmia ex. gr. ornata, Phoretodagmia ephemerophila, Tetsimulium kerisorum, Cnephia undecimata*. Although these species are not so abundant as the dominants, their absolute numbers are fairly large (over 20 larvae and pupae per 10 cm^2 of stone area).

Phoretodagmia ephemerophila occurs frequently along the middle reach of the main channel on mayfly nymphs and pupae and on stones.

Very different is the blackfly species composition in the upper reach zone (2000–3670 m). In this zone the monocyclic species *Sulcicnephia jankovskiae* and *Metacnephia kirjanovae* are common, making up the main bulk of blackflies.

In the affluents of the upper Akbura I–III stages of the *Tetisimulium alajensis* larvae prevail (90%). Because of the difficulties in their identification and absence of adult stages, the taxonomic affinity was determined approximately on the basis of similarity with the larvae which had been collected in the middle and lower reaches. In the upper river other related species or genera may occur.

There is no doubt, however, that in the Akbura River and its affluents the *Tetisimulium* blackflies are widespread and represent a quantitatively prevailing group.

They are encountered all year round and hibernate in the larval stage. Species of this genus inhabit mostly the mountain and high-mountain rivers and brooks in which they prevail numerically over other groups. They also occur in lowland rivers and in the submontane zone where they are regular bloodsuckers. At present the blackflies of this synantropic, rapidly evolving genus are typified by ecological and seasonal subspecies, forms and hardly identifiable species.

It is very likely that it is because of the inadequate knowledge that some of related species and subspecies inhabiting the mountain regions of Middle Asia can not yet be differentiated to a species.

4.3. Family *Chironomidae*

a. Species composition

Of the family *Chironomidae*,[31] the subfamily *Orthocladiinae*[32] is most typical of fast flowing waterbodies and what we have said about many other

---

[31] The author is very thankful to V. Ya. Pankratova for the discussion of this section.

[32] In her "Key", Pankratova (1970, p. 4), following Brundin, combines *Orthocladiinae* and *Diamesiinae*, which earlier were considered as separate subfamilies.

hymarobionts is also relevant here: "The *Orthocladiinae* fauna in many geographical regions is very little known, if at all." (Pankratova, 1970, p. 5); "most forms in the subfamily *Orthocladiinae* live mainly on stony substrata in cold springs, brooks, rivers. . . or lakes. . . , the nearest relatives of the *Orthocladiinae* are the larvae of the subfamily *Diamesinae*. . . , almost all members of this subfamily also are coldwater litho-rheo-oxyphyls. . . , the characteristic complex of the *Chironomidae* in lotic waters at elevations of 1100–1200 m also consists of the cold-water lithorheophylous *Orthocladiinae* and *Diamesinae*" (Pankratova, 1950, p. 196). The *Orthocladiinae* are found up to the highest elevations in the mountain waterbodies and were encountered by us as the only representatives of the lithorheophylous fauna in brooks flowing from the glaciers and even on the glacier itself. It may be assumed that in the Tien Shan they ascend up to 3800–4000 m and higher. We have already noted the poor and uneven state of knowledge of *Orthocladiinae*. Thus Pankratova (1970) described and named for the mountain rivers and torrents 28 larvae (species), among them 27 were from the Western Pamirs and the Gorno-Badakhshan! The Tienshanian fauna of these midges is still almost *terra incognita*.

Some of the *Chironomidae* lists available in the literature include the forms which are not yet identified to a species (*e.g.* for the Chu River; Ovchinnikov, 1936) or whose identification cannot be relied upon (for example, abundance of "European" species among the *Chironomidae* from the Turgen River; Kurmangalieva, 1974). The list of *Chironomidae* for the Zeravshan River (which is somewhat conventionally assigned by us to the Tien Shan) has been compiled by Sibirtzeva (1961); it covers the main channel of the Zeravshan and its affluents, the Magiandarya and Kshtut-darya. Not all sections of these rivers, particularly of the main channel, can be considered to be of the torrential type. It is not possible to separate the species on the basis of their ecological conditions, therefore the list combines ecologically different forms of the *Chironomidae*. A more reliable list was kindly made available to our use by Kustareva of the Issyk-kul Biological Station of the Kirghiz Academy of Sciences, and we should like to acknowledge her contribution.[33] The list includes more uniform ecological data and gives an idea of the species composition of the Chironomid larvae from the rivers of the Issyk-kul depression: the Aksu, Karasu and the Cholpon-Ata Rivers:

Heptagia accomodata Pankr.,
Syndiamesa sp.,
Diamesa sp. (similar to *hygropetrica*),
D. thienemanni Kieff.,
D. latitarsis Geotg.,
D. insignipes Kieff.,
D. quinquesetosa Pankr.,
D. pseudostylata Tschern.,

[33] The forms were identified by Kustareva and verified by Pankratova at the Institute of Zoology, the USSR Academy of Sciences.

Odontomesa fulva Pag.,
Eukiefferiella longicalcar (Kieff.),
E. communis Pankr.,
Eukiefferiella sp. (similar to *similis*)
E. bavarica Goetg.,
E. masordariensis Pankr.,
E. tschernovskii Pankr.,
E. discoloripes Goetg.,
E. atrofasciata Gortg.,
E. sellata Pankr.,
Eukiefferiella sp. (similar to *calvescens*),
Eukiefferiella sp. (similar to *popovae*),
Eukiefferiella sp. (similar to *oxiana*),
Eukiefferiella sp.,
Orthocladius rivicola Kieff.,
O. thienemanni Kieff.,
O. frigidus (Zetterst),
O. rivulorum Kieff.,
Orthocladius ex. gr. olivaceus Kieff.,
Orthocladius sp. 1,
Orthocladius sp. 2,
Paratrichocladius sp.,
Corynoneura scutellata Winn.,
Thienemaniella clavicornis Kieff.,
Orthocladiinae gen. sp.?
Ablabesmyia ex. gr. tetrasticta,
Cryptochironomus camptolabis,
Micropsectra sp.

In 1976 the Institute of Zoology organized a survey of the fauna in the Tiup river (an affluent of the Issyk Lake). Pankratova has treated the *Chironomidae* collected from this river (Alimov *et al.*, 1977). Below is the list of *Chironomidae* which were collected in the Tiup River from stones (at depths of 20–30 cm) in swift waters:

Syndiamesa orientalis Tshern., *Syndiamesa* sp. 1, *Syndiamesa* sp. 2, *Pothastia gaedi* (Mg.), *Diamesa insignipes* Kieff., *Diamesa* similar to *insignis* Kieff., *D. thienemanni* Kieff., *D. hygropetrica* Kieff., *Odontomesa fulva* Pag., *Eukiefferiella longicalcar* Kief., *E. bavarica* Goetg., *E. masordariensis* Pankr., *E. tschernovskii* Pankr., *E. discoloripes* Geotg., *E. calvesens* Edw., *E. oxiana* Pankr., *E. longipes* Tschern., *E. atrofasciata* Geotg., *E. quadridentata* Tshern., *E. hospita* Edw., *E. coerulescens* Edw., *Eukiefferiella* sp., *Orthocladius rivulorum* Kieff., *O. thienemanni* Kieff., *O. rivicola* Kieff., *O. saxicola* Kieff., *O. oblidens* Walk.?, *Orthocladius* similar to *frigidus* Zett., *O. fuscimanus* Kieff.?, *Orthocladius* sp. 1, *Orthocladius* sp. 2, *Orthocladius* sp. 3, *Orthocladius* sp. 4, *Cricotopus biformis* Edw., *C. algarum* Kieff., *C. trifasciatus* Pan., *C. holsatus* (Geotg.), *C. ephippium* Zett., *C. atritarsis*

Kieff., *C. silvestris* F., *cricotopus* sp., *Acricotopus lucidus* Staeg., *Paratrichocladius inserpens* (Walk.), *Limnophyes distrophilus* Tshern.?, *Limnophyes* sp., *Synorthocladius*? sp., *Microcricotopus*? sp., *Lapposmittia*? sp., *Thienemannia*? sp., *Psectrocladius simulans* Joh., *Corynoneura scutellata* Winn., *Thienemaniella* sp., *Orthocladiinae* gen. spp., *Microdentipes* sp., *Ablabesmyia* sp.

The presence of a number of "European" species in the list raises some doubts and, at least, indicates inadequacy of the existing *Chironomidae* systematics for the Tienshanian torrents.

The paper cited above gives a short list of *Chironomidae* from the Kara-Kol River which looks more like a mountain torrent than the middle Tiup River: *Syndiamesa* sp. 1, *Syndiamesa* sp. 2, *Eukiefferiella* sp. similar to *E. popovae* Tshern., *E. tshernovskii* Pankr., *E. masordariensis* Pankr., *Orthocladius* similar to *O. rivulorum* Kieff., *O. fuscimanus* Kieff.?, *O. saxicola* Kieff.

b. Notes on adaptations in Orthocladiinae

Adaptations in the larvae and pupae of *Orthocladiinae* to the conditions of fast current are very diverse and composed of a whole system of different morphological formations. First of all, the workers point out the small size of larvae which allows them to use the roughness of stones, rocks, etc. and, of course use the slower current in the boundary layer. Larvae of some species have web glands to attach themselves with the web threads. Some others build a case-cover which is firmly attached by one end to the substrate. Still others have much shortened pushing prolegs which form a kind of sucker (Fig. 80) for tight attachment to the stones (Thienemann, 1936; Pankratova, 1950). In her guide, Pankratova again writes about the "suckers" in *Orthocladiinae*: "The forms adapted to life on the nearshore stones in rough torrents (*Heptagia*) have tremendously shortened pushing

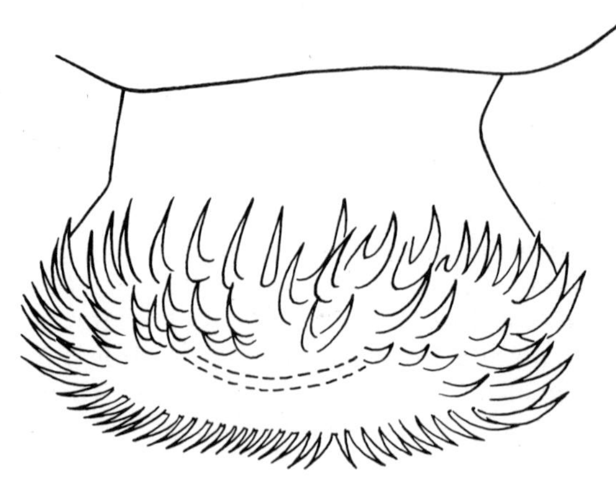

Fig. 80. "Pushing" proleg in larva of chironomid *Heptagia*. (From Thienemann, 1954, after Pankratova, 1970.)

prolegs with their tops, rounded by hooks, modified into suckers" (1970, p. 15). The use by specialists in *Chironomidae* of the term "suckers" for modified pushing prolegs in *Heptagia*, of course, cannot be justified, since we deal in this instance with a false sucker or, more correctly, with a structure which is similar to the sucker only by analogy but which is not a true sucker.

c. Ecological data on the preimaginal phase of Chironomidae

Inadequate knowledge of the systematics of *Orthocladiinae* prevents us from discussing the longitudinal distribution of *Orthocladiinae* in the Issyk and Akbura Rivers. However, certain ecological observations made by us in the Issyk River are of some interest, as they relate to a fairly limited group of species of this subfamily of *Chironomidae* which inhabit only typical mountain torrents with ecological conditions preferred by true torrential dwellers.

The Issyk River is inhabited nearly exclusively by the subfamily *Orthocladiinae*. The occurrence of this group is very high over the whole length of the torrent, and almost without interruption. The preferential substrate is diverse, mainly stones, pebbles, rocks and moss. The size of inhabited stones varies from large pebbles to sharp-angular rocks. In deep waters the larvae and pupae were not found. They were recovered chiefly from the nearshore zone where very small-sized forms of *Orthocladiinae* occurred also on projected stones, if they were constantly watered by waves and spray. Therefore, these forms, like *Orphnephilidae* and *Dixidae*, may be assigned to the hygropetric fauna.

Adults of *Orthocladiinae* can be observed in large numbers on nearshore stones or, in the evening, when they swarm in the air. Small Orthocladiinae were a usual entomological catch on the riverside vegetation, and bridges over the water, *etc.* Among various forms of the Orthocladiin larvae from the Issyk River, of interest are those in silty tubes, which seem to tend to be widespread farther downstream and to play a certain role in the fauna of irrigation canals. But the pupae were in slimy capsules. In moss the Orthocladiin larvae, especially the small-sized ones, were found in mass and in larger numbers than on stones; from eleven sites in moss 2,222 individuals (97.8%) were counted but on stones 50 (2.2%).

In the Issyk River, the larvae and pupae of *Orthocladiinae* were most numerous in places having the following general features: (i) where the current was roughest, thus in most typical sections of mountain torrent and cascades; (ii) more or less developed moss growths on stones and rocks; (iii) presence of water spray forming a zone which is most suitable to the hygropetric fauna development; (iv) large percentage of foam (up to 70–100%) on the water surface, which is indicative of fast and rapid currents (turbulent flow over the bed with numerous obstacles).

Where the Orthocladiin larvae and pupae were most numerous, the observer noticed an abundance of moss-living forms and an absence of any forms which prefer more tranquil current. In places of most abundant *Orthocladiinae*, the mayfly nymphs are entirely absent, except for larvae and nymphs of the *Baetis*. In such places there were neither larvae nor pupae of the stoneflies and caddisflies. Instead, blackfly larvae accompany the *Orthocladiinae* in large numbers.

Therefore the suitable conditions (on the one hand, extremely rough current, availability of a wetted zone on the nearshore rocks and, probably, on the other the absence of enemies) enable the Orthocladiin larvae to inhabit the cascade zone's very extreme conditions. Of course, the two requirements are interrelated and again we must give credit not only to the abiotic factors, but also to the biocenotic aspects, and it is hard to say which of them should be placed first.

5. Some other families of the *Diptera*

The list of families in the *Diptera* order is far from exhausted by the *Blepharoceridae, Deuterophlebiidae, Simuliidae* and *Chironomidae*. One might also mention the aquatic phases of a few species from some other dipteran families inhabiting fast-flowing watercourses: *Tipulidae, Limoniidae, Psychodidae, Dixidae, Ceratopogonidae, Stratiomyiidae, Tabanidae, Rhagonidae, Empididae, Tetanoceridae, Ephydridae, Anthomyiidae* and others. However, information about their aquatic phases is still too scarce (excluding the biting midges) for their description to be adequate. The four previously described families (*Blepharoceridae, Deuterophlebiidae, Simuliidae* and *Chironomidae*) are typical of mountain torrents and their aquatic phases comprise a major dipteran complex in this type of watercourses.

Sampling from the riverside vegetation on the Issyk river gave a number of forms in the imaginal stage. Some of them are not aquatic insects, but we give here the complete list for characterization of the riverside entomofauna (preliminary identification was made by the well-known dipterologist A. A. Stakelberg).

Culicidae
    Culex sp.
    Aedes (*Ochlerotatus* sp.)
Dixidae
    Dixa maculata Mgn.
Simuliidae
    Simulium (*Odagmia* sp.)
Tipulidae and *Limoniidae*
    Tipula irrorata Schumm.
    Symplecta punctipennis Meig.
    Limonia spp. (two species)
    Dicranomyia sp.
    Malophilus propinquus Egg.
    Pachyrrhina cornicina L.
    Pericoma sp.

Syrphidae
    Platychirus pellatus Meig.
    Paragus bicolor
    Syritta sp.
    Syritta pipiens L.
Chironomidae
    g. sp. (a number of forms)
Sciaridae
    Sciara sp.
Lonchopteridae
    Lonchoptera furcata Fall.
Phoridae
    Phora sp.
Drosophilidae
    Drosophila sp.

Empididae
  Clinocera spp. (two species)
  Bicellaria sp.
  Hemerodromia raptoria Mgn.
Dolichopodidae
  Hydrophorus sp.
  Syntormon pumillum Mgn.
  Xanthochlerus tonellus Wd.
  Dolichopus sp.
Cecidomyidae
  Thuraia sp.
Psychodidae
  Pericoma palustris Meig.
Ephydridae
  Coenia pallustris Fall.

Parydra littoralis Meig.
Scatella lutosa Meig.
Pelina aenea Fall.
Cordylluridae
  Hydromyza sp.
  Amaurozoma sp.
  Scatophaga squalida Mgn.
Muscidae
  Chortophila aestiva Meig.
  Stomoxys calcitrans L.
Sepsidae
  Sepsis sp.
Chlorophidae
  Oscinella sp

Besides the above *Diptera* were found *Liancalus* and *Atherix* in the adult stage and *Calliophrys* in the adult, larval and pupal stages. The hygropetric fauna of wet stones on the torrent shores was represented by imagoes, larvae and pupae of the *Dixa* and some *Chironomidae*. Among adults collected from the riverside vegetation were species which do not hatch in the torrent, *e.g. Mycetophilidae* and *Sciaridae*. Thienemann characterizes their aquatic phase as inhabitants of springs. He also assigns *Cecidomyiidae* to this group. There is no doubt that many adults collected on the Issyk River shores have hatched in the wet zone, under leaves, in mushrooms, *etc.*

Larvae of various flies were seldom found in the torrent but exclusively in moss; it seems impossible to assign them to the lithorheophylous torrential fauna.

Among *Limoniidae* in large numbers were found adults of the *Limonia* and in more limited numbers their larvae and pupae. The larvae were fairly large, white-coloured, with small knobs across the abdominal side and completely covered with chitinous spines. The pupae were in capsules made from web threads with sand grains adhered to them; the capsules were open from their lower side and attached to the stone. The pupa has an elongated cylindrical shape, with numerous appendages at the anterior end of body where they form a real bunch. The pupae were of dirty brown colour. All the larvae and pupae were found exclusively on stones where they lay in cavities on the lower, upper and lateral sides hidden in their capsules firmly attached to the stone. This form occurs exclusively in the lower Issyk River, not higher than 1540 m above sea level, at water temperature 10.4°. Only in one site, at absolute elevation of 1900 m and water temperature 9.0° seven larvae and pupae of this form were found.

Both sexes of the *Limonia* may be encountered on stones and rocks above the roughest places of the torrent – waterfalls, cascades and similar sites. The insects place themselves so closely to the water surface that they are constantly sprayed. The ruffled water frequently washes them away, but it does no real harm to these dipterans which are surprisingly well adapted to such conditions. When washed away, the adults once more ascend over the rock or, after a short flight, take their old seats – to be driven away again by the rough water or abundant spray. In this respect, the *Limonia* adults resemble the winged stages of *Blepharoceridae* which also choose similar places. The aggregation of the *Limonia* adults above cascades and rough sites of the torrent may be so large that a log of wood stuck in a waterfall sometimes seems of grey colour on its lower side, just above the water, where most of them usually sit. The bridges across the Issyk River are favorite breeding sites of the *Limonia*. Everywhere, on boards and abutments of the bridges, large numbers of males and females sit just above the rapids. On the boards they copulate and deposit their eggs onto the wetted side of the board. Sometimes, the copulating pair found itself under water but quickly showed up again above the water surface; there was no instance when the *Limonia* adults could be transported away by the current. This form did not demonstrate the soaring flight over the water observed in adults of *Blepharoceridae*. Catching with the entomological net above the water surface provided large numbers of the *Blepharocera asiatica* adults and only single specimens of the *Limonia* which had been probably disturbed with the net on the bridge abutments. The seating movement in the *Limonia* adults differs from that in the *Blepharocera* adults, as we mentioned earlier.

When the female deposits her eggs, she bends her abdomen and constantly touches the rough and wetted surface of the board with her elongated and rigid ovipositor; the eggs are deposited either at the water level or a little above it.

The *Limonia* adults, larvae and pupae were encountered in the same river sites. Copulation in the *Limonia* lasts from June to late August.

In the favourite locations of the *Limonia* adults, large numbers of dipterans of the *Empididae, viz. Clinocera* and *Hilara* also occur, as well as single individuals of the *Dolichopodidae, viz. Liancalus*. All of them sit on the bridge boards in a close proximity to the *Limonia*, but a little higher, beyond the zone of spray and waves. On the stones one could observe a specific distribution of their adults. The *Limonia* prefers the upstream surface of the stone, but the *Clinocera* occupies the downstream surface and dwells in its cavities (Fig. 81).[34]

The *Atherix* adults, whose larvae, many authors believe, are moss dwellers of lotic waterbodies, were found in the lower sites of the torrent, especially on bridges. In the Issyk River we often observed a phenomenon which is described in the literature but whose significance is unclear. On the lower side of a bridge board, above the water, one can see numerous dead adults glued together and mixed with clusters of deposited eggs. No young larvae were seen, while the dead bodies of flies were dry and empty envelopes. It is hardly conceivable that the larvae feed only on the chitinous envelopes of the maternal organisms, as Grünberg (1910) suggested. All the

[34] An interesting account of the Caucasian *Clinocera* was published by B. Kovalev (1973) in the journal of "Molodoi naturalist" (Young naturalist).

190

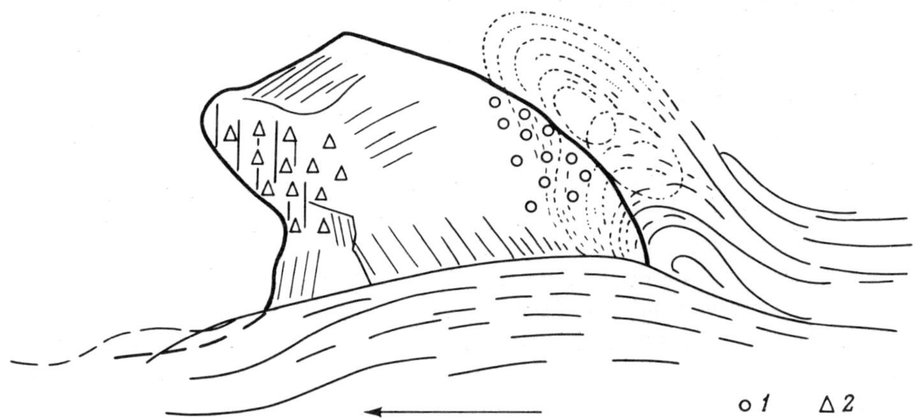

Fig. 81. Schematic diagram of adult dipteran distribution over stone with respect to flow direction (arrow).
1, *Limonia* sp., 2, *Clinocera* sp. (Original.)

aggregation of eggs and dead flies was webbed and the spiders seem to have contributed greatly to the demolition of the flesh of the dead flies. All this aggregation has an appearance of a broad pancake, which could by no means drop into the water. Therefore the possibility that all the mass drops into the water and serves as food for the *Atherix* larvae must be ruled out.

To conclude our observations of the *Diptera*, it can be suggested that some *Tipulidae* and *Limoniidae*, at least those from the *Limonia* genus, or, more correctly, their aquatic phases, should be assigned to torrential dwellers.

6. Notes on other invertebrate groups from the mountain torrent

Here we confine ourselves to some comments on insects (bugs, beetles), water mites, crustaceans, molluscs, aquatic earthworms, roundworms and turbellarians. Knowledge of these groups from the Tienshanian mountain torrents is extremely incomplete and irregular, so there is a possibility that some species of the above groups of invertebrates may also be true hymarobionts. For example, some water mites, beetles and turbellarians are well adapted to life in fast-flowing watercourses. But even so, the total mass of these groups is much inferior in abundance to the mayflies, caddisflies, *Blepharoceridae* and even stoneflies. That is why we give less attention to the groups which are considered in this section, and we do so not because of their inadequate knowledge, but because of their little significance in the biocenosis of lithorheophylous fauna of the Tienshanian torrents.

No bugs (*Hemiptera*) were found in the Issyk and Akbura Rivers. As is known, only one form, the *Aphelochirus*, is typical of torrents, but, having been found in Europe, this form has not yet been discovered in torrents on the Tien Shan. In a small waterbody connected with the Issyk River, one specimen of the *Acanthia arenicola* Scholtz (in Kirichenko's determination) was caught, and also a water scorpion *Nepa cinerea* L. at the end of the river, in a small mire. It goes without saying that these species have no respect to the lithorheophylous fauna of the torrent and they are mentioned here to stress that if the *Aphelochirus* were in the Issyk River, it would be found.

The beetles (*Coleoptera*) are rather common in the Issyk River, but only in its lower reach and in the irrigative canals, where the whirligig beetles are especially numerous. In the upper and middle river they are rarely encountered and in small numbers. The rare representatives of this order are by no means characteristic inhabitants of torrents. Their presence is explained by availability of sites with slow and quiet current like pools, the nearshore zone, *etc.*

Let us give some examples of the forms found in such sites, but with the serious reservation that their identification is only preliminary and, to be frank, of low reliability for reasons explained below. Thus the Issyk River yielded *Agabus femoralis?*, *Ochtebius exculptus?*, *Rhysodidae* gen.?, *Limnius* sp., *Colymbetini* gen. sp.?, *Gyrinus colymbus?*. The *Agabus femoralis?* was recovered, with larvae of *Ameletus* mayfly, in standing waters (in a small pool flooded at an increased water level). The *Ochtebius* and *Rhysodidae* were found in moss in the river, in one site. The whirligig beetles in the torrent are, of course, a random phenomenon, as they are numerous only in the backwaters, pools and slack brooks. In the Zeravshan River, the *Helmis* sp. and *Gravelens rioloidea* Reitt. were found (Sibirtzeva *et al.*, 1961). In addition to this list, one more form of aquatic beetles occurs in middle Asia – *Helmis quadricollis* (Shadin, 1950c). A number of species is pointed out for Middle Asia by Reinhardt and Ogloblin (1940), but the lack of accurate ecological data does not permit identification of the typical torrent dwellers.

Besides the Issyk and Akbura Rivers, abundant water beetles were often found in other torrents of the Tien Shan, *e.g.* in a river in the northern Alai Mountains, the Abshirsai River, where adult beetles of the *Dryopidae* family were observed. Its general character was very peculiar in that it had a more laminar current than the Issyk or Akbura; however the reasons for the beetles' preference for certain types of waters remains unclear.

Among the larvae of aquatic beetles are many well-adapted to life in fast-flowing watercourses (*e.g.* the families *Psephenidae*, *Dryopidae* and *Elmidae*), but the beetle fauna in torrents of the Tien Shan still represents a gap in our knowledge of the invertebrate fauna from this kind of waterbodies. The situation is better with aquatic beetles from swift waters of the Caucasus. Thanks to the studies by F. A. Zaitzev and some other coleopterologists, the beetle composition from rivers, brooks and torrents in this region is more or less complete. We are informed by a well-known specialist in Middle Asiatic beetles, O. L. Kryjanovski, that even for the Himalaya there has been a number of papers on aquatic beetles from swift waters; as to the Tien Shan, the work on this fauna has not yet started, and for this reason all identifications of beetles from the Tienshanian streams should be taken *cum grano salis*, *i.e.* very critically. One may expect that this group of torrential inhabitants of Middle Asia will be investigated before long both taxonomically and ecologically.[35]

From a preliminary identification, about 10 forms of water mite *Hydracarina* found in the Issyk River are typical rapid water dwellers: their legs have lost swimming hairs, but are provided only with short setae; they have a flattened body in a rigid armour. It is known that in mites of the genus *Lebertia*, the number of hairs on the legs is inversely proportional to the flow velocity (Schwoerbel, 1961).

In the Issyk River, most of the *Sperchon* mites (13 specimens) were found at depths from 20 to 90 cm, water temperature 11.5° and flow velocity over 3 m/sec. The river bottom at the sampling site is composed of rock fragments and large stones overgrown with mossy turfs near the banks. The mites of the genera *Lebertia* and *Megapus* were found in similar environmental conditions. Some mites were collected from the stones in the nearshore zone.

In the Chu River, the mites (*Hydrobates* (*Rhabdotobatus*) *intricatus* Sok. and *H. kirgisicus* Sok.) were found only within the backwaters. There is a hope that special investigation of the torrential mite fauna will considerably contribute to our knowledge of its composition; however, most of the mites are likely to be assigned to moss and vegetational growth dwellers of the nearshore zone of the torrent rather than to the lithorheophylous fauna.

Sampling in the Tiup River (the Issyk-kul Valley) made by A. I. Yankovskaia provides the following list of water mites (Alimov *et al.*, 1977): *Feltria* sp., *F. minuta* Koen., *Aturus* sp., *A. duplex* S. Thor., *Atractides nodipalpis* ssp. *fonticolis* Viets., *Sperchon* sp., *S. glandulosus* Koen., *S. g.* ssp. *cubanicus* (Sokolov), *S.* (*Hispidosperchon*) sp., *S. clupifer* Piers., *S.* (*Mixosperchon*) sp., *S. brevirostris* Koen., *Lebertia* sp., *L.* (*Lebertia*) *rivulorum* Viets., *L.* (*Pseudolebertia*) sp., *L.* (*P.*) *glabra* Thor., *L.* (*P.*) *tuberosa* Thor., *Kongsbergia materna* S. Thor., *Hydrobates foreli* Lebert.

7. *Crustacea*

In the Issyk River, *Harpacticidae* (*Canthocamptus* sp.) were found in moss together with rotifers and water bears. Undoubtedly, special investigation will reveal a rich moss fauna of this group. Below the Issyk Lake, *Calanoida* and *Cladocera* were found with a plankton net. Their origin leaves no doubt – they had been transported down from the lake. It is of interest that although these crustaceans were exhausted (they could hardly move) they had survived a 500 m travel downstream of very rough current, cascades, waterfalls, *etc.* Besides *Crustacea*, the "plankton" contained small numbers of very young mayfly larvae (*Ameletus* and *Baetis*) and the *Chironomidae* in the silty tubes.

[35] Recently a guide to the world fauna of aquatic beetles has been published, but only with reference to larvae and pupae (Bertrand, 1972). From the guide it is clear that the fauna of aquatic beetles from Central and Middle Asia, India and other regions of continental Asia is almost entirely unknown, while from mountain waterbodies of Middle Asia is completely *terra incognita*.

In eight sampling sites along the river, chiefly in moss, young side-swimmers *Gammarus spinulatus* Mart. were encountered (Martynov, 1935). They were recovered both from brooks and from the littoral zone of lakes. In brooks the *Gammarus* were so numerous that, if sufficient care had not been taken, they penetrated into the insectary and consumed the entire population there. Probably these animals prefer brooks of spring origin. The *Gammarus* occurred at elevation only up to 1790 m at water temperature of 8.5–10.5°. Judging from the Issyk fauna it must be suggested that the side-swimmers are accidental dwellers in mountain torrents but typical for springs, brooks and rivers.

For the Chirchik River *Gammarus* sp. was noted, however, not in the main channel, but in an affluent (Sibirtzeva, 1966). In the Chu River, *Rivulogammarus pulex* f.? was found in the main channel (the lower reach?); its biotope was not pointed out (Ovchinnikov, 1936). In ponds of the river basin the same species was encountered, while in the mountain river Ala-Medyn *Riv. angustatus* Mart. According to our own observations and those made by workers of the Issyk-kul Biological Station, the side-swimmers are not representative dwellers of the typical mountain torrent but common in springs, slow brooks and various subordinate waterbodies of the river. They are so abundant there that, *e.g.* in springs, they constitute a significant forage component for the Sevan trout fry which has been acclimatized in the Issyk Lake.

Perhaps, in other mountain regions the rheophylous side-swimmers play some role in the biocenoses of rapid waters, but not in the Tien Shan, and this group can by no means be assigned to hymarobionts.

8. *Mollusca*

No *Mollusca* were found in the Issyk and Akbura Rivers. This is not easy to explain. Muttkowsky did not find *Mollusca* in Rocky Mountain torrents in North America either and wrote in this connection: "To my surprise, I did not find any snails or clams represented in any of the mountain waters. This is contrary to the findings of Steinmann for Switzerland and of Thienemann for Germany" (Muttkowsky, 1929). When sampling the torrential fauna of the north Caucasus and Adjaria, we did not find *Mollusca* either, except for large numbers of the *Ancylus* which is entirely absent from torrents and small brooks on the Tien Shan, where the *Limnaea* was encountered. Leaving aside the *Ancylus*, the absence of *Mollusca* from torrents may be explained by the effect of high flow power and low calcium content. As for Steinmann's and Thienemann's findings, these were restricted to small mountain brooks. The same phenomenon was reported by Hubault who did not find *Mollusca* in mountain rivers or streams in the Vosges (Hubault, 1927). Information from Hora (1923) about his finding *Mollusca* should be regarded with caution, as he did not differentiate between the types of fast-flowing watercourses.

It is very remarkable that in the *Mollusca* list for the Chu River and its

194

subsidiary waterbodies none of the 39 species was described as typical of the "main channel" (Ovchinnikov, 1936)! All of them are restricted to the following waterbodies: the river arms and springs, brooks and also other slack waterbodies. Of curiosity is the finding of *Ancylus lacustris* L. noted for the "slack waterbodies" of the Chu River basin. This information raises doubts.

In the Chirchik River three species of *Mollusca* were found (two species of the *Radix* and one of the *Pisidium* genus), but they were again recorded in a brook, spring and only one species in the Magian-darya River (Sibirtzeva *et al.*, 1961). According to Ya. I. Starobogatov's identification, the Akbura River has only three species of *Mollusca* (*Physa acuta, Radix lagotis* and *Limnaea auricularia persica*), but all of them were solitary and encountered in the lowermost river site where pools, slack places with vegetation and similar biotopes occur.

Certainly in this case the same general conclusion applies: that *Mollusca* are not typical but accidental for the torrents of the Tien Shan and that *Ancylus*, in contrast to the European and Caucasian torrents, are absent from Middle Asian torrents.

9. *Oligochaeta* (aquatic earthworms)

Aquatic earthworms are rare in the Tienshanian torrents. This is probably due to the effect of fast current which eliminates forms lacking any special organs for attachment and makes the waterbody extremely oligotrophic. In the Issyk River there are *Nais bretscheri* (with or without the eye-spot) in growths on stones and in moss; *Nais pseudoobtusa, Lumbriculus* sp., *Stylaria lacustris, Eiseniella tetraeda* and *Enchytreoides* sp. were also found. Special investigation of the Oligochaet fauna could reveal some other forms in the river, but it is hardly conceivable that the fauna of these worms is as rich as that from rivers with more or less slow current. In this respect, the Chu River and its middle and, especially, lower waters are remarkably different from mountain torrents; it was noted that the Worm Type is well represented in the Chu River and its relative waterbodies by *Oligochaeta* and *Leeches* (Ovchinnikov, 1936).

There are few publications on the faunistics of *Oligochaeta* from Middle Asian torrents. For the Issyk River there is Chernosvitov (1930) who gave a list of 23 species, but only eight of them at most may be considered dwellers of fast-flowing watercourses. These eight species are for the most part the same as those cited for the Issyk River. Some are partially adapted to life in fast current, *e.g.* the *Nais bretscheri*, which Lastochkin (1949) described as "capable of living in algal growths on stones over the channel bars owing to its gigantic setae" (p. 126). Another work on the Oligochaet fauna from Middle Asia (Grib, 1950) is devoted to waterbodies of the Western Pamirs, not the Tien Shan (some data from this article are given in Chapter 7). We believe that *Oligochaeta* should not be included in the hymarobiont community nor even assigned to the lithorheophylous fauna.

10. *Rotatoria* and *Tardigrada* (rotifers and water bears)

The same conclusions can probably be applied to these groups as to the aquatic earthworms. The rotifers are inhabitants of either standing waterbodies or slack rivers. They are mentioned here only because seven species are reported by Sibirtzeva (1964) on the general faunistical list for the Chirchik River. But this is because the list includes not only the rotifers from the main river channel, but also from subsidiary waterbodies and some sites of the lower reach where the Chirchik River loses its character of a mountain torrent. The water bears are known to be inhabitants of moss and other vegetation and they can not be assigned to the lithorheophylous fauna.

11. *Nematodes* (roundworms)

For the Issyk River were reported *Dorylaimus crassus*, *Prismatolaimus dolichurus* (from mucuous growths on stones in diatoms and in *Nostoc parmeloides* colony) and *D. filiformis* from moss. But the *Nematodes*, even if taking into account not only the above listed, but also a number of undetermined immature ones from the upper and middle reaches of the river, are very infrequent. They are more abundant in small brooks where they correlate with detritus content in the river. The gordian worm was found at an elevation of 2190 m. There have been no special investigations of *Mermithoidea*, although they seem to be abundant in the river, as it follows from the records published by Rubzov (1974).

12. *Tricladida*

The group is a significant component in the brook fauna, but in torrents it is found only within the nearshore zone. Here as many as ten or more individual triclad turbellarians are encountered per stone. Their species composition is unknown for waters on the Tien Shan, where one may expect to discover a peculiar and endemical fauna.[36]
  At least a few forms are reported for the Issyk River: one form from the Middle Asiatic genus *Sorocelis* and three species of the *Crenobia* genus. The *Sorocelis* are widespread in the Issyk River. Besides the nearshore zone, they were encountered in small affluents different from the main river channel in the ecological respect. In the latter, the triclads are smaller and light-coloured. In the Issyk River proper the triclads from the *Crenobia* genus were found only in three sites and in small numbers. Our observations of the waterbodies on the Northern, Central and Southern Tien Shan suggest a certain preference of the triclads for a certain type of rapid waters. Like the water beetles, they mainly preferred brooks with slower current than torrents, and the bottom of rounded boulders and pebbles rather than large fragments. These waterbodies contained only a few *Blepharoceridae* and specific species of mayflies from the *Iron* genus.

[36] Other regions of the USSR are more lucky in this respect; see, *e.g.* Zabusova-Shdanova (1970).

According to Beklemishev (1949, p. 29), from mountain regions of Middle Asia are known *Polycelis stummeri* Seidl., *P. sabussovi* Seidl., *P. gracilis* Seidl., *P. lactea* Seidl., *P. eburnea* Muth.

13. Comparison of the composition of most representative lithorheophylous invertebrate fauna from Tienshanian torrents (including Zeravshan river)

On the basis of evidence collected for torrents and mountain rivers by L. K. Sibirtzeva, E. O. Omorov, K. A. Brodsky, L. A. Kustareva, E. O. Konurbaev, A. O. Konurbaev, L. A. Madjar and others, let us compare the composition of mayflies, caddisflies, stoneflies and *Diptera* (*Blepharoceridae* and *Deuterophlebiidae*) from 15 Tienshanian torrents. The groups are listed not in a systematic order, but in a sequence we have accepted for the presentation of the data on the species composition and ecology of these groups, *i.e.* mostly on the basis of weight of the respective group in the torrential lithorheophylous fauna. In other words, we discuss only those groups which contain hymarobionts (Table 36) and are known from several rivers (see Table 35 for blackflies and pages 184,185 for *Chironomidae*).

There is no need to stress here the incomplete knowledge of the lithorheophylous fauna in respect to its composition, but we believe however, that due to special attention to the most important groups the list of representatives of torrential insects does contain their most typical members. We also believe that the backbone of the mass forms is known, to a certain degree, for the Tienshanian torrents and can give an idea of their rheophylous fauna. Undoubtedly, further investigation may greatly detailize the genera, *i.e.* a number of new species will be described, but the general character of the fauna will be hardly changed. Moreover, we think that the character of the torrential fauna is the same in the Western Pamirs (on this see below) and in the mountain regions adjacent to Middle Asia (the Kuen Lun, the Chinese Tien Shan, Hindu Kush, Karakoram and the Himalaya), at least with respect to the generic composition.

Table 35 gives the blackfly species only for the Akbura River, but it does not mean that the blackflies are absent from other rivers, because they simply have not yet been examined.

We have often emphasized the unsatisfactory condition of the systematics of the groups in question, but we have not drawn the readers attention to the fact that this is due not only to the inadequate investigation of hymarobiont systematics, but because this work cannot be done without settling a number of complex evolutionary problems, in particular, the different evolution of the air and the water phases of hymarobionts (from aquatic insects). This problem concerns most of the hymaro-groups and, of course, it is of great value to biology in general.

With the study of the ecology, systematics and, on this basis the phylogeny of the caddisflies, especially the *Rhyacophilidae* family which includes a number of typical hymarobiontic species, this problem is most closely

Table 36. Mayflies, caddisflies, stoneflies, net-winged midges and mountain midges from some rivers and torrents on the Tien Shan.

Name of form	Akbura R. (Alai Mts.)	Koshchan R. (Alai Mts.)	Chiilisai R. (Alai Mts.)	Apshirsai R. (Alai Mts.)	Gulcha R. (Alai Mts.)	Tar R. (Ferghanskii Range)	Jassy R. (Ferghanskii Range)	Aksu R. (Terskei Alatau)	Kizyl-Ungur R. (Ferghanskii Range)	Karasu R. (Alai Mts.)	Chechmesai R. (Alai Mts.)	Ahangaran R. (Chatkalskii Range)	Issyk R. (Zailiiskii Alatau)	Chirchik R. (Chatkalskii Range)	Zeravshan R. (Zeravshanskii Range)
*Ephemeroptera**															
Iron montanus Brod.	+	+	+	+	+	+	+	+	+	+	+	−	+	−?	−?
Ir. rheophilus Brod.	+	+	+	+	+	+	+	+	+	+	+	−	+	−?	−?
Ir. nigromaculatus Brod.	+	+	+	−	+	+	+	+	+	−	−	−	+	+	+
Iron sp. 1	+	−	−	−	−	−	−	−	−	−	−	−	−	+	+
? *Epeorus* sp.	−	−	−	−	−	−	−	−	−	−	−	−	−	+	−
Rhithrogena tianschanica Brod.	+	+	+	+	+	+	+	+	+	+	+	+	+	+	+
Rhithrogena sp. 1	+	?	?	?	?	?	?	?	?	?	?	?	?	?	?
Ecdyonurus rubrofasciatus Brod.	+	+	+	+	−	−	+	+	−	+	+	+	−	+	+
Ecd. snojko Tschern.?	+	−	−	−	−	−	−	−	−	−	−	−	−	−	−
Ecdyonurus sp. 1	+	−	−	−	−	−	−	−	−	−	−	−	+	?	?
Ecdyonurus sp. 2	−	−	−	−	−	−	−	−	−	−	−	−	+	?	?
Heptagenia perflava Brod.	+	+	−	−	−	−	−	+	−	+	+	+	−	+	+
Cinygma sp.	+	−	−	−	−	−	−	−	−	−	−	+	−	+	+
Ephemerella submontana Brod.	+	+	+	+	+	+	+	+	+	+	+	+	+	+	+
Ephemerella sp.	+	−	−	−	−	−	−	−	−	−	−	+	−	−	−
Ordella macrura Steph.	−	−	−	−	−	−	−	−	−	−	−	+	−	+	+
Ameletus alexandrae Brod.	+	−	−	−	+	−	+	+	−	+	−	+	+	+	+
Baetis transiliensis Brod.	?	?	?	?	?	?	?	?	?	?	?	?	+	?	?
B. heptopotamicus Brod.	+	−	+	−	+	−	+	+	+	+	−	−	+	−	−
B. issyksuvensis Brod.	?	?	?	?	?	?	?	?	?	?	?	?	+	−	−

Baetis sp.	+	+	−	+	+	+	+	+	+	+	+	+	+	+	+
Pseudocloeon sp.	+	+	−	−	−	−	−	+	−	−	+	−	−	−	−
Caenis sp.	+	−	−	−	−	−	−	−	−	−	−	−	−	−	−
Cloeon sp.	−	−	−	−	−	−	−	−	−	−	−	−	−	+	+
Trichoptera															
Rhyacophila extensa Mart.	+	−	−	−	−	−	+	−	−	−	−	−	+	−	−
Rhyacophila "larva prae-branchiata" Lepn.	+	−	−	−	−	−	−	−	−	−	−	−	−	+	−
Rhyacophila obscura Mart.	+	−	+	+	+	+	+	+	+	+	−	−	+	+	+
Himalopsyche gigantea Mart.	+	−	−	−	−	−	−	−	−	−	−	−	+	+	+
Himalopsyche "larva hoplura" Lepn.	+	−	−	−	−	−	−	−	−	−	−	−	−	+	+
Glossosoma dentatum McL,	+	−	+	+	+	−	+	+	−	+	+	+	+	+	−
Glossosoma sp.	−	−	−	−	−	−	−	−	−	−	−	−	−	+	+
Agapetus kirgisorum Mart.	+	+	−	−	−	+	+	+	−	−	+	+	+	+	+
A. tridens McL.	+	+	+	−	+	−	+	+	+	+	+	−	+	−	−
Hydroptila insignis Mart.	−	−	−	−	−	−	−	−	−	−	−	+	−	+	+
Hydroptila sp.	+	−	−	−	−	−	−	−	−	−	−	−	−	+	−
Oxyethira sp.	−	−	−	−	−	−	−	−	−	−	−	+	−	+	−
Agraylea sp.	−	−	−	−	−	−	−	−	−	−	−	+	−	+	−
Stactobia sp.	−	−	−	−	−	−	−	−	−	−	−	−	−	+	−
Psilopterna pevzovi Mart.	−	−	−	−	−	−	−	−	−	−	−	−	−	+	−
Dolophilodes ornata Mart.	+	−	+	−	−	−	+	−	−	−	−	+	+	+	−
Arctopsyche sp.	+	+	−	+	−	+	+	+	+	+	−	−	−	−	−
Hydropsyche sp. 1	+	+	+	+	+	+	+	+	+	+	+	−	+	+	+
Hydrosyche sp. 2	+	−	−	−	−	−	−	−	−	−	+	−	−	−	−
H. gracilis Mart.	+	−	−	−	−	−	−	−	−	−	+	−	−	−	−
H. stimulans McL.	+	+	+	+	+	+	+	−	+	+	−	+	−	+	−
Apatania copiosa McL.	+	−	+	−	−	−	+	+	−	+	−	+	+	+	+
Dinarthrum pugnax McL.	+	−	−	−	−	+	+	−	−	+	+	+	+	−	+
Dinarthrum reductum Mart.	+	+	−	+	−	−	+	+	−	−	−	−	+	+	+
Brachycentrus maracandicus Mart.	+	+	−	+	+	−	−	−	−	+	+	+	−	+	+
B. montanus Klap.	−	−	−	−	−	−	−	−	−	−	−	−	+	−	−
Oligoplectrodes potanini Mart.	−	−	−	−	−	−	−	−	−	−	−	−	+	−	−

Table 36. (continued). Mayflies, caddisflies, stoneflies, net-winged midges and mountain midges from some rivers and torrents on the Tien Shan.

Name of form	Akbura R. (Alai Mts.)	Koshchan R. (Alai Mts.)	Chiilisai R. (Alai Mts.)	Apshirsai R. (Alai Mts.)	Gulcha R. (Alai Mts.)	Tar R. (Ferghanskii Range)	Jassy R. (Ferghanskii Range)	Aksu R. (Terskei Alatau)	Kizyl-Ungur R. (Ferghanskii Range)	Karasu R. (Alai Mts.)	Chechmesai R. (Alai Mts.)	Ahangaran R. (Chatkalskii Range)	Issyk R. (Zailiiskii Alatau)	Chirchik R. (Chatkalskii Range)	Zeravshan R. (Zeravshanskii Range)
Plecoptera															
Amphinemura crenata Kop.	+	–	–	+	+	+	+	+	+	+	+	+	+	+	–
Nemoura sp.	+	–	–	–	+	–	+	+	–	+	–	–	+	+	+
Phasganophora sp.	+	–	–	–	–	+	–	–	–	–	–	–	–	–	–
Mesoperlina pecircai Klap.	+	–	–	–	–	–	–	–	–	–	–	–	–	–	–
M. ochracea Klap.	+	–	–	–	+	+	+	+	+	+	+	–	–	–	–
M. capnoptera McL.	–	–	–	–	–	–	–	–	–	–	–	–	+	–	–
Chloroperla curta McL.	+	–	–	–	–	–	–	–	–	–	–	–	–	–	–
Chloroperla sp. (*tripunctata* Scop.)	–	–	–	–	–	–	–	–	–	–	–	+	–	+	–
? *Isogenus* sp.	–	–	–	–	–	–	–	–	–	–	–	–	+	–	–
Capnia prolongata Zhiltz.	–	–	–	–	–	–	+	–	–	–	–	+	+	–	–
Capnia sp.	–	–	–	–	–	–	+	–	–	–	–	+	+	–	–
Diptera															
Blepharoceridae															
Blepharocera asiatica Brod.	+	+	+	+	+	+	+	+	+	–	+	–	+	+	+
Tianschanella monstruosa Brod.	+	–	–	–	+	+	+	–	+	–	–	+	+	+	+
Asioreas tianschanica Brod.	+	–	–	–	–	+	+	–	–	–	–	+	+	+	+
A. nivia Brod.	+	–	–	–	–	+	+	–	–	–	–	+	+	+	+
A. turkestanica Brod.	–	–	–	–	–	–	–	–	–	–	–	–	–	–	+
Philorus asiaticus Brod.	–	–	–	–	–	–	–	–	–	–	–	–	–	+	–
Deuterophlebiidae															
Deuterophlebia mirabilis Edw.	+	–	–	–	+	+	+	+	+	–	–	+	+	+	+

* To complete the list of mayflies from Middle Asian torrents, we should mention *Notacanthurus zhiltzovi* TCHERN. from Talasskii Alatau Range; *Ephemerella karasuensis* KUST. and *Rhithrogena brodskii* KUST. from rivers in the Issyk-kul Valley; *Rh. minima* SIN. and *Rh. asiatica* SIN. from

200

approached by Lepneva (1964). She has clearly outlined different evolution-ary ways of the air and the water phases in caddisflies and it is worthwhile citing her on this question: "Such is the *Rhyacophilidae* family, whose adults approach, in their morphology and way of life, the concept of phylogenetic relics,[37] whereas the intricately specialized and progressively developing larvae have ensured the way for high species diversity of this family and its wide geographical distribution... in *Rhyacophilidae* the progressive evolu-tion has involved only larvae, not adults. The evolution of the adult stage, as a relict, has been delayed, and this has permitted to distinguish, within the large subfamily of recent *Rhyacophilinae*, only two large primitive genera – the *Rhyacophila* and *Himalopsyche* – whereas the diversification and mor-phological specialization in the *Rhyacophila* larvae, which have developed in the dynamic conditions of the torrent, are more pronounced than the differences of characters on the generic or even subfamily level in other groups: varying form of the gills, and the posterior legs and highly differen-tiated primary arrangement of setae. In the *Rhyacophilidae* family, against the background of its totally primitive character caused by peculiarities of the adult insects, the larvae display a very high specialization which accounts for the current prosperity and wide geographical occurrence of the family".

Different rates (and intensity) of evolution are peculiar not only to caddisflies as torrential dwellers, but also to mayflies, blackflies (which are also torrential species), *Blepharoceridae* and, probably, many other hymarobionts. The aquatic phases in these groups display a significant diversification of their morphology and ecology, which is a reflection of their geographical and ecological variability, whereas the adults, if compared with preimaginal phases, are much more monotonic (uniform), both in their structure and ecology. In the *Blepharoceridae* we count for the whole of Middle Asia only six species of adults, but with their larvae we can "easily" identify 25–30 species. The same probably applies to the torrential black-flies. We have observed a similar phenomenon in the mayfly adults and larvae of the *Iron* genus.

Therefore the main problem in the systematics of hymarobionts is not to describe new species, but to reveal a species volume with reference to both the air and aquatic phases. It cannot be resolved without studying this variability (geographical and ecological) in the larvae, pupae and adults. Of importance is the rearing of adults from larvae and pupae or nymphs, *i.e.* the substantiated continuity of the generative phases. This necessitates station-ary investigations. That is why many forms of hymarobionts do not have a species name, but are labeled with letters or numbers. However, the solution of specific taxonomic problems related to the hymarobiontic groups is, at the same time, a solution of the general biological problem of different evolutionary rates in the various phases; here, in addition to the study of variability, the comparative functional-and-morphological method of in-terpretation becomes of great importance.

[37] On phylogenetic relics see in Rodendorff (1959).

7 Short faunistical review of torrents of the Western Pamirs – a region adjacent to the Tien Shan

In Chapter 6 we discussed specific groups of lithorheophylous fauna within the Tien Shan (and the Zeravshan River bordering on the Tien Shan), but did not touch upon the fauna of torrents of the Alai-Pamirs. This is because the two mountain regions of Middle Asia – the Tien Shan and the Alai-Pamirs – are considered by some authors as individual areas with their own peculiar features of geomorphology and climate (Gerasimov *et al.*, 1964; Geller and Ranzman, 1968; Murzaev, 1968). However, from the point of view of biogeography of torrential inhabitants such a strict discrimination between the Tien Shan and the Alai-Pamirs seems to be untimely, since their faunas are still inadequately known. So far the biogeographical regionalization has been made only for the fauna of standing waters, mainly lakes (Gurvitsch, 1966, 1971). Perhaps, taking into account the hydrological features of the river systems, one could consider the torrential fauna by separate basins, *i.e.* the Ili River basin which includes the Issyk and the Chu Rivers, the Syrdarya basin with the rivers Chirchik and Akbura, the Zeravshan basin considered by some authors as a basin in its own right (Schultz, 1965), and, lastly, the Amudarya basin which, by its many parameters, cannot be included into the Tienshanian physiographical region, but it is situated just at the border of the Tien Shan, *i.e.* more closely to the Tien Shan than the Hindu Kuch, the Karakoram, the north-eastern Himalaya or the Tibet.

However, the discussion of torrential faunas by river basins could not be justified biogeographically. That is why we have preferred a completely conventional faunistic subdivision of Middle Asia into the Tien Shan and the Western Pamirs. It is possible to speak at present about a certain specificity of the torrential faunas in these two regions, having in mind the predominance of the mediterranian and boreal components in the Tienshanian fauna and of Middle Asiatic and Indo-Himalayan ones in the Alai-Pamirs fauna. On the faunistic grounds, we include the Zeravshan River basin in the Tien Shan, but not in the Alai-Pamirs as do geomorphologists. From the geomorphological Alai-Pamirs we consider only the so-called Western Pamirs. It should be noted that, judging by the torrential fauna, the northern slopes of the Alai Mountains pertain to the Tien Shan. Thus the faunistic border between the Tien Shan and the Alai-Pamirs runs either along the southern slopes of the Alai or even along the Alai Valley.

Now we shall consider shortly the fauna of swift waters of the upper Amudarya with its source on the Pamirs (in the broad sense of the word). It is known that the Pamirs is heterogenous in the physiographical respect[1] and we therefore exclude from consideration the watercourses of the eastern and the central Pamirs in view of the specificity of their waters which do not correspond in their characters to the mountain torrent type (Yankovskaia, 1950; Shultz, 1965; Middle Asia, 1968). In Yankovskaia's work, which is devoted to the aquatic fauna of the Central and the Eastern Pamirs, we read: "The fauna of these rivers also differs from that of typical mountain rivers. Rheophylous forms are rare among the aquatic organisms: here predominate the forms characteristic of slow waters, and the river plankton becomes widespread" (p. 47). The same follows from the materials of the expedition for the Pamirs in 1928 (The expedition accounts: Zoology). According to V. F. Gurvitsch's observations, *Deuterophlebia* is absent from waters of these regions (*in litt.*). Thus, from the Alai-Pamirs as an area adjacent to the Tien Shan, we shall consider only the part which is known under the name of Western Pamirs, or the Gorno-Badakshan, or the Western Pamirs Foreland.

The occurrence of mountain torrents here is due to the topography of the area. "The Badakshan mountains... are represented by deep valleys and narrow and high ranges... The mountain ranges of Badakshan descend gradually towards east and pass... into less elevated ranges of the Eastern Pamirs, and the valleys... give place to broad flat valleys" (Geller and Ranzman, 1968, p. 57).

Because of insufficient data on the lithorheophylous invertebrate fauna from the Western Pamirs, and taking into account the differences in the geomorphology, vegetation and such faunistic units as ichthyofauna (Berg, 1949), we consider the watercourses of the Tien Shan and the Western Pamirs separately. Streams and rivers of the latter area pertain to the Amudarya basin.

The basin borders on the Alaiskii, Turkestanskii and Nuratinskii Mountains in the north, the Hindu Kush in the south and the Sarykolskii Range in the east. Despite the more southern position and the high snow line (3800–5250 m above sea level), the mountain area of the Amudarya basin is characterized by an exclusively wide field of the perrenial snow and glaciation: over a thousand glaciers, including the Earth's largest glacier the Fedchenko Glacier (77 km long), and the total glaciation area (glaciers and firn fields) as large as 10,00 km^2 (Shultz, 1965). The Amudarya headwaters: the Wahdjir River, starting from the Hindu Kush at an elevation of over 5000 m, confluences with the Pamir River, taking the name of the Punj (Pundsch) River (or "the Five Rivers", according to the number of its main affluents: the Wachandarya, the Pamir, the Gunt, the Bartang and the

[1] As pointed out earlier, the faunistic regionalization with respect to the aquatic organisms has been made only for lakes (Gurvitsch, 1966, 1971). The border of the Tibetan province, according to Gurvitsch, coincides with the Zaalaiskii Range. North of this range, *i.e.* within the Alai Valley, the Pamirian species are not found. However, some species which are widespread on the Tien Shan are also encountered in the waterbodies of the Alai Valley and on the northern slopes of the Zaalaiskii Range, but are absent south of it, *i.e.* on the Pamirs (private communication, Gurvitsch 1969).

Wantsch). The rivers Pamir, Gunt, Bartang, Jasgulam and Wantsch have their sources on the Pamirs, running down the Western Pamirs. The upper Kizylsu River (the last right-side affluent of the Punj River) lies at a lower absolute height of about 3000 m (Shultz, 1965).

We have listed the major affluents of the Amudarya to draw the reader's attention to the numerosity of watercourses on the Western Pamirs (or the Gorno-Badakshan), some of which should be certainly assigned to the type of "true" mountain torrents. The study of their lithorheophylous fauna is definitely of ecological interest and also for the understanding of the biogeographical problem of faunistic interrelations on the Tien Shan, the Badakshan, the Hindu Kush, the Karakoram and the Himalaya.

Unfortunately, very little is known about these faunas (except for the Zeravshan River described earlier). The only faunistic investigation for the upper Amudarya River was made by the Institute of Zoology, Academy of Sciences, USSR, in 1942–45. Its results were published in a topical collection of the Institute in 1950 and in "The Kondara Gorge" (1951). The same data are discussed in the "Freshwater Life" (vol. III, 1950). They are also summarized in the book by Shadin and Gerd (1961). In fact, these are the only records available on the lithorheophylous fauna from torrents on the Western Pamirs.

In recent years some data have been published on different faunistic groups, in particular, on stoneflies (Zhiltzova, 1964, 1970, 1971a,b 1972a,b; Gritzai and Zhiltzova, 1973) and *Blepharoceridae* (Brodsky, 1972a–c; Brodsky and Omorov, 1972b). More complete data on the composition, distribution and ecology of larvae of the stoneflies, *Orthocladiinae* and blackflies can be found in the collected edition "The Fauna of the USSR" and in the "Keys issued by the Institute of Zoology" (Rubzov, 1956; Lepneva, 1964, 1966; Pankratova, 1970).

When comparing the ecological and faunistical investigations in the Tienshanian torrents (including the Zeravshan and in the upper Amudarya, it should be noted with regret that in the latter case there have been no systematic surveys from the headwaters down to the inflow into a large aquatic artery or to the disappearance in the desert (as for the Issyk, Akbura, Zeravshan and Chirchik Rivers). There are scattered observations and samplings (occasional or seasonal) made for single sites of watercourses, few of which can be assigned to the type of glacial or snowy torrent. More complete investigation has been made of the Kondara River which can hardly be compared with glacial torrents as it is of the spring origin, but the available data cannot be ignored in view of our inadequate knowledge of streams in the Western Pamirs. However, we shall try to elucidate as completely as possible the data on the lithorheophylous fauna from the upper Amudarya River despite the serious difficulties arising because, for instance, the dominant torrential group, the mayflies, is identified only to genera, which makes impossible a faunistic comparison of different areas of Middle Asia. The stoneflies sampled in 1942–45 were not treated and only in recent years have they been described in some detail by Zhiltzova, but separately for several genera and families. Some other

204

groups, on the other hand, *e.g.* the chironomid midges and aquatic earth-worms, are known more exhaustively than those from the Tienshanian watercourses.

From the upper Amudarya and its affluents, observations and samplings made by workers of the Institute of Zoology concern the following swift watercourses: the Mazordarya, the Kolondue (an affluent of the Warzob River), the Kafirnigan River and its affluent, the Hanaka, the Luchobe, the Obi-Sangou, the Dushanbinka and the Kondara Rivers. All these lie within the Western Pamirs.

Not all of these streams can be assigned with respect to their origin, flow velocity, water temperature and other conditions, to the type of mountain torrent. However, the streams for which the general faunistic accounts are available (Shadin *et al.*, 1951) are very close to this type, *e.g.* the Warzob River. Shadin briefly characterizes it as follows: a rough mountain river with huge granite boulders, stones and pebbles on the bottom, the flow velocity is very fast. The annual range of water temperature is 14°, the diurnal variations are within 0.5–1.5° in winter, 2.3–4.6° in spring and about 4.5° in summer. On stones there occur growths of the green *Prasiola fluviatilis*, the chrysomonad *Hydrurus foetidus* and diatoms (over 70 species: see Kiselev and Vozjennikova, 1950). The Warzob River has its source in glaciers and snow-fields, *i.e.* it is of the mixed, glacial-snowy alimentation. The headwaters of this powerful right-side affluent of the Kafirnigan River lie at high altitudes on the southern Hissarskii Range. Its upper reach is called Zidda, but after the confluence with the Luchob River it has the name of Dushanbinka River. Its length is 63 km and mean annual velocity over 1.5 m/sec. In May the river carries a large quantity of suspended loads. The water temperature in January is 3.7°, in August 13° (maximum). Algal sampling was made at elevations of 1400, 1300, 1100, 1050 and 1000 m (Kiselev and Vozjennikova, 1950).

The list of lithorheophylous fauna given below for the Warzob River from Shadin *et al.* (1951) is derived from works of systematizers of the respective faunistic groups (Grib, 1950; Shadin, 1950a; Pankratova, 1950; Lepneva, 1951; Rubzov, 1951a–c). Identification of some forms was somewhat refined later (Rubzov, 1956; Lepneva, 1964, 1966; Pankratova, 1970), but even in this most preliminary state the list gives a certain idea of the lithorheophyl-ous fauna from the Warzob River:

Ephemeroptera
  Iron
  Rhithrogena
  Cinygma?
  Baetis
  Ephemerella
Trichoptera
  Glossosoma sp.
  Apatelia sp.? (*Copiosa?*)
  Hydropsyche sp.

Rhyacophila obscura larva *pallidiceps*
Plecoptera, not determined
Diptera
  Blepharocera asiatica
  Deuterophlebia mirabilis
  Simulium (Astega) ovtschinnikovi
  S. (Odagmia) alajensis
  S. (Wilchelmia) veltistschevi
  Heptagia larva *accomodata*
  Orthocladiinae g.? larva *principata*
  Orthocladiinae g.? larva *carbonaria*
  Orthocladiinae g.? larva *sellata*
  Orthocladiinae g.? larva *fracta*
  Lukiefferiella larva *oxiana*
  lauterbornia
  Corynoneura
  Tanytarsus ex. gr. *exignus*
  Tendipedini g.? larva *duclis*
  Tipulidae
  Atherix
Hydracarina
  Sperchon plumifer
  S. glandulosus cubanicus
  Megapus sp.
Turbellaria
  Polycelidia receptaculosa
Oligochaeta
  Lumbriculus variegatus
  Nais pardalis
  N. bretscheri
  Pristina rosea

The cited work by Shadin and others contains also some quantitative data. For example, the biocenoses of the stony substrate occupy 95% of total bottom area. There occur from 203 to 806 animals per $0.1 \, \text{m}^2$ of the stony bottom, or 580 per $0.1 \, \text{m}^2$ in average. Here caddisfly larvae and pupae predominate, comprising from 74 to 94% of the entire population, the remaining 6% being divided among the mayfly larvae, the water mites and the worms. One can see that the proportions of different groups in the Warzob River are not the same as in the typical watercourses of the Tien Shan where the mayflies predominate. This seems to be because of the quantitative underestimation of the mayfly larvae which have not been treated taxonomically.

A general preliminary list of the lithorheophylous fauna is available also for a river of different character, the Kondara (Shadin *et al.*, 1951), which is "a typical mountain river about 10 km long. The flood occurs in the period of rains and snowmelting. Flow velocity is 0.2–0.3, in some places 0.8–1.0 m/sec and greater. The water is warmer than in other rivers" (Grib, 1951, p. 203, 204). The river takes its origin at an elevation of 1860 m from springs, its point of confluence with the Warzob River lies at 1100 m. Stony bottom, water temperatures in summer up to 20.5°, supersaturated with oxygen up to 101.8%, pH 7.7–8.0. On the submerged stones there are growths of blue-green *Phormidium uncinatum, Ph. autumnalis f. minor*, at places *Stratonostoc verrucosum, Spirogyra, Mougeotia, Zygnema* and various diatoms (Kiselev and Vozjennikova, 1950). Shadin and co-workers (1950) noted a large slope of the river bed, 15 m/km; the flow velocity is high only in spring, during the flood time, and is 0.4–1.0 m/sec at low water.

"In view of absence of the summer-round snow fields and the low power of the Kondara River at its source, there are no conditions suitable for the high-altitude insects (caddisflies)" (Lepneva, 1951, p. 153); "a specific feature of the caddisflies from the Kondara area is the absence... of high-altitudinal forms... this negative feature typifies the local caddisfly fauna as being peculiar to the middle altitudes (1100–1850 m), *i.e.* different from the caddisfly fauna of mountain valleys on more elevated areas of the Hodja-Obi-Garm (1500–2000 m) and the Zidda (2200–2800 m) regions" (p. 156).

From this one can see that the Kondara River cannot be assigned in any of its characteristics to the glacial-snowy type of mountain torrent. However we present here a preliminary list of the lithorheophylous fauna in the river (Shadin *et al.*, 1951) to illustrate the fauna of fast-flowing watercourses originating from springs and having relatively high water temperature:

Ephemeroptera (identified by O. A. Tschernova)
   Rhithrogena
   Ecdyonurus
   Heptagenia
   Ephemerella
   Ordella
   Baetis
   Pseudocloeon?
Trichoptera (identified by S. G. Lepneva)
   Rhyacophila obscura
   Glossosoma dentatum
   Agapetus sp. sp.
   Dolophilodes larva *dolichocephala*
   Hydropsyche sp.
   Dinarthrum pugnax
   Apatelia copiosa
   Leptoceridae g.? sp.?
Plecoptera (unidentified)
Diptera
   Blepharocera asiatica
   Simulium (Odagmia) ornatum var. *caucasicum*

S. (O.) humerosum
S. (Obuchovia) albellum
S. (Simulium) multistriatum
S. (S.) subornatoides
Tendipedini g.? larva *dulcis*
Tanytarsus ex. gr. *exignus*
Microspectra
Lauterbornia
Ablabesmyia
Corynoneura
Orthocladiini g. larva *condarensis*
O. g. larva *principata*
O. g. larva *incurva*
O. g. larva *oxiana*
Eukiefferiella
Heptagia larva *accomodata*
Aatopynia
Dixa
Bezzia
Coleoptera
Staphylinidae (*Lestera fasciata*?)
Ochtebius sp.
Hydracarina
Sperchon plumifer
S. glandulosus cubanicus
Atrictides turkestanicus
Calonyx montanus
Dartia longispora
Megapus sp.
Lebertia sp.
Mollusca
Limnaea (*Galba*) *truncatula*
Oligochaeta
Lumbriculus variegatus
Nais pardalis
N. bretscheri
Pristina rosea
Chaetogaster sp.
Turbellaria
Polycelidia receptaculosa
Seidlia sp.

The following quantitative data characterize the lithorheophylous fauna in the Kondara River. The nubers of animals per 0.1 m² are: 773 in summer (July), 1069 in autumn (October), 934 as annual average. The mudflows decrease these numbers to 200–206 animals per 0.1 m².

The predominant form is caddisfly larvae which constitute 61% in March,

86% in July and 47% in October of the total amount of aquatic animals. In a sample taken after the mudflow the caddisflies made up 62%. The next most abundant group is dipteran larvae (mostly *Chironomidae*), 8–19% per sample. Third or, sometimes, second in abundance is the mayfly larvae (from 1.2 to 29%). Other animals (stoneflies, water mites, aquatic earthworms, turbellarians, molluscs) "commonly play an insignificant role here" (Shadin *et al.*, 1951, p. 251).

One's attention is drawn to the composition of the groups, even unidentified to a species, which differs from that in a typical torrent. The absence of such a typical torrential genus as the *Iron* of mayflies is remarkable; it is replaced here by the *Heptagenia* which is not typical for mountain torrents, but rather for the lower reaches of rivers or streams with slow current. No less remarkable is the absence of the caddisfly genus *Himalopsyche, etc.*

Analysis of the above data on the fauna from the Warzob and Kondara Rivers shows that the former is closer to the torrential type than the latter.

For other rivers investigated by the Institute of Zoology in the Western Pamirs no general faunistic lists are available, except some data on single groups.

Ephemeroptera (mayflies)

No data are available on this major group of the lithorheophylous fauna of the Western Pamirs. There are more data for the Himalaya and the Hindu Kush, as pointed out above. Althought the treatment of mayfly samples was made by O. A. Tschernova, a specialist in this group, the identification is only on the generic level, which does not permit a comparison with the Tienshanian fauna. However, judging by mayfly fauna from the Akbura and Zeravshan Rivers (which are "terminal" rivers for the Southern Tien Shan), it may be suggested that the mayflies from the Western Pamirs and the Gorno-Badakshan are similar in their species composition to those inhabiting the Tien Shan.

Trichoptera (caddisflies)

The additions to the lists for the Warzob and Kondara Rivers from a paper by Lepneva (1951) are few: only two *Hydropsyche* species from the Warzob River – *H. gracilis* Mart. and *H. exocellata* McL. A general list of torrential caddisfles (Lepneva, 1964, p. 147) was given above, which indicates a significant similarity of the caddisfly faunas from the Tien Shan and the Western Pamirs. It goes without saying however that differences between them will emerge and grow as the records on the mayflies from the Tien Shan and the Pamirs accumulate. It is of interest that S. G. Lepneva discovered from a brook (a left-side affluent of the Warzob River) the Indian species *Apsilochorema indicum* Ulm. subsp. *turanicum* Mart. (family *Rhyacophilidae*), whose main form is widespread in the Punjab.

Plecoptera (stoneflies)

A paper by Grizai and Zhiltzova (1973) gives a general account of the stonefly fauna from Tadjikistan on the basis of samplings made in the Turkestanskii, Alaiskii, Zeravshanskii, Hissarskii, Karateginskii Ranges, the Western Pamirs, Gorno-Badakshan, and the Murgab River area on the Eastern Pamirs (the latter is of no interest to us).

As the authors point out, the stonefly fauna from Tadjikistan (as well as from Middle Asia in general) is poor in species composition, has a high degree of endemism, lacks common palaearctic species and a connection with the Central Asiatic fauna (the general list of stoneflies from the Tien Shan and the Pamirs is given in Table 30). Even the few widespread species discovered in other areas of Middle Asia are absent from Tadjikistan. Here the percentage of species which are endemic to Middle Asia is 65% and only 10 out of 30 species are not endemic but described from Pakistan, China, Afghanistan and Kashmir. If compared with the Tien Shan, the stonefly fauna on the Pamirs is poorer: 13 species found on the Tien Shan are absent here. Especially poor is the fauna of the Badakshan with only a half (about 20) species recorded for the Tien Shan. Thus the family *Perlidae* is absent and the genus *Mesoperlina* so typical of the Tien Shan is represented by only one species (*M. ochracea*). Although more than 10 species recorded for the Tien Shan are absent from the Badakshan, there are a number of species similar to those from the Hindu Kush, the Karakoram and the Himalaya, and yet not found north of the Badakshan: *Perlodes* (Scobeleva) *cashemirica* and *Xanthoperla kishanganga*. Some species common in the Central Asia are widespread in the Badakshan and penetrate as far as the Hissarskii and adjacent ranges (*Nemoura vaillanti, Capnia pedestris*) and even the Tien Shan (*Eucapnopsis stigmatica transversa, Filchneria mongolica, Mesoperlina ochracea, M. pecirkai*).

Table 37 gives the percentage of species which are present both on the Tien Shan and the Pamirs and individually for each region. The data are derived from Table 30.

Diptera

Family Blepharoceridae

At present six species (with imagoes, larvae and pupae) are known from Middle Asia (Brodsky, 1972) and were encountered both on the Tien Shan and the Pamirs. There are some differences in their distribution, however. Thus the common western pamirian genus *Philorus* was found only in the Western Tien Shan (on the Chatkalskii Range – *Ph. asiaticus*). The other species occur throughout the entire mountainous Middle Asia, except the Kopet-dagh, whose Blepharocerid fauna is unknown.

210

Table 37. Distribution of Middle Asiatic stoneflies.

Locality	Number of species	% %
On the Tien Shan and the Pamirs	15	32.5
Only on the Tien Shan	17	37.0
Only on the Pamirs	14	30.5
Total	46	100.0

Family Deuterophlebiidae

Throughout the Tien Shan and the Western Pamirs, as well as in the Altai, Kashmir, the Hindu Kush, the Karakoram and the Himalaya the species *Deuterophlebia mirabilis* is widespread, but there are grounds to suggest that it is not the only species of this genus present in these regions.

Family Simuliidae (blackflies)

For the Kondara and Warzob Rivers and their small affluents and brooks, Rubzov (1951a) reported an extensive list of species (16) from which only three species refer to the Warzob and five to the Kondara River (Shadin *et al.*, 1951). It appears that this selection was made with the aim of showing only the representative forms in these streams. We cited earlier the blackfly list for the Tien Shan (Konurbaev *et al.*, 1972; see Table 35), and now we present the proportions of species found either in the Tien Shan or the Pamirs, as well as in both regions. These data are derived from calculation of all species reported by Rubzov in his monograph published in the series "Fauna of the Soviet Union" (1956). No matter how these species were described – with reference to the larva, pupa, imago or all stages taken together – their selection was made on the ecological basis. Only those forms whose aquatic phases were said to inhabit small springs were excluded. Where the data on the biology of the aquatic phases were absent, the species were not eliminated from the list. We are of course aware that our data concerning species distribution show only the order of magnitude. They were also affected by the different knowledge of the blackflies from various regions of Middle Asia (Table 38).

The blackfly distribution data in Table 38 show that the number of species restricted only to the Tien Shan or to the Pamirs is larger than the number of species which are common to both regions, but, generally speaking, the values are quite close.

Family Chironomidae

In her monograph on the larvae and pupae of the subfamily *Orthocladiinae* representative for swift waters, Pankratova (1970) described a number of species. Those inhabiting mountain rivers and brooks made up 28 species,

Table 38. Distribution of Middle Asiatic blackflies.

Locality	Number of species	%%
On the Tien Shan and the Pamirs	13	25.2
Only on the Tien Shan	19	37.4
Only on the Pamirs	19	37.4
Total	51	100.0

but 27 among them were reported for Tadjikistan, which is indicative of the inadequate and, chiefly, irregular knowledge of the chironomid fauna in Middle Asia. It would be naive to assume that many species found in the Gorno-Badakhan (and in the Western Pamirs in general) will never be discovered in the Tienshanian torrents. Unfortunately, it is impossible at present to compare the faunas of these two mountain regions.

In her work on the fauna of chironomid larvae from the Amudarya basin, Pankratova (1950) gives some examples of species representative for different ecological environments. We include in these examples also the species inhabiting brooks since it is difficult to distinguish between a "brook" with large discharge and a "torrent" with small discharge.

In brooks at absolute heights of 1800–2200 m and water temperature of 6–11° there occurred on stony substrate and in moss *Diamesa* l. *adumbriata* Pankr. and *Metriocnemis* l. *incompletus* Pankr.; in silty places *Orthocladiinae* g.? l. *curvata* Pankr. Among the forms typical of rivers of Tadjikistan was *Orth.* g.? i. *carbonaria* Pankr. In brooks at an absolute height of 1100–1200 m on stones were *Orth.* g.? i. *principata* Pankr., *Eukiefferiella* l. *sellata* Pankr., and the common forms like *Corynoneura*, *Tanytarsus* ex. gr. *exiguus* Joh., *Ablabesmyia*.

In the Saradjou, Mazordarya and Kolondue Rivers originating from snowfields at absolute heights of 1800–2200 m and the water temperature near the source 2–3°, up to 13.5° farther downstream were found larvae of 13 *Chironomidae* species. Most abundant on stones were the species *Eukiefferiella* and *Orth.* g.? l. *dissimylata* Pankr., others being solitary. At absolute heights of 1100–1200 m on stones in the rivers live mostly species of the *Orthocladiinae* subfamily. Larvae of the same subfamily predominate also in rivers of the subarid region (the submontane zone), but here also occur the forms inhabiting rivers, brooks and springs of the humid (mountainous) region (*Heptagia accomodata*, *Orth.* g.? l. *nana*, *Orth.* g.? l. *proxima* and others). "In contrast to the rivers of the humid region, a considerable area here is invaded by the common forms (known in Europe)" (Pankratova, 1950, p. 182).

Other dipteran families

The outstanding systematizer, the dipterologist A. A. Shtakelberg, took part in the investigation of the Kondara River and treated the winged mature

Diptera. The species list and the observations he made are of great interest also to our work, since among the imagoes sampled on the shores of different watercourses and in the Kondara River there are many forms with the aquatic preimaginal phase, some of which refer to the lithorheophylous fauna. Unfortunately, not all species are known in this phase and we present from the Shtakelberg's list (1951) only those species which were labelled as having been found on the stream shores (Table 39). We recognize that some species which should have been included in the list are absent. Of interest is the abundance of species (five) of the genus *Dixa*, whose larvae are very common in torrents. A number of forms, in particular, from the genera *Limonia* and *Clinocera*, were often found on the shores or in the river channel itself on stones constantly watered with spray.

Table 39. Dipteran imagoes from the Kondara River region (after Shtakelberg, 1951).

Family	Species	Note
Limoniidae	*Dicranomyia modesta* WD.	—
	Limonia inusta MG.	
	Antocha turkestanika MEIJ.	Mountain and submontane rivers in Middle Asia and Semirechie; a typical member of "fauna hydropetrica"
Dixidae	*Dixa christophersi* EDW.	Tibet, M. Asia (Hissarskii, Darwazskii Ranges); on stones on the shores of mountain torrents
	D. lepnevae STACK.	Tadjikistan (Hissarskii Range). On shores of mountain torrents
	D. nebulosa MG.	Northern and Central Europe, M. Asia (Hissarskii Range). On shores
	D. platystyloides STACK.	Hissarskii and Darwazskii Ranges. On shores of mountain torrents
	D. zernovi STACK.	Tadjikistan (Hissarskii Range). On shores of mountain torrents
Blepharoceridae	*Blepharocera fasciata asiatica* BROD. (= *B. asiatica*)	Common species in Tadjikistan
Stratiomyidae (included only one species from 14 reported)	*Adoxomyia cinerascens* LW.	Turkmenistan, Tadjikistan (Hissar and Darwaz Ranges, W. Pamirs, Iran). On stones on the shores of mountain torrents
Tabanidae (8 species)	—	—
Dolichopodidae (15 species)	*Hericostomus leptocercus* STACK.	Medium altitudes of Hissar and Darwaz Ranges. On stones on the shores of mountain torrents.
	H. ogloblini STACK.	Hissarskii Range. On shores of mountain torrents
	Tachytrechus gussakowskii STACK.	Hissar and Darwaz Ranges. Near the water.
	Liancalus virens SCOP.	Most of the Palaearctic. Near waterfalls.
	Syntormon cilitibia STACK.	W. Pamirs. At the water edge.

The following quotation from the above work by Shtakelberg is a vivid description of his observations of the dipteran imagoes from the Kondara River: "There is much interesting about the *Diptera* in the Kondara River itself, a rough mountain torrent which allows the existence of unusually rich fauna (fauna *hygropetrica*). On large stones, which lie either in the Kondarinka channel or in the zone of water spray, there live large numbers of delicate long-legged crane flies of the *Limoniidae* family – the *Antocha turkestanica*; either decreasing or increasing in numbers, they occur in this environment almost the year round. Together with the *Antocha*, on the same stones occur delicate carnivorous flies – *Atalanta* of the family *Empididae* – hunting small insects (*Collembola*). In the Kondara River this genus is represented by several still undescribed species. The Kondara River is the habitat of. . . midges of the family *Dixidae* which sometimes occur en masse on vertical stony walls descending into the river, or on vegetation. Towards evening in late autumn, one can observe a massive flight of *Dixidae* over the water surface where they make up small swarms. *Dixidae* larvae live at the water edge. Of interest is. . . that. . . the Kondara species, *Dixa christophersi* Edw., is widespread on the uplands of Central Asia (Tibet) and in the mountains of Northern India (Kashmir)" (p. 130).

Mollusca

One can say little about the *Mollusca*. In a special paper devoted to molluscs Shadin (1950, p. 68) notes: "The area of the drainage formation is practically devoid of molluscs. Only rarely in the nearshore zone with slower current may one find *Limnaea truncatula*. All the *Mollusca* fauna is concentrated in springs." We mentioned earlier the causes of absence of molluscs from mountain torrents; the above author explained this as due to high flow velocity and water turbidity in the drainage dispersion area.

Oligochaeta (aquatic earthworms)

A comprehensive paper on the aquatic earthworms from the swift waters of the Gorno-Badakshan was published in a topical collection of papers by the Institute of Zoology (Grib, 1950). This paper gives a good idea of the distribution of aquatic earthworms in different types of watercourses (Table 40). As a general conclusion, Grib states the following: "In mountain brooks and streams with rough current (the Mazordarya, Kolondue, Sarydjou Rivers) the oligochaet fauna is almost absent. Only near the banks, at slow current, on the stony substrate are solitary rheophylous *Nais pardalis*, *Nais* sp. In rivers with slower current – the Kondara, the Dushanbinka, the Luchob, the Hanaka – there are aquatic earthworms mostly near the banks, where the flow current does not exceed 1 m/sec and there is no loess silting of the substrate in summer. These are *Nais bretscheri* v. *bidentata*, *N. pardalis*, *N. variabilis*. With loess silting, the oligochaets occur further away from the banks" (p. 237).

214

Table 40. Aquatic earthworms in rivers of the humid zone of the Gorno-Badakshan (after Grib, 1950).

Name of river	Ecological station	Species of *Oligochaeta*
Kondara R.	Stony substrate, flow velocity about 1 m/sec	*Nais bretscheri* var. *bidentata* *N. pardalis* (up to 1,311 sp./m^2)
	At slower current	*Chaetogaster diastrophus* *Pristina rosea*
	Sand and slightly silted substrate	*Nais variabilis*
Dushanbinka R.	Stony substrate, only near banks, flow velocity about 1 m/sec	*Nais pardalis* *N. bretscheri* var. *bidentata* *Nais* sp.
	Flow velocity more than 1 m/sec	No oligochaets
Kafirnigan R.	In main channel	Only solitary *Enchytraeidae*
	In river arms, at velocity of 1 m/sec, near banks	*Nais bretscheri* (8–10 sp./m^2) *N. paradalis* (12 sp./m^2)
	Affluent the Hanaka R., near banks, flow velocity less than 1 m/sec, stony substrate covered with algae	*Nais bretscheri* var. *bidentata* (up to 275 specimens per m^2) *N. variabilis* (up to 57 sp./m^2)

From the evidence available in the literature we have briefly reviewed the lithorheophylous fauna of invertebrates from the swift waters of the Gorno-Badakshan and, partly, the Western Pamirs. Unfortunately, there have been no monographs on this region, which would consider the torrents over their entire length. But even present knowledge indicates a great similarity between the torrential fauna of this mountain region and that of the Tien Shan.[2] The Zeravshan River and, to some degree, the Akbura River represent a sort of transitional zone between the Tien Shan and the Pamirs. However, as we pointed out earlier, it seems to be premature to discuss the biogeographical regionalization of the fauna as has been done with respect to the lakes. For this, the systematics of some, even major groups of torrential invertebrates like mayflies, caddisflies, *etc.* is much too inadequately developed. Certainly, this is an urgent problem for the future.

[2] As Mani concludes in his monograph on the high altitude insects (1968, p. 267), "Biogeographically the Pamirs, the Tien Shan and the Northwest Himalaya are closely related". Savchenko (1974) traces an interrelation between faunas of the longicorn dipterans (families *Tanipodidae, Tipulidae, Limoniidae* and others) from Tadjikistan and the Himalaya.

215

8 The effect of ecological factors on the distribution of lithorheophylous fauna in torrent

There are two viewpoints in the literature on the significance of ecological factors for the distribution of torrential organisms. The first one is held by most of the authors mentioned in the bibliography at the end of this essay. They claim that the distribution of organisms is affected by this or that individual factor. Also, the characterization of torrential organisms is made using the classical ecological terminology which emphasizes the relation of the organisms to one such factor: to the dissolved oxygen – oxyphils, oxybionts; to the current in general – rheobionts, rheophils, rheoxens; to the water temperature – stenotherms, eurytherms; *etc.* The other viewpoint, which we favour, consists in the evaluation of the ecological factors (abiotic and biotic) not individually, but mutually interrelated in a complex which affect the distribution of organisms as a holistic system. We expressed this viewpoint on the significance of ecological factors as early as 1935 (Brodsky, 1935) and mention it again in the chapters devoted to the physical and chemical characterization of torrents. Later, the same opinion was expressed by Mani (1968) in his monograph on the high altitude insects of the world. Considering different ecological approaches of the authors writing on the inhabitants (insects) of mountain torrents, he notes: "Most workers have, however, overlooked the fact that the biota of the mountain torrents are governed not only by these factors (temperature of water and amount of dissolved oxygen), but also by a complex set of other factors, including the nature and character of the bank, the stream bottom, the width and the depth of the stream, the current velocity, the total water discharge per second, *etc.* Brodskii (1935) is perhaps the most important worker to realize the complexity of ecological factors in the high altitude torrents" (p. 96).

Recognition of the great importance of the complex of factors, rather than any one of them, taken in isolation, raises also the question on a specific ecological terminology. Of principal value for this is the terminology proposed by Martynov (1929) in respect to inhabitants of the mountain torrents. In contrast to the ecological terms which designate a relation of torrential organisms to separate ecological factors (rheobionts, oxybionts, stenothers, *etc.*), Martynov proposed the terms like hymarobionts, hymarophils and hymaroxens, which designate a relation of torrential organisms not to a single factor, but to their certain set typical of the rough mountain torrent (the terms were produced from the Greek word χειμαρον, *i.e.* the rough mountain torrent). Unfortunately, Martynov's terminology have not become generally accepted (except for the papers of the author of this essay), because the hydrobiologists, ecologists, zoologists and botanists engaged in the study of the fauna and flora of fast-flowing watercourses do not

216

distinguish any types of streams by their strength which entails a change of the whole system of factors. We stressed this in the essay when trying to illustrate the specificity of the mountain torrent proper and its fauna. Thus we believe that Martynov's terminology deserves careful attention, because it emphasizes the need for a different approach to the swift waters and their faunas. As we have already tried to show, the matter is not the different flow velocities, different depths and other individual factors, but their whole system which determines the type of watercourse, the composition, abundance and distribution of its fauna and flora. The hymarobionts are dwellers of the mountain torrents; they are specific to just this type of swift waters and are not capable of changing the mountain torrent for other types of waters, in particularly, for the fast-flowing watercourse of the lowlands. This point is well illustrated by the change of the species composition along the longitudinal profile of a stream.

Thus, speaking about a complex (or a self-consistent system) of factors, we distinguish among the fast-flowing watercourses such types as the mountain brook, the torrent and the mountain river.

However, within each type we can also trace the effect of individual factors on the distribution of animals, which can be detected only if one of the factors deviates greatly from its usual state in the system. It should be stressed that such an analysis of the influence of individual factors is purely conventional, as we temporarily dismiss the interrelationship of factors within the total system of a mountain torrent.

Let us consider a few examples in which an individual factor may affect the distribution of organisms in a torrent: (i) the flow velocity which depends on the bed slope, the bottom substrate, the discharge, the wetted perimeter – the width and the depth of the torrent, *etc.*; (ii) the water discharge which depends on the same conditions plus the precipitation, the substrate porosity, *etc.*; (iii) the substrate which depends on the flow velocity, the rocks underlying the bottom, *etc.*; (iv) the water temperature which depends, in turn, on the flow velocity, the discharge, elevation, *etc.*; (v) the salt composition and dissolved oxygen amount, both of which are dependent on the above factors and some others like the nature of the rocks in the channel, the drainage, the water discharge, the contribution of springs and snowfields to the alimentation, *etc.*

As we said, the differential evaluation of the significance of one factor in the torrential system is an artificial procedure, nothing more than a way of approach, since all the factors are acting within the torrent, not beyond it.

Flow velocity. The flow velocity is, undoubtedly, a very important condition in the torrent, as it forms the total ecological situation in which the aquatic organisms live. It regulates the transportation of nutrients, the saturation of water with dissolved oxygen, the thermal differences in the water; it sorts out the substrate and creates a specific biocenotic background. Here we shall speak only about its direct influence on the distribution of aquatic animals in the torrent, *i.e.* the flow velocity as a mechanical force.

The effect of the mechanical force of fast-flowing waters on organisms may be considered in two aspects: the influence on the distribution of individual organisms or on the entire fauna of the torrent. As to the first aspect, there are extensive data in the literature and our own observations of the Tienshanian torrents. The data concerning the second aspect are not so rich, but the vertical zonality of the Tienshanian torrent is based on the original information (see the previous chapter on the ecological and faunistic zonality of the torrent), and here we shall confine ourselves to a few examples.

Returning to the first question, a work by Percival and Whitehead (1929) must be mentioned in which the change in abundance of different members of the lithorheophylous fauna was shown to depend on the flow velocity (Table 41).

There are also other data, including experimental evidence. Thus *Simulium ornatum* prefers a flow velocity between 50–120 cm/sec; most blackfly larvae are encountered at the velocity of 80 to 90 cm/sec (Phillipson, 1956). According to Wu (1931), the velocity range for the larvae of the same dipteran family is 17–84 cm/sec. Dittmar (1955) reported the most current-resistent organisms: the blepharocerids *Liponeurs* (from Europe) and *Simulium*, which can withstand the flow velocity over 240 cm/sec. Then come the organisms inhabiting the upper surface of stones: the mayfly *Epeorus*, the caddisflies *Rhyacophila*, *Brachycentrus* and others, 100–130 cm/sec; then the organisms which live on or under stones: the mayflies *Rhithrogena* and *Baetis*, the caddisflies *Agapetus*, *Glossosoma* and others, 80–100 cm/sec. The next is a group of species of some stonefly genera *Amphinemura*, *Isoperla*, *Protonemura*, the mayfly *Ecdyonurus* and the caddisfly *Apatania*, 48–77 cm/sec.

Experimental observations of the transport of organisms by the water in a flume gave a maximum flow velocity at which they can resist to the current (*i.e.* before they are washed away), being capable of moving against the current. These results are presented in Table 42 as the averages of 10 trials.

The same genera as those in Table 42 are known in torrents of Middle Asia, but they are represented here by other species. Similar values were also obtained from observations on the natural occurrences of the North American torrential insects at certain flow velocities. Thus, according to

Table 41. Percentage of animals at various flow velocities (after Percival and Whitehead, 1929).

	Flow velocity (m/sec)	
Organisms	3.6	2.3
Mayfly nymphs:		
Rhithrogena	8.8	1.7
Baetis	8.8	10.7
Caddisfly larvae:		
Agapetus	25.0	35.0
Chironomid larvae	20.5	25.0

Linduska (1942), in the Rattlesnake Creek, Montana, the upper side of a boulder of 50 cm^2 was inhabited at a surface velocity of 240 cm/sec only by the mayfly nymphs *Ironopsis* and by two species *Baetis* and the lateral sides by *Iron longimanus* and *Ephemerella doddsi*. *Rhithrogena* was found only at velocity lower than 60 cm/sec.

These data, as well as some original evidences presented below, show with certainty the significance of the study of microbiotopes, primarily, of the micro-zonal distribution of the flow velocity, for the understanding of the ecological conditions in which the torrential organisms live.

Observations in the Issyk River made on the distribution of different torrential dwellers show a considerably wider range of velocities at which these species occur and their stronger resistivity to the mechanical force of the current. All these data are summarized in Table 44. For many species of the Table the upper limit of velocity is as high as 4 m/sec and even higher.

It may be explained by the fact that the species studied by the European investigators were not true hymarobionts, but were dwellers of brooks and springs. The mountain torrent of the Tien Shan with faster current is inhabited by true hymarobionts.

Of course, the question arises of the concrete microbiotopes of these organisms. We shall discuss it somewhat later, and now we shall satisfy ourselves by saying that any organism inhabiting a microbiotope with slow current will be affected, some time or other, by a faster current. In other

Table 42. Minimal flow velocity (cm/sec) resisted by animals (after Dittmar, 1955, and Dorrier and Vaillant, 1955; from Hynes, 1970, abridged).

Organisms	Dittmar, 1953	Dorrier and Vaillant, 1955		
	A	B	C	D
*Liponeura**	>300	>240	>240	220
Simulium	280	240	117	114
*Epeorus**	124	>240	230	109
Rhyacophila	122	200	100	125
*Ancylus**	118	240	109	24
*Rhithrogena**	96	182	125	—
*Agapetus**	94	89	36	36
Baetis	84	117	99	—
*Ecdyonurus**	57	154	99	—
Protonemura	53	198	147	145

Note: A and B, maximal flow velocity in a trial incapable of animal removal from the substrate; C, maximal flow velocity at which the animals are capable of moving against the current; D, maximal flow velocity observed in nature and resisted by the animals.

* Indicates that the same species were studied.

words, periodically or constantly, the organism will be exposed to an appreciable mechanical force of water which it must be adapted to resist.

As to the selective effect of the flow velocity on the torrential fauna rather than on single species, we shall give some examples for a number of sites in the Issyk River which have fairly similar ecological conditions except for the flow velocity. The conventionality of the isolation of a single factor from the interactive complex should be stressed again, and the examples below are very rare in nature. It is only in an experiment that a parameter can be changed arbitrarily (Ambühl, 1959; Zimmerman, 1961; Pleskot, 1962; Bournaud, 1972).

The first example is the comparison of the lithorheophylous fauna from two points (58 and 1, from the middle Issyk River) which differ in flow velocity:

Point 58	Point 1
Abs. height 1220 m, water temperature 14.3°, substrate of small pebbles, flow velocity 1.53 m/sec	Abs. height 1763 m, water temperature 14.0°, substrate of small pebbles, flow velocity 0.56 m/sec

Point 58

Deuterophlebia mirabilis
Blepharocera asiatica
Tianschanella monstruosa
Iron rheophilus
I. nigromaculatus (abundant)
Ephemerella submontana
Baetis issyksuvensis
Ecdyonurus sp.
Brachycentrus montanus
Apatelia copiosa
Agapetus tridens
A. kirgisorum
Dinarthrum sp.
Limnius sp.
Limonia sp.
Chironomidae
Simuliidae

Point 1

Iron montanus
Baetis sp.
Brachycentrum montanus (abundant)
Oligoplectrodes potanini
Dinarthrum sp.
Dolophilodes sp.
Leuctra sp.
Simuliidae

The faunal differences in the two sites are easily recognizable, although the hydrological data are rather similar, *i.e.* the substrate (small pebbles), the salt content and the amount of dissolved oxygen. From this example one can easily see the significance of flow velocity as a mechanical factor. Point 1 with slower current entirely lacks the members of the family *Blepharoceridae, Deuterophlebia mirabilis* and the mayflies *Iron rheophilus, Ecdyonurus* sp. and *Ephemerella submontana*.

Let us compare two other points, 11 and 56, lying at the upper half of the lower Issyk River: they are similar in many conditions except the flow velocity.

220

<table>
<tr><td>

Point 11

Abs. height 980 m, water temperature 17.5°, pebbles, flow velocity 2.5 m/sec

Blepharocera asiatica
Iron nigromaculatus (abundant)
Rhithrogena sp.
Ephemerella submontana
Baetis issyksuvensis
Caenis sp.
Dinarthrum sp.
Oligoplectrodes potanini
Apatelia copiosa
Agapetus tridens
A. kirgisorum
Rhyacophila extensa
Brachycentrus montanus
Chironomidae
Simuliidae

</td><td>

Point 56

Abs. height 1100 m, water temperature 14.6°, silty substrate, flow velocity 0.37 m/sec

Baetis issyksuvensis
Apatelia sp.
Colymbetini
Chironomidae
Simuliidae

</td></tr>
</table>

The comparison of these points shows great faunistic differences. From point 56 with slow current such forms as *Blepharocera asiatica, Iron nigromaculatus, Rhithrogena* sp., *Ephemerella submontana, Ephemerella* sp. and some others are completely absent. It can hardly be accounted for by the difference in the water temperatures because it is small; at the point of faster current the temperature was even higher. The principal difference between the points is only in the flow velocity and the substrate. But these factors cannot be separated in natural environments. As the flow velocity decreases, the fine loads begin to precipitate; as a result, the lithorheophylous organisms to which most rheobionts belong disappear.

When discussing the effect of flow velocity on the distribution of torrential animals, one should mention the so-called "organic drift". This question has been discussed extensively in the literature (Ambühl, 1959, 1962; Waters, 1961, 1962b, 1964; Illies, 1962; Pleskot, 1962; Anderson and Lehmkuhl, 1968; Lehmkuhl and Anderson, 1972; Reisen and Prins, 1972; Steine, 1972, and others). This drift, *i.e.* the downstream transportation of many organisms by swift waters, was considered to be of great importance, but in more recent years some observations have become available showing that the importance of the drift has been overestimated and that this phenomenon can hardly occur on a large scale, as it was thought earlier (Bishop and Hynes, 1969). It seems more likely that the drift was caused by abrupt and great change in the flow velocity and water discharge in the torrent, as observable, for example, during the flash-floods produced by the snow and ice melting due to high insolation or heavy rains. In fact, not all torrential organisms will be transported downstream, but only some. Thus in the lower reach of a river in the southern Kungei-Alatau, after a flood caused by a three-day rainfall, we examined the water pools and wet sand in the previously dry channel after the water had infiltered in the substrate. The pools and the sand were covered with a continuous layer of the *Iron* mayfly nymphs. Usually, such a phenomenon is not observed on the bottom of irrigative canals after the flood has abated.

Even without the drift caused by the rains or fast snowmelting in the mountains, the downstream transport of torrential organisms does take place all the time in a greater or smaller degree. Subjected to the downstream transportation are the eggs deposited by mayflies and caddisflies, the imagoes hatched from pupae or nymphs, the young larvae, etc. But at the same time there are compensatory adaptations to maintain the constant position of the organisms in the torrent (Madsen *et al.*, 1973). Otherwise the whole mass of organisms would be swept away downstream where they would perish from the foreign ecological conditions. Among such compensatory adaptations are the flight of imagoes upstream (for a very short time), the active movement of the preimaginal forms upstream or closer to the bank, into the slower waters. Many organisms (the hymarobionts) are not transported by the current but stick to their constant place in the torrent.

We are discussing the drift problem because it is of principal and universal significance. However different from the ocean the torrents may be, here too, the life of organisms is greatly affected by the water movement. The continuous water movement in streams, the currents in the ocean and the tides at the coasts undoubtedly affect the distribution of organisms. But these currents are not merely a means of transport or a device for the drift of organisms, as some authors believe. If we assumed a passive transportation of organisms by water, then it would be reasonable to assume also that the organisms represent inanimate objects, having no activity of their own, incapable of resisting the current. It is out of place to discuss this question here, but we shall recollect that even the passive oceanic plankton has a number of adaptations enabling the planktonic animals and, partly the plants, to maintain their constant place in the vast spaces of the ocean and sea (vertical seasonal and diurnal migrations, active locomotion, *etc.* – Brodsky, 1972c). The hymarobionts occupy a constant place in the torrent and the drift affects mostly such components of the fauna which are causal or foreign to the mountain torrents (*e.g.* the favourite object of study for the drift investigators is the side-swimmers, typical dwellers of springs, not torrents).

One of the main compensatory mechanisms allowing drift-resistance is the adaptations which serve for maintaining the torrential organism on the substrate under the conditions of fast current.

The morphological adaptations were described when we discussed the mayflies, the caddisflies, the net-winged and mountain midges and blackflies. Here we give very briefly the general classification of adaptations made for the organisms of running waters by Hynes (1970), which we shall illustrate by several examples from the fauna of the Tienshanian torrents.

Commencing with the work of Steinmann (1907), extensive data have been published on the adaptations, which were summarized by Hynes in his

222

monograph. He distinguishes between the morphological, the ethological, and the physiological adaptations. The morphological adaptations are the following: (i) flattening of the body (we have shown this with reference to the Middle Asiatic mayflies of the genera *Iron* and *Rhithrogena*); (ii) streamlined shape of the body: the width of an ideally streamlined body is about 36% of its length (Bournaud, 1963) (examples were given earlier – the mayfly nymphs of *Ameletus* and *Baetis* which are capable of swimming even in very fast current: their tail cerci are directed downstream, the outer filaments have hairs only on the inner side and the central filament on both sides); (iii) suckers: only in the Blepharocerid larvae (the leeches are not typical for torrents); (iv) reduced projecting structures: the gills in *Baetis* nymphs are reduced and many mayfly nymphs have lost the central tail (*Iron, Baetis*); for *Baetis* see Berner (1950); (v) small size of the body, due to which the width of some organisms is within the boundary water layer, 2–3 mm: small larvae of *Chironomidae*, mayflies, *etc.*; (vi) silk and sticky secretions: the web glands and the use of the web threads is observed in blackflies, caddisflies, *Psychodidae* and *Chironomidae*; the web threads are used for attachment of the "ballast" to the caddisfly cases and their attachment to the stone; the sticky secretion firmly attaches the pupae of *Blepharoceridae, Deuterophlebia*, blackflies, *Chironomidae* and some others to the substrate and serves for fixation of the mayfly eggs on stones; (vii) ballast: in many larval and pupal caddisflies; (viii) adaptations for living among torrential vegetation (moss): some stonefly and mayfly nymphs living in moss have specialized spines to hold on the vegetation.

From others, non-morphological adaptations, the reduction of the powers of flight must be mentioned. Imagoes of *Deuterophlebiidae* (Kennedy, 1958), stoneflies and mayflies are confined to a limited section of the torrent where the oviposition occurs.

The oviposition by aerial insects was mentioned earlier and now we note only that if the oviposition takes place on stones projecting from the water, then it is a problem for the youngest larvae to creep into the water (as in some *Blepharoceridae* and mayflies).

The ethological adaptations include the distribution of organisms between microbiotopes, the oviposition in the period of low water, *etc.*

So far we have discussed the distribution of the organisms as a function of a general (mean) flow velocity, however, stressing repeatedly that, in contrast to the mean velocities, of great significance for the torrential dwellers is the "micro-pattern" of velocity in different microbiotopes selected by the organisms themselves, which differ for different species or stages of each species.

Lately, much attention has been given to the study of microbiotopes, in particular, of the influence of water jets with different velocities. Careful investigations of microbiotopes in the Tienshanian torrents (Brodsky, 1935, and some other works) have shown that the generally accepted view of the mountain torrent as a biotope where the dwellers are subjected to high pressure of water jets is erroneous. We discussed it in this essay in the section on the physical and chemical characteristics of the Tienshanian torrents. Similar ideas, however in a very general form and without quantitative data, were expressed with respect to the Himalayan (Hora, 1930) and the Vosgesan torrents (Hubault, 1927). At present a number of papers are available where the question is discussed in detail, in particular, in papers by Ambühl (1959, 1962).

What is the reason for the complicated adaptations of the organisms to the life in waters with high flow velocity?

The variability of the flow velocity in torrents observed during the flood, at low water or caused by heavy rains, fast snowmelting, diurnal variations of the water level – all this necessitates the adaptation to life not at a minimal flow velocity in certain microbiotopes, but to maximal velocities which may arise any time of the day, in different seasons and with sudden variations of the discharge. Otherwise the organisms exposed even for a short time to a strong water jet would be eliminated from the torrent. As was previously noted, the hymarobionts sooner or later find themselves in the conditions of maximal velocity in their microbiotopes. Therefore, the microbiotopes are not stable in their parameters, including the current, but very variable.

Water discharge – the strength of the torrent. What do we mean by the term the "torrent strength"? – It is an annual mean water discharge per second, *i.e.* all the mass of the water running during a certain unit of time through the channel. The role of the water mass in the organic life of fast-flowing watercourses is certainly great. That is why we recognized a special type of swift waters, the mountain torrent, with a large mass of water as differing from the brook (Gebirgsbach, according to Steinmann) with small water discharge. The essential role of the water mass in the distribution of organisms is manifested not in the direct mechanical action on the organisms (water pressure, *etc.*), but in the change of other physical and chemical factors. Above we noted the significance of the water mass for the variation of the water temperature and flow velocity, for the formation of the river bed, the transportation of the loads, *etc.* Of course, its effect on the biocenotic relations should not be neglected.

There are two controversial opinions about the effect of the discharge on the composition and distribution of the torrential inhabitants. The most common is the view that the total mass of water in the torrent has no direct effect on its organisms. This idea was also supported by Hynes, the author of a general account of the ecology of running waters: "Discharge itself is, however, of little direct interest in most biological studies; biologists are usually much more interested in the rate of flow of the water where animals and plants actually live" (Hynes, 1970, p. 5). We hold a different viewpoint, believing that the water volume is of considerable significance to the torrential dwellers. A similar opinion of the important role of the torrential power (discharge) for organisms can be found in a paper of Baker (1928) on the distribution of molluscs in torrents. He writes that the life of large watercourses differs from that of small ones. The faunistic differences which correspond to the volume of a stream must be regarded as a universal rule. In the literature one can also find some data on the role of the water volume

224

in the distribution of organisms. Thus Muttkowsky reported that large rivers
are poor in the flatworms. The best places to be inhabited by them are the
medium-size streams. Some stonefly species (from the genus *Acroneura*) are
most abundant in small rivers and near the shores of mountain lakes, but
they are less common in large rivers.

A good example of the effect of the stream strength at different sites is the
distribution of flatworms of the genus *Sorocelis* in the Issyk River cross-
section. For instance, in the river arms and the subsidiary channels at shallow
depths with a uniform substrate in the bed, the flatworms occur over the
whole length of the cross-section; in the channel with deeper waters the
flatworms are found only in such places where the nearshore shallow zone is
well developed; they disappear from the deeper waters towards the midstream. The diagrams of the cross-sections in the channel of varying depth
and width and the sites of occurrence of flatworms are shown in Fig. 82.

A similar distribution pattern in the transverse profile of the Issyk River
as a function of the water volume can be stated with respect to some
mayflies like *Baetis issyksuvensis*, *Ecdyonurus* sp., *Ephemerella submontana*
and the caddisflies *Brachycentrus montanus*, *Agapetus tridens*, *A. kirgisorum*.

In the Issyk River the youngest larvae of all species live in the nearshore
zone and move in deeper waters only when they reach a certain age. In the
relatively quiet upper Issyk River the larvae of *Baetis* mayflies do not
appear in the midstream, although they occur at a considerable distance

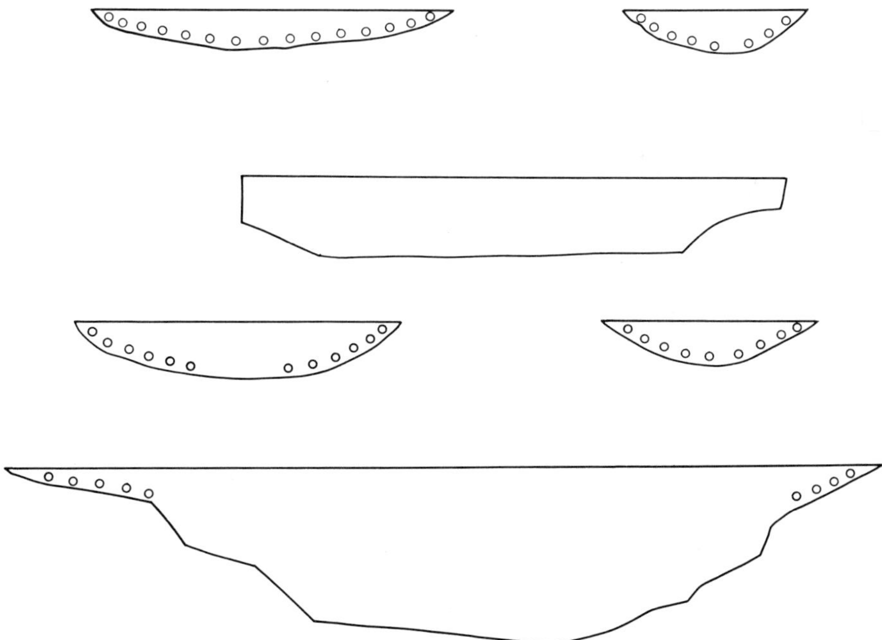

Fig. 82. Transverse distribution of flatworms (circles) over some cross-sections in the Issyk
River.

from the bank (Fig. 83a). With the greater strength in the next cross-section which is situated a little downstream from the former, the nymphs of the same genus occupy only a narrow belt just near the bank (Fig. 83b).

The transverse pattern of distribution in the Issyk River for mayfly nymphs of the genera *Rhithrogena, Iron, Ecdyonurus* and *Baetis* is as follows. In the nearshore zone there is the *Baetis* and farther from the bank *Ecdyonurus* sometimes overlapped by the *Baetis*. Still farther from the nearshore zone and nearer to the midstream is *Rhithrogena*. Almost in the same zone are also nymphs of the *Iron* genus, situated somewhat deeper (nearer to the midstream) than the *Rhithrogena* (Fig. 83c). Downstream in the shallower midstream there are mayfly nymphs of the genera *Iron, Rhithrogena, Ephemerella* and *Baetis*. Nymphs of the latter genus are nearly always the nearest to the bank. Partly overlapping the *Baetis* belt and somewhat deeper there are numerous nymphs of *Ephemerella* and at the midstream (depths of 35–40 cm) there are nymphs of the *Iron* genus. *Rhithrogena* has wedged itself between the *Iron* belt and that abundant of *Ephemerella* nymphs (Fig. 83d).

Let us recall here our assertion (Brodsky, 1935) that at the same flow velocity (V), but different channel cross-section (F), *i.e.* different water discharge ($VF = Q$ m^3/sec), the faunas are different. A certain contribution to the distribution of the fauna is made by the value of the hydraulic radius (R) which is inversely proportional to the wetted perimeter (P): $F/P = R$. The larger the R, the deeper the torrent and the poorer is the population near the midstream, especially of the aquatic insects since their larvae and

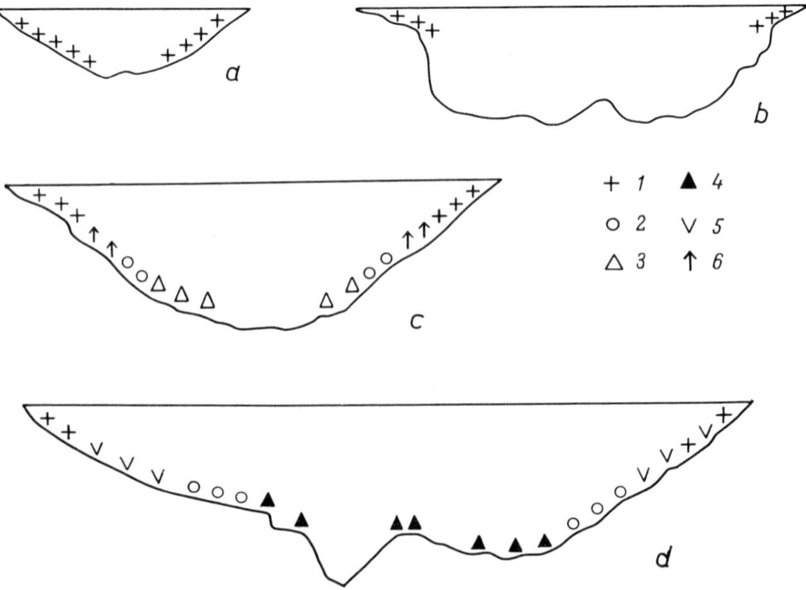

Fig. 83. Transverse distribution of mayfly nymphs over some cross-sections in the Issyk River. 1 – *Baetis* spp., 2 – *Rhithrogena tainschanica*, 3 – *Iron montanus*, 4 – *Ir. rheophilus*, 5 – *Ephemerella submontana*, 6 – *Ecdyonurus* sp. 1. For explanations see the text.

226

pupae prefer shallower waters. Abundance of insects near the banks and nearly complete absence of them from the midstream was reported by us for many streams in Middle Asia. The same is reported by Sibirtzeva (1961) for the invertebrate fauna from the Zeravshan River basin. For instance, in the main channel with very rough current and large water volume, the bentos was concentrated mainly near the bank or on the channel bars. The bentic organisms inhabit immobile stones, backwaters and river arms. In affluents with quieter current and considerably smaller water volume the aquatic organisms extend their range of distribution, and in small affluents and brooks they occupy all the bottom.

Other examples can be given of the negative role of the large water volume as a mechanical factor which affects the distribution of torrential organisms. As we noted previously, pupae of the aquatic insects cannot live in deep waters where the flight of imagoes would be difficult. For example, pupae of *Blepharocera asiatica* and *Tianchanella monstruosa* were always found near the water surface; in deeper torrents which had no rocks or large boulders in the bed, the pupae were encountered only near the bank. Torrents with greater discharge have poorer population than their affluents. Thus in the Zeravshan River were found 30 species, while in its two affluents 47 and 53 species (Sibirtzeva *et al.*, 1961). When studying the Chirchik River basin, she reported less abundant fauna in the main channel than that in the affluents. In the main channel the biomass of bentic invertebrates varied from 0.6 to 3.1 g/m^2 at abundance of 113–2426 individuals per m^2, but in the affluents it was 4.5 g/m^2 at 2925 individuals per m^2 (Sibirtzeva, 1964).

Therefore, the water discharge in torrents (strength, or water volume) with the same channel width considerably affects the transverse distribution of lithorheophylous fauna and its composition. In the former instance, the transverse distribution of preimaginal stages of aquatic insects depends on their ecological characteristics (adaptations!). However, this distribution varies with the volume of water: the larger the discharge in a torrent (in a long period of flood or low water stage), the less the population in the midstream, and, alternatively, with decreased discharge all the cross-section becomes inhabited. This applies both to species and to their stages (larval and pupal ages).

Let us recall that we have discussed the effect of the strength with reference only to one kind of swift waters, namely, the mountain torrent. It is no use comparing torrents, brooks and rivers by their strength, since they are different types characterized not only by the water discharge, but by other conditions.

It is worthwhile mentioning the effect of the catastrophic floods, *i.e.* the discharge of the drastically and abruptly increased mass of water, on the torrential dwellers.

Many authors report a quantitative and qualitative alteration of the torrential fauna after periods of high water. Thus the disastrous effect of flood was observed in a brook in the Vienna Woods when, after a storm, the discharge in the brook increased 15,000 times (Pomeisl, 1953a,b). Heavy rains in Utah on 7 August 1934 produced the first major flood in a river for fifty years, which completely altered the stream bed. On 20 August no animals were found there. Only in late September did the recovery of the fauna begin and this continued into November. At first, the organisms with short life cycles reappeared (*Chironomidae* and blackflies), then mayflies, stoneflies and caddisflies. Similar observations were made in other rivers in Utah (Moffett, 1936) and in other states. In Wales a July flood cut down the fauna of a river from 613 and 1091 animals per square foot to 445, and this was further reduced to 48 (Jones' 1951). Sometimes the flood effect may be different: the flood may increase the population owing to its drift from the upstream sites.

Besides the normal reduction of the fauna during the summer flood, another important factor characteristic of the Tien Shan and Middle Asia in general – the disastrous mudflows – may remarkably alter the composition and abundance of the lithorheophylous fauna. For example, for the Kondara River (the Western Pamirs) Shadin reports the following decrease of the population: before mudflows 994 animals per 0.1 m^2 of the stony bottom in average, after mudflows this number decreased down to 200–206 (Shadin *et al.*, 1951). However, the quantitative estimates for the faunal alterations due to the mudflows in Middle Asia are almost absent, which makes it impossible to characterize this phenomenon. That animals are washed away from the river headwaters by the catastrophic floods was pointed out when discussing the organic drift, but we believe that this concerns only few organisms and does not concern the Blepharocerid larvae, some caddisflies or their pupae.

We think that the faunal decrease during catastrophic floods is accounted for not only by the increased discharge but by the disturbed river bed (if it is made up of fine components), by higher flow velocity, increased turbidity (due to the bank washout and destruction of the channel), *etc.*

Substrate. The composition and disposition of the substrate are so closely related to the flow velocity that it is very difficult to select for investigation such points of a torrent which would differ only by substrates but not by flow velocities. This can be done only with immobile substrates, in other cases such differentiation is nearly impossible.

The most comprehensive works devoted to the distribution of aquatic animals versus the substrate composition are those of Percival and Whitehead (1929, 1930). In their work of 1929 they classified substrates and reported animals which are confined to certain substrates (they presented both quantitative and qualitative data).

1. Loose stones. Most abundant on this substrate are *Rhithrogena, Ecdyonurus, Agapetus* and *Chironomidae*, about 70% of the entire population.
2. Stones embedded in the bottom: *Ancylus, Ephemerella, Agapetus, Glossosoma, Psychomyia, Chironomidae.*
3. Small stones: *Baetis, Agapetus, Helminthinae, Chironomidae.*

228

4. Stones inhabited by *Cladophora*: *Ephemerella, Chironomidae, Naididae*.
5. Loose moss. By this term the authors mean moss which does not form thick growths, permitting the water passage: *Ephemerella, Baetis, Chironomidae*.
6. Thick moss: *Ephemerella, Hydropsyche, Chironomidae, Naididae*.

The numbers of animals per square decimetre of the substrates are as follows:

Loose stones	33	*Cladophora* on stones	444
Embedded stones	46	*Potamogeton* on stones	2,440
Small stones and pebbles	34	Thick moss	4,319
		Loose moss	798

With reference to some species, the authors give concrete examples of their relation to the substrate.

Rhithrogena is most common in sections with fast current, on the substrate from rounded, smoothed stones without any visible vegetation, which indicates mobility of the substrate.

Ecdyonurus venosus appears on less mobile substrates replacing *Rhithrogena* which disappears with increased substrate stabilization.

Baetis spp. are widespread in all stony places and disappear as soon as the stones become covered by fine sand or other such material. These nymphs are most common.

To illustrate the significance of the substrate for the distribution of torrential animals, we compare below the faunal composition of two pairs of points from the lower Issyk River, which differ in the substrate:

Point 30

Flow velocity 0.36 m/sec, water discharge 0.019 m³/sec, water temperature 18°, substrate of small pebbles and gravel

Blepharocera asiatica
Baetis issyksuvensis
Agapetus tridens
A. kirgisorum
Simuliidae
Chironomidae
Gyrinus colymbus

Point 56

Flow velocity 0.37 m/sec, water discharge 0.046 m³/sec, water temperature 14.6°, substrate of silt

Baetis issyksuvensis
Apatelia copiosa
Culicoides sp.
Colymbetini
Oligochaeta

Point 52

Flow velocity 0.66 m/sec, water discharge 0.104 m³/sec, water temperature 18.3°, substrate of small pebbles and gravel

Point 28

Flow velocity 0.66 m/sec, water discharge 0.133 m³/sec, water temperature 17.0°, substrate of small clay particles
No torrential animals, only aquatic earthworms

Baetis issyksuvensis
Baetis sp.
A. kirgisorum
Amphinemura sp.
Limonia sp.

Apatelia copiosa
Agapetus tridens
Chironomidae
Simuliidae (abundant)
Gyrinus colymbus

At point 30 there are typical torrential forms: *Blepharocera asiatica*, *Agapetus tridens* and *A. kirgisorum,* and *Simuliidae* which are not present at point 56 despite the higher water temperature at point 30 (18.0° against 14.6°). Although there is the thermal difference and a slight difference in the water masses (0.046 and 0.019 m³/sec), the effect of the substrate on the fauna formation is very remarkable. Salt content and dissolved oxygen are nearly the same at both points. Two other points (52 and 28) nearly coincide in their parameters, such as the flow velocity, the water discharge, the salt content, dissolved oxygen and water temperature (in the latter case it is somewhat higher, 18.3° against 17.0°). But changes in the substrate entail changes in the fauna. Absence of *Chironomidae* from point 28 is due to the predominance of the *Orthocladiinae* group, since it prefers moss and does not usually live on a silty substrate.

From the lists of animals found at points with silty substrate one can see clearly that the silty biocenoses from the lower waters of a mountain torrent do not contain typical torrential stenobionts. We find only *Oligochaeta*, their inherent group of *Chironomidae* which does not occur within the stony biocenoses, and eurybiont organisms like *Baetis issyksuvensis*. The pools with silty bottom and most downstream sections of irrigative canals are also populated by organisms which are generally not common to the mountain torrent (*Dytiscidae* and *Nepa cinerea*).

The general conclusion from the examples of the relation between the substrate and the fauna is the same as the one we made elsewhere in this essay when we described the torrential fauna. We invariably stress that we deal with the lithorheophylous fauna, whose substrate is the stone in all its modifications (rocks, fragments, boulders, pebbles). Silting of the stony substrate with loess-like loads is very frequent in the lower waters and unfavourable for the hymarobionts.

Stability of the substrate is a favourable factor; however, among the hymarobionts there are some forms (mostly, mayfly larvae and nymphs) which easily settle down on movable stony substrate. This explains why the loose boulders are very poor in caddisflies and dipteran pupae, whereas mayflies are always easily detected.

Therefore, other conditions being equal, the substrate taken within one type of swift waters – the mountain torrent – is a factor affecting the distribution of organisms.

Water temperature. The effect of water temperature on the distribution of organisms in swift waters is given much attention in the literature, and the thermal conditions are often considered to be of paramount importance among all other factors. Steinmann (1907) noted that the water temperature

230

in mountain brooks is very low. This leaves a deep mark on the organic life of the brook. For example, the distribution of flatworms is related to the water temperature as a factor related in turn to the sexual process, whereas other factors (the degree of the water saturation with dissolved oxygen, the concentration of calcium compounds, etc.) may act only as minor causes affecting the distribution of the triclads. Dodds and Hisaw (1925) considered the water temperature distribution to be the major cause of the vertical zonality in stoneflies, mayflies and caddisflies. But other authors who had examined the life of mountain torrents, *e.g.* Hubault (1927), reported the great significance of dissolved oxygen for torrential animals, while Powers (1929) asserted that the distribution of the trout can be due not only to the water temperature.

Let us compare the number of species at the study sites with the water temperature over the whole length of the Issyk River, from its glacial headwaters to the end of the stream (Fig. 91). It turns out that the number of species gradually increase from one, at the foot of the glacier, to a maximum at the beginning of the debris cone on the outlet from the gorge, and then drops down to the initial number of one. The water temperature increases down to the point of the maximum number of species, after which it rises rapidly while the number of species as rapidly drops. One may suggest that the strong effect of water temperature on the number of species takes place only to a certain limit, which creates a false impression that the main factor affecting the distribution of organisms in the middle river is the temperature. It can be illustrated by examples of distribution of individual species. Thus *Himalopsyche gigantea* may be found both in lotic and in nearly standing waters, but always at low temperature. On the other hand, *Blepharocera asiatica* inhabiting the waters with a temperature range 10.5 to 21.5° was never found in slow or standing waters. The same applies to *Deuterophlebia mirabilis* found in a thermal range from 5.0 to 17.5°, but only in running, mostly, swift waters. We can find many examples of the thermal range "broadening" by the stenotherms. But our aim is not to discredit the thermal factor, but to show that it does not overwhelm any other factor and that its role varies with the rest of the environmental conditions.

Having determined the lowest absolute heights (with maximum water temperatures) for each faunal component of the Issyk River and irrigative canals, we shall arrange these temperatures in the increasing order to get a histogram (Fig. 84) which shows clearly that the lowest distribution limits (downstream) of nearly all organisms in the canals are determined by higher temperatures than in the river channel. For example, *Deuterophlebia mirabilis* in the main channel is found only up to 11.5°, after which a careful survey could reveal none of this species in the river. In the irrigative system this species, although solitary, was found at 17.5°, i.e. here it resists the additional 6° of warmer water – a considerable value for a stenotherm. From the mayflies, *Iron nigromaculatus* occurs in the main channel only at temperatures up to 15.5° while in the irrigative canals up to 21.6°, i.e. the difference is again 6°. Zero thermal difference between the main channel

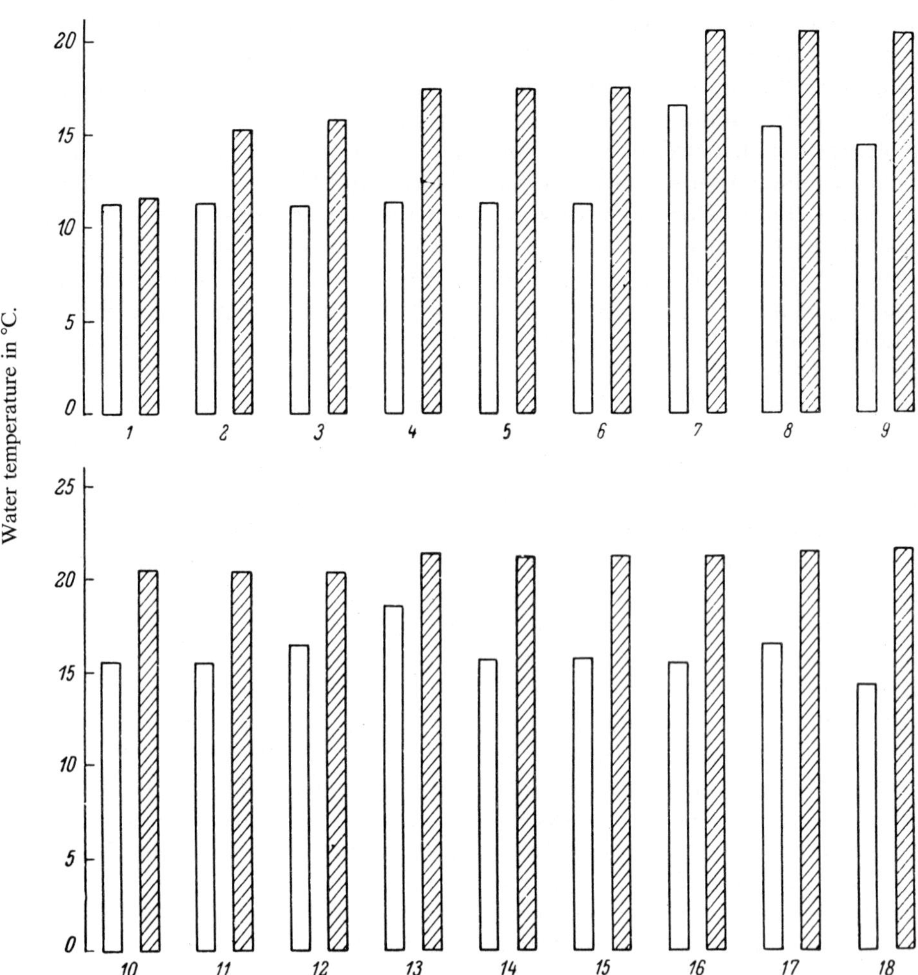

Fig. 84. Water temperature variation at the lower distribution limit for lithorheophylous invertebrates in the Issyk River channel and in the irrigative canals (shaded). The species are arranged with increasing water temperature:

1 – *Iron montanus*, 2 – *Ir. rheophilus*, 3 – *Ecdyonurus* sp. 1, 4 – *Deuterophlebia mirabilis*, 5 – *Tianschanella monstruosa*, 6 – *Rhyacophila extensa*, 7 – *Ephemerella submontana*, 8 – *Rhithrogena tianschanica*, 9 – *Apatelia* sp., 10 – *Brachycentrus montanus*, 11 – *Caenis* sp., 12 – *Ecdyonurus* sp. 1, 13 – *Blepharocera asiatica*, 14 – *Ameletus alexandrae*, 15 – *Iron nigromaculatus*, 16 – *Ephemerella* sp. 1, 17 – *Rhithrogena* sp. 1, 18 – *Agapetus* sp.

and the canals is found for the following forms: *Hydropsyche* sp., *Simulium* spp., *Baetis issyksuvensis* and *Baetis* sp. The listed organisms are likely to be assigned to relative eurybionts.

What conclusion should be made from this discussion? Most organisms in the main channel of the Issyk River occur only up to the water temperature of 16°, the flow velocity of 2 m/sec and the water discharge 3 m³/sec. In the canals the border of distribution lies at 22°, flow velocity 0.5 m/sec and water discharge 0.01 m^3/sec. One can conclude that the thermal factor does not have very great significance for the distribution of organisms, as it is sometimes assumed. Were the thermal factor the dominant one, we should expect the same thermal limits both in the main channel and in the irrigative canals. In this instance, the effect of the thermal factor is weakened by some other factors. Perhaps, this is due to great instability of the river channel in the lower reach (the channel wandering) and, therefore, instability of the substrate, which does not occur in the canals.

The general conclusion about the significance of the water temperature for the distribution of hymarobionts in the torrent is similar to the one we made when discussing other factors. The water temperature is as inherent component of the system of factors acting in the torrent as the flow velocity, the strength of torrent, the substrate, *etc.*

It is the daily and annual ranges of temperature which is of importance to torrential organisms, rather than the temperature at a given moment. These ranges in mountain torrents in spite of the relatively thin water layer and, therefore, the possibility of considerable warming and cooling, are small (see Chapter 3 containing a general account of the Tienshanian mountain torrent). Due to high flow velocities and the glacial and snow-glacial origin, the mountain torrent may be named a low temperature thermostat. As a result of these thermal peculiarities, most hymarobionts are cold-water stenotherms.

Salt composition of the water. Relatively low mineralization of water in the Tienshanian torrents impedes discussion of the effect of this factor on the fauna distribution. Some difference in the dissolved calcium content can be found in the Issyk River but only in a few of its affluents, not in the main channel.

The significance of varying calcium content has received special attention in studies of European swift waters. Some authors tend to consider high calcium content as a factor unfavourable for faunal abundance in rivers and brooks (Steinmann, 1907). According to Muttkowsky (1929), the scanty mollusc fauna in the rivers of Yellowstone National Park is due to low calcium content in these waters. Some authors attributed a great role to calcium content in the distribution of some groups of organisms (Carpenter, 1928, for the flatworms), but others argued against this relation (Hubault, 1927). There is a classification of rivers based on calcium content in the water (Ohle, 1937).

To illustrate the effect of calcium on the fauna composition in the Issyk River basin, let us consider the following examples (Table 43). Point 53: CaO 0.084 mg/l, water temperature 14.2°, flow velocity 0.40 m/sec, water discharge 0.87 m^3/sec, substrate of rock fragments. The point lies at an affluent of the Issyk River, which runs down from low mountains and inflows in the river in the vicinity of an irrigative canal. Point 54: CaO 0.01 mg/l, water temperature 12.0°, flow velocity 1.00 m/sec, water discharge 0.416 m^3/sec, substrate of pebbles and stones. Point 58: CaO 0.01 mg/l,

water temperature 14.3°, flow velocity 1.58 m/sec, water discharge 1.763 m³/sec, substrate of small pebbles.
What conclusions can be made from the data in Table 43?

1. According to simultaneous sampling at three points, the total number of individuals at point 53, where the calcium content is highest, is not smaller than those at points 54 and 58, but even remarkably larger. The number of species is nearly the same at the three points.

2. Relatively calciphilous organisms can be revealed, like *Ecdyonurus* sp. 1, *Rhithrogena* sp. 1, *Baetis* sp., *Limnius* sp. The "true" calciphilous form is only *Ecdyonurus* sp. 1 found at one point (53) with high calcium content in strikingly large numbers.

3. One can find organisms which suffer deterioration due to high calcium content in the water (except the forms not found at flow velocities under 0.40 m/sec and water discharge less than 0.087 m³/sec) – *Blepharocera asiatica* and *Iron nigromaculatus*.

As to the other forms not found at point 53, one can suggest from the investigations in other torrents of the Tien Shan that *Deuterophlebia mirabilis, Tianschanella monstruosa, Iron rheophilus, Ephemerella submontana, Ecdyonurus* sp. 2 and *Sorocelis* spp. are calciphobic.

Clearly, relatively few true dwellers of the Tienshanian mountain torrents inhabit waters with rich calcium salts (the absence of molluscs from the Tienshanian torrents is likely to be caused by low calcium); those to be assigned to calciphils are from the biocenoses of the lower torrent and of irrigative canals. Higher calcium content in water seems to result in a change of the species composition, but does not decrease the totals of animals as suggested by Steinmann, Thienemann and Martynov.

Dissolved oxygen. The oxygenic factor in torrents never occurs at a minimum and the winter-kills are never observed (torrents are free from ice in winter, except their headwaters near the perrenial snows). The high content of dissolved oxygen is due to the fast and turbulent current, *i.e.* a physical factor which depends but little on the photosynthetic activity of algae. Torrential waters are usually saturated and supersaturated with oxygen (mostly over 100% saturation).

The minimal oxygen requirements in many torrential animals are considerably lower than the oxygen content in the water. Thus for trout this is 2.5 to 3.5 cm³/l, for stoneflies (*Perla*) 4.9 cm³/l, for other genera of the same order (*Perlodes, Taeniopteryx, Nephelopteryx, Protonemura, Amphinemura*) 6.0 cm³/l (Schoenemund, 1925). In mayflies from mountain torrents, in *Blepharoceridae* and *Deuterophlebia* these requirements are still higher, but not beyond the values observed in the water.

Because of the specific conditions in mountain torrents, the saturation and supersaturation of water with oxygen, the thin water layer, the daily and seasonal fluctuations of the water level, the organisms often find themselves in aquatic and aerial environments alternatively. Such an alteration of environment caused them to develop, in the process of evolution, very

234

Table 43. Numbers of animals per sample collected at three points (53, 54 and 58) of the Issyk R. system with different calcium contents (mg/l).

Name of animal	53	54	58
		CaO	
	0.08	0.01	0.01
Ephemeroptera			
Iron rheophilus	0	1	11
I. nigromaculatus	0	0	3
Rhithrogena sp. A	109	0	0
Ecdyonurus sp. A	186	0	0
Ecdyonurus sp. B	0	1	2
Ephemerella submontana	0	30	39
Baetis issyksuvensis	7	0	7
Baetis sp. A	48	0	0
Caenis sp. A	0	4	0
Ameletus alexandrae	1	0	0
Trichoptera			
Rhyacophila extensa	0	1	0
Agapedus tridens and			
kirgisorum	80	9	40
Apatelia copiosa	7	12	4
Brachycentrus montanus	20	3	2
Dinarthrum reductum	4	15	3
Hydropsyche sp.	2	1	0
Dolophilodes sp.	0	2	0
Plecoptera			
Amphinemura sp.	32	1	0
Neoperlina capnoptera	1	0	0
Diptera			
Blepharocera asiatica	0	73	95
Tianschanella monstruosa	0	5	25
Deuterophlebia mirabilis	0	1	10
Simuliidae	36	48	31
Chironomidae	2	2	5
Limonia sp.	8	4	1
Coleoptera			
Limnius sp.	8	4	1
Crustacea			
Gammarus spinulatus	11	1	0
Turbellaria			
Sorocelis sp. sp.	0	3	0
Crenobia sp.	4	1	0
Total, individuals	572	218	279
Total, species	18	21	16

peculiar and specific respiratory organs. Some are adapted equally well to the absorption of oxygen from the water and the air (the tracheal gills in some dipteran pupae, in particular, in *Blepharoceridae*). In mayfly nymphs the high oxygen content and flow velocities caused a reduced area of gills and their functional modification when the gill fringes become thickened to form a sucker like in nymphs of *Iron, Rhithrogena* and others (Dodds and Hisaw, 1924a,b).

Therefore, in contrast to the standing and slow waters, the oxygen in torrents is in a satisfactory state and does not inhibit the organic life, but as a component of the "mountain torrent system" it is very specific.

"Biocenotic factors". Strictly speaking, this is not a factor but a whole intricate system of animalistic relations, *viz.* interspecific and intraspecific interactions, the structure of biocenoses, *etc.*

Unfortunately, this realm of ecology is almost entirely unknown in relation to the torrential dwellers and we must confine ourselves only to general considerations and assumptions. One thing is apparent that the "biocenotic factor" is of great importance to hymarobionts. We have repeatedly emphasized that this factor causes the distribution of torrential animals to a large degree. Among other things, the refuge to the extreme conditions, as it is observed in some typical hymarobionts, may be caused by the carnivors pressure and the competition. This phenomenon, *i.e.* the developed resistivity to unfavourable conditions with the aim to avoid the competitors, is of course of universal importance (Brodsky, 1957).

Some data are available on the feeding in hymarobionts, but their trophology has been much less examined than that of the standing water dwellers. Most hymarobionts are known to be algivorous animals feeding on diatoms which are abundant in growths on the stones, rocks and boulders. To scrape their food from the stones, larvae of torrential insects have developed specialized devices by modifying their mouth parts, as described for caddisflies (Lepneva, 1964), blackflies (Rubzov, 1956), mayflies (Strenger, 1953), *Blepharoceridae* (Anthon and Lyneborg, 1968) and *Deuterophlebia* (Brodsky, 1930b).

The main feature of the specialized adaptations in the mouth apparatus of these hymarobionts are the strong and resilient brushes of bristles (on mandibles, maxillae and labial and maxillary palps) acting like steel scrapers.

Some hymarobionts are carnivorous (some stonefly nymphs and caddisworms), but the feeding type is insufficiently known and one may expect surprises. For example, blackfly larvae which feed by particle catching with a large fan may be also cannibals as they are capable of swallowing their relatives of smaller size (Konurbaev, private communication).

Examples of parasitism (the *Mermithidae* infection) and commensalism (blackflies and *Chironomidae* on mayflies), *etc.* are also known among torrential dwellers.

But now we must confine ourselves only to the statement of the need to study the biocenotic relations among torrential dwellers.

All ecological factors discussed are of significance for the distribution of

236

the torrential fauna. There seems to be no "indifferent" factors. The physical and chemical factors must be supplemented by a number of factors of biocenotic nature such as the absence or availability of food, its character, presence or absence of enemies and parasities, the character of vegetation, *etc.* Of great importance to the torrential dwellers, we believe, is the elimination of competition, which makes it possible to flourish for those organisms which are well-adapted to the "extremal" conditions.

All the factors affect the organisms as a complex but not individually. Any factor may be, in certain conditions, more important than others, and the isolation of one or two most important factors is possible only when the other factors are in a certain combination. We have attempted to do this with the reservation that it is an artificial approach which can rarely be effective and only when it is applied to the same type of fast-flowing watercourses.

9 General review of the invertebrate fauna from the mountain torrents of the Tien Shan

Ecologically, the fauna of the fast-flowing watercourses may be referred to as association, since there are different biocenoses in a torrent ("the mosaic of biocenoses"; Illies and Botosaneanu, 1963). But in a mountain torrent with its 95% of stony bottom in the channel there predominates the lithorheophylous biocenosis, *i.e.* biocenosis of stones, rocks, boulders and pebbles. The biocenoses of sand and silty substrate occupy a negligible area; one exception is the biocenosis of moss which is present in torrents at places, but it occupies a smaller area than the lithorheophylous one.

We have tried to gather as much evidence as possible on the fauna of Middle Asian torrents. Some of these data were not yet published and we are very grateful to E. O. Omorov, N. K. Sibirtzeva, E. O. Konurbaev and L. A. Kustareva for providing them. Some of the data were received from the Biological Station, Academy of Science, Kirghiz SSR, and the Hydrobiological Department of the State University, Tashkent, thanks to the authorities of these institutions V. F. Gurvitsch, A. O. Konurbaev, and L. A. Folian, whose assistance we acknowledge.

To reveal the routine list of faunal composition for invertebrates of the Tienshanian mountain torrent it is appropriate to use the data for the Issyk River, with reference to which we characterize ecologically the widespread forms inhabiting the entire length of the river. (As the details of the species identification for various torrents of the Tien Shan are different, see also Table 30 in addition to the species list for the Issyk River in Table 44.)

The lithorheophylous fauna list for the Issyk River contains several groups, including some unidentified species (*e.g. Chironomidae* and black-flies in Table 44). The upper limit of the flow velocity for these species and groups is 4.0 m/sec. This value is very approximate, since the flow velocity was measured with the Pitot-Darcy tube which operates correctly only up to 3.3 m/sec and, therefore, only the lower limit of flow velocity in the site of the animal occurrence may be accurately known. Higher velocities were measured by the float method which is less accurate. Although the limits of parameters are given in Table 44 only for the Issyk River, their ranges are very wide for many organisms. This is a result of the recording by the presence or absence of animals without any quantitative estimates of their abundance. Such an approach gives a pattern of the maximum ecological amplitude, *i.e.* it reveals the limits at which the animals in the Table may be encountered even as single individuals. Some quantitative data will be given later for the torrential animals, but now, taking into account the above considerations and the data presented in Chapter 6, we single out the ecological groups on the fauna list for the Issyk River and the dominant group of torrential animals, the hymarobionts.

Do the hymarobionts actually exist? Even a glance at the torrential fauna is enough to see many *Deuterophlebia mirabilis, Tianschanella monstruosa, Asioreas nivia, A. tianschanica* (in torrents other than the Issyk River), *Rhithrogena tianschanica, Iron montanus, I. rheophilus* and others. But if scrutinized, many of them appear to occur also in relatively slow waters with rather high temperature, *i.e.* they may be encountered in different types of running waters, *e.g.* in brooks. From this a hasty conclusion may be deduced that no animals exist which inhabit exclusively the mountain torrents. But from the comparison of the abundances of a given species in different types of watercourses, the distribution of organisms in the waterbody, their habits and, last but not least, their adaptations one can find the true dwellers of the mountain torrent, or the hymarobionts. For example, *Deuterophlebia mirabilis* occurs both in torrents and brooks. But in torrents it is numerous and confined to the typical torrential biocenosis, that of the stony substrate, and lives mostly near the banks, while in brooks it is solitary and occurs over the whole cross-section of the watercourse.

As can be understood from the above arguments, the data of Table 44 are not sufficient in themselves to characterize the animals. Only on the strength of all the evidence like the frequency of occurrence, the biomass, the distribution in the fast-flowing watercourses, the life cycle and the adaptation, one can judge correctly about the ecological pattern of the aquatic forms.

Table 44 shows the numerical dominance of hymarobionts over other ecological groups of animals. The relative eurybionts live only in the lower reach of the torrent. Some animals are also typical of other types of watercourse like brooks, springs and alpine lakes.

One will see later that the hymarobionts play an important role not only in the Issyk River, but also in other typical mountain torrents, but their significance varies with the type of watercourse.

Before discussing the faunal composition of single groups on the basis of the numbers or the biomass per square metre or both, we quote the percentage data on the generic composition of the invertebrate fauna from different types of swift waters. Little evidence is available on this question, but the fauna of the Issyk River may be compared in this respect with that of some European watercourses in France, Great Britain and Germany from the works by Bornhauser (1912), Carpenter (1927) and Hubault (1927):

Issyk River	32	genera of invertebrates
England, short streams	57	genera of invertebrates
France, lotic springs (emissaires des sources)	35	genera of invertebrates
France, cascades in rivers (barrages sur les riviéres)	42	genera of invertebrates
France, brooks (ruisseaux)	73	genera of invertebrates
Germany, springs (Quellen)	106	genera of invertebrates

The greatest variety of genera is observed in springs; it is much less in torrents and cascades, but this fact does not indicate the faunal scarcity or

Table 44. The extreme physical parameters for the Issyk River.

Name of organism	Absolute height, m	Water temperature in °C	Flow velocity in m/sec	Water discharge in m^3/sec.	Substrate
Ephemeroptera					
*Iron montanus	1250–2230	3.8–11.8	1.66–4.00	1.95–18.0	Rocks, pebbles
I. nigromaculatus	900–1020	11.2–21.2	0.34–2.50	0.01–18.0	Pebbles, gravel
*I. rheophilus	1050–1439	11.5–15.5	0.71–4.00	0.10–18.0	Rocks, pebbles
*Rhithrogena tianschanica	1100–2230	3.8–20.5	0.85–4.00	0.03–18.0	Rocks, pebbles
Rhithrogena sp.	900–1180	11.5–21.5	0.52–4.00	0.01–18.0	Pebbles, gravel
*Ecdyonurus sp. A	1100–2230	5.0–15.8	0.71–4.00	0.10–18.0	Rock fragments, pebbles
Ecdyonurus sp. B	1280	14.2	0.40	0.087	
+Ecdyonurus sp. C	830–1120	12.0–20.5	0.71–2.30	0.03–10.0	Pebbles
*Ephemerella submontana	830–1558	10.5–20.5	0.71–4.00	0.10–18.0	Rock fragments, pebbles
Ephemerella sp.	900–950	15.8–21.2	0.34–2.30	0.02–10.0	Pebbles, gravel
Baetis issyksuvensis	690–2230	5.0–30.0	0.05–4.00	0.01–18.0	Pebbles, silt
Baetis sp.	610–1300	10.2–30.0	0.05–4.00	0.01–18.0	Pebbles, gravel
Caenis sp.	950–1280	11.5–20.5	0.85–4.00	0.03–18.0	Pebbles, sand
+Ameletus alexandrae	850–2230	5.0–21.2	0.34–4.00	0.01–18.0	Silt
+Ameletus sp.	2900	2.0	?	?	Pebbles
Trichoptera					
*Hymalopsyche gigantea	1737–2210	5.5–8.0	4.00	18.0	Rocks, pebbles
*Rhyacophila extensa	950–1805	7.5–17.5	1.00–4.00	0.29–18.0	Rocks, pebbles
Rhyacophila sp.	2900	2.0	?	?	Pebbles
*Oligoplectrodes potanini	980–1763	14.0–14.5	0.56–2.50	0.03–6.0	Rocks, pebbles
*Dolophilodes sp.	1020–1780	7.5–13.8	0.71–4.00	0.26–18.0	Rocks, pebbles
*Dinarthrum reductum	830–1737	8.0–21.2	0.34–4.00	0.02–18.0	Rocks, pebbles
Brachycentrus montanus	950–1338	10.6–20.5	0.71–4.00	0.03–18.0	Rocks, pebbles
*Agapetus kirgisorum	980–1737	8.0–21.5	0.36–4.00	0.02–18.0	Rocks, pebbles
*A. tridens	980–1737	8.0–21.5	0.36–4.00	0.02–18.0	Rocks, pebbles
Apatalia copiosa	980–2230	5.5–20.5	0.37–4.00	0.03–18.0	Rocks, pebbles
Hydropsyche sp.	950–1020	10.8–15.8	1.00–2.50	0.40–6.0	Pebbles, silt

	1020	11.8	3.33	8.0	
Hydroptila sp.	1020	11.8	3.33	8.0	Pebbles
Plecoptera					
*Neoperlina capnoptera	950–1982	5.7–15.8	0.40–4.00	0.09–18.0	Rock fragments, pebbles
*Nemoura sp.	1780–1982	5.7–8.0	4.00	18.0	Rocks, pebbles
Amphinemura sp.	820–1558	10.6–20.5	0.66–4.00	0.11–18.0	Rocks, pebbles
Leuctra sp. A	1763–2230	7.0–14.0	0.56–4.00	0.03–18.0	
*Leuctra sp. B	2900	2.0	?	?	Pebbles
Diptera					
*Tianschanella monstruosa	1050–1554	11.0–17.5	1.00–4.00	0.29–18.0	Rocks, pebbles
*Blepharocera asiatica	900–1439	11.0–21.2	0.36–4.00	0.01–18.0	Rocks, gravel
*Deuterophlebia mirabilis	1050–2230	5.0–17.5	0.71–4.00	0.10–18.0	Rocks, pebbles
Simuliidae	720–2500	3.0–29.0	0.05–0.40	0.01–18.0	Rocks, pebbles
Chironomidae	690–4000	0.4–30.0	0.06–4.00	0.01–18.0	Rocks, pebbles, silt
*Limonia sp.	820–2021	7.0–18.3	0.66–4.00	0.11–18.0	Rocks, pebbles
Coleoptera					
Gyrinus colymbus	690–1400	11.5–30.0	—	—	Water surface
Limnius sp.	1220–1280	14.2–14.3	0.40–1.53	0.09–1.76	Rock fragments, fine pebbles
Tricladida					
+Sorocelis sp. sp.	950–2230	5.5–12.8	1.00–4.00	0.40–18.0	Rocks, pebbles
+Crenobia sp.	1120–2083	5.8–14.2	0.40–4.00	0.08–18.0	Rocks, pebbles

Note: The asterisk indicates hymarobionts; other forms are relative eurybionts; + stands for organisms typical of brooks, springs and alpine lakes.

abundance. We wrote earlier that only few species (or genera) may produce large populations under extreme conditions.

In literature the percentage data are available for species numbers in different animalistic groups from the Chirchik and Zeravshan Rivers (Sibirtzeva, 1961; Sibirtzeva *et al.*, 1961). These data show clearly the important role of aquatic insects in the fauna of the Tienshanian swift waters:

Chirchik River, the species percentages:

Insects	95.5
among them	
mayflies	7.9
caddisflies	11.6
stoneflies	5.8
Diptera:	
blackflies	16.7
Chironomidae	35.6
other dipterans	16.7
dragonflies	0.7
beetles	5.0
Mites	4.1
Molluscs	1.3
Worms	3.2

Zeravshan River, the species percentages

Insects	92.0
Crustacea	1.7
Molluscs	2.7
Worms	3.6

We find it useful to present here the initial field data taken when the fauna of the Tienshanian swift waters was quantitatively sampled. Although tedious work like the picking of animals from the measuring frame, identification of species and counting and weighing of the animals gives somewhat random data (because of the dynamism of the torrent and its fauna, the change of the quantitative relations between the faunal components due to an occasional short-time flood, the random installation of the frame in a site which is either abundant or poor in animals, *etc.*) the quantitative aspect of the faunal composition can be revealed, *i.e.* one gets an idea of "the order of magnitude" which helps understand the fauna structure better than merely the recording of the presence or absence of a certain species. The quantitative data for the Tienshanian torrential fauna are few and we give all of them, expecting them to be useful for further studies. Some annual means for the abundance and biomass are known for rivers of the Turgen River basin, but these, unfortunately, are not classified by the animalistic groups (Kurmangalieva, 1976a). We again warn the reader against "the hypnotism of the figures": the data in the tables show nothing more than "the order of magnitude"!

Let us compare the invertebrate faunas (not only the lithorheophylous one, but all the quantitatively sampled) by large, not detailized groups from the mountain torrent Aksu River (Table 45) and the springs situated in the river basin (Table 46). The Aksu River runs in the north-western part of the Terskei Alatau Range; it is a large affluent of the Djergalan River running into the eastmost part of the Issyk-kul Lake.[1]

Tables 45 and 46 show that the samples were taken at seven different dates in the river and at five dates in the springs, and for this reason the above random changes in the fauna tend to be somewhat smoothed out.

The differences in the abundance of invertebrate groups in the river and springs are still more evident when the data are averaged for every group (Table 47). The biomass is not shown in the table because it was given in the previous tables or is unknown for some (not abundant) groups. But the biomasses in the river and springs correlate with the numbers of animals.

The comparison of the data in Table 47 and in Fig. 85 shows that the invertebrate fauna in the springs is more diverse than in the mountain torrents and rivers. In the springs predominate such groups as blackflies, *Chironomidae*, beetles, mites, side-swimmers, *Ostracoda*, *Oligochaeta* and *Nematodes*. Some groups are not represented in the river – the biting midges, the molluscs and the flatworms. Earlier we noted that single members of these groups may occur in the torrent, but only by chance, not as a typical component. Thus, the flatworms may be found in a torrent where the current is slower but are much less abundant here than in brooks and springs.

The higher abundance of mayflies in the springs is due to the higher species diversity of the larval and nymphal forms compared to the torrents. We emphasize this fact, since the balance between caddisflies and mayflies determines the type of mountain stream: in the "true" mountain torrent there predominate mayflies, rather than caddisflies. In springs from the lower Akterek River (the northern slope of the Terskei Alatau Range) there were no mayflies of the genus *Iron* and *Rhithrogena tianschanica* and no *Blepharoceridae*, but these were however encountered in the Akterek River (Kustareva, 1976a).

Tables 48 and 49 give the data for comparison of the invertebrate faunas from mountain rivers (and torrents) and springs.

The mean values are shown in Table 50 with a difference from the previous example: in Table 47 the data were averaged for the Akbura River and the adjacent springs, while now the data used are for springs and for four rivers from different sites of the same region of the Tien Shan, the Issyk-kul Valley.

Analysis of Table 50 reveals a smaller difference between the faunas of rivers and springs than it followed from the comparison of the fauna from the Akbura River alone and that from its springs (see Table 48). The smaller difference results from the fact that Table 50 gives the fauna lists not for

---

[1] Data on the Aksu River fauna are included in some tables which follow. They were taken in different years and are somewhat different numerically, but the quantitative relations between the animalistic groups are the same. The repeated sampling of one and the same river may show the quantitative data to be valid.

Table 45. Numbers of animals per 1 m² (numerator) and biomasses in mg/m² (denominator) of invertebrate groups in the Aksu River, an affluent of the Djergalan, the northern Terskei-Alatau Range. Sampled in 1967–69 (after Ivanova, 1973).

Name of group	17 V 1967	18 VI 1967	19 III 1967	19 VII 1967, at Djergalan R.	29 VIII 1967	31 X 1969	11 VII 1969, after flood
Ephemeroptera	$\frac{170}{1502}$	$\frac{112}{462}$	$\frac{112}{222}$	$\frac{486}{1.492}$	$\frac{305}{3.87}$	$\frac{292}{131}$	$\frac{188}{146}$
Trichoptera	$\frac{104}{850}$	$\frac{424}{1.451}$	$\frac{9}{10}$	$\frac{16}{-}$	$\frac{8}{2}$	$\frac{144}{60}$	$\frac{4}{26}$
Plecoptera	$\frac{228}{596}$	$\frac{119}{412}$	$\frac{2}{14}$	$\frac{18}{78}$	$\frac{48}{12}$	$\frac{588}{922}$	$\frac{16}{6}$
Diptera:							
Simuliidae	$\frac{2}{-}$	—	$\frac{2}{-}$	—	—	$\frac{672}{308}$	—
Chironomidae	$\frac{530}{282}$	$\frac{125}{55}$	$\frac{227}{107}$	$\frac{190}{110}$	$\frac{21}{186}$	$\frac{1144}{132}$	$\frac{218}{40}$
Blepharoceridae and other	$\frac{24}{222}$	$\frac{18}{418}$	$\frac{36}{388}$	$\frac{10}{80}$	$\frac{14}{127}$	$\frac{76}{73.4}$	—
Diptera							
Odonata	—	—	—	—	—	$\frac{4}{?}$	—
Rhynchota	$\frac{2}{?}$	—	$\frac{35}{10}$	$\frac{52}{29}$	$\frac{2}{?}$	—	$\frac{6}{1.3}$

Coleoptera	$\frac{2}{30}$	—	—	—	$\frac{2}{1.0}$	—	—
Hydracarina	$\frac{56}{22}$	$\frac{62}{34}$	$\frac{10}{1.0}$	$\frac{10}{4.0}$	$\frac{3}{?}$	$\frac{44}{2.0}$	$\frac{20}{4.8}$
Amphipoda	$\frac{8}{?}$	$\frac{8}{114.0}$	—	$\frac{6}{114.0}$	—	$\frac{4}{1.5}$	—
Cladocera	—	—	—	—	—	$\frac{16}{?}$	—
Copepoda	—	$\frac{2}{?}$	$\frac{4}{?}$	—	—	$\frac{16}{?}$	—
Ostracoda	—	$\frac{1}{?}$	$\frac{2}{?}$	$\frac{2}{?}$	—	$\frac{44}{?}$	—
Mollusca	$\frac{2}{?}$	—	—	—	—	—	—
Oligochaeta	—	$\frac{5}{2.0}$	$\frac{14}{4.0}$	—	—	$\frac{186}{8.2}$	$\frac{20}{1.1}$
Nematodes	—	—	—	—	—	$\frac{76}{?}$	—
Turbellaria	—	—	—	—	—	$\frac{4}{16.0}$	—

Table 46. Number of animals per 1 m^2 (numerator) and biomasses in mg/m^2 (denominator) of invertebrate groups in springs along the Aksu River, the northern slope of Terskei-Alatau Range. Sampled in 1967–69 (after Ivanova, 1973).

Name of group	19 V 1967	20 VI 1967	19 VII 1969	28 VIII 1968	31 X 1969
Ephemeroptera	$\frac{362}{4436}$	$\frac{1110}{2728}$	$\frac{218}{674}$	—	$\frac{1506}{274}$
Trichoptera	$\frac{330}{2324}$	$\frac{1252}{5089}$	$\frac{870}{7104}$	$\frac{20}{144}$	$\frac{1538}{4789}$
Plecoptera	$\frac{464}{2595}$	$\frac{38}{248}$	—	$\frac{144}{22}$	$\frac{20}{?}$
Diptera:					
Simuliidae	—	$\frac{1576}{798}$	$\frac{2}{?}$	—	2118
Chironomidae	$\frac{224}{74}$	$\frac{356}{189}$	$\frac{330}{290}$	$\frac{112}{32}$	$\frac{7422}{1997}$
Heleidae	—	—	—	—	$\frac{748}{4.5}$
Blepharoceridae and other *Diptera*	$\frac{60}{588}$	$\frac{45}{1347}$	$\frac{132}{582}$	$\frac{48}{594}$	$\frac{64}{74}$
Rhynchota	$\frac{13}{10}$	$\frac{50}{10}$	$\frac{12}{2.0}$	$\frac{4}{104}$	—
Coleoptera	—	$\frac{9}{1.0}$	$\frac{398}{727}$	—	$\frac{8}{?}$
Hydracarina	$\frac{94}{52}$	$\frac{417}{279}$	$\frac{560}{192}$	$\frac{4}{4.0}$	$\frac{20}{2.2}$
Amphipoda	$\frac{2}{0.75}$	$\frac{883}{3.357}$	$\frac{48}{1.360}$	$\frac{212}{20}$	$\frac{20}{668}$
Cladocera	—	—	—	—	—
Copepoda	—	$\frac{2}{?}$	$\frac{26}{?}$	—	$\frac{50}{?}$
Ostracoda	$\frac{6}{0.75}$	$\frac{2254}{126}$	$\frac{114}{4.0}$	$\frac{212}{20}$	$\frac{478}{?}$
Mollusca	$\frac{3}{?}$	$\frac{40}{36}$	$\frac{14}{19}$	$\frac{4}{28}$	$\frac{78}{46}$
Oligochaeta	$\frac{20}{948}$	$\frac{40}{598}$	$\frac{22}{2448}$	$\frac{48}{1.760}$	$\frac{268}{94.2}$
Nematodes	—	—	—	—	$\frac{166}{?}$
Turbellaria	$\frac{20}{22}$	$\frac{28}{74}$	—	$\frac{104}{220}$	$\frac{104}{101}$

"typical" mountain torrents but for mountain rivers and even for those their sections which are under a strong influence of the inflowing springs (*e.g.* the Karasu River). Especially non-typical of the torrent is the fauna from the offing area of the Tosor River with larger numbers of *Chironomidae*, water mites and *Oligochaeta* but a smaller caddisfly population.

Although this fauna is more specific to mountain rivers than to true

Table 47. Averaged numbers of animals per 1 m² in the Aksu River and its springs (at population density less than one animal per m², the group was not included in the table).

| | Animals per m² | | |
Name of group	Aksu R.	Springs	Note
1 *Ephemerpoptera*	238	635	Various species
2 *Trichoptera*	105	803	Various species
3 *Plecoptera*	144	132	?
Diptera:			
4 *Simuliidae*	97	734	Various species
5 *Chironomidae*	352	1750	Various species
6 *Heleidae*	—	147	Various species
7 *Blepharoceridae* and other *Diptera*	25	69	In river this group includes *Deuterophlebia*
8 *Rhynchota*	14	16	Various species
9 *Coleoptera*	—	83	Various species
10 *Hydracarina*	29	218	Various species
11 *Amphipoda*	4	253	Various species
12 *Copepoda*	3	16	For the river occasional occurrence, comes from springs
13 *Ostracoda*	7	615	The same
14 *Mollusca*	—	28	The same
15 *Oligochaeta*	32	79	Various species
16 *Nematodes*	11	33	Various species
17 *Turbellaria*	—	83	Various species
Total	961	5704	

Note: The serial numbers are as in Fig. 85.

mountain torrents, it differs from the fauna of springs quite notably. Thus in rivers, stoneflies predominate, while caddisflies, blackflies, bugs, beetles and flatworms are in smaller numbers.

After the faunal comparison between the mountain rivers and the springs, let us consider the composition of the invertebrate groups from a typical mountain torrent, the Cholpon–Ata River. The river runs down from the northern slopes of the Kungei-Alatau Range and therefore it also is within the Issyk-kul Valley. Through the courtesy of workers of the Issyk-kul Lake Biological Station, Academy of Sciences, Kirghiz SSR, we have at our disposal quantitative data for the fauna of this river from its single points (Table 51) or from the upper, the middle, and the lower reaches (Table 52).

At first, let us consider Table 51 which contains the data for four points of study, three of them being from the middle and one from the lower reach. The data show a significant predominance of mayflies over caddisflies (in 33.5 times), the abundance of *Blepharoceridae* larvae, which make up nearly a half of the blackfly number, and negligible numbers of such groups as bugs, beetles, water mites and round-worms. From the studied sections of the Cholpon-Ata River flatworms were absent. A very clear picture of the

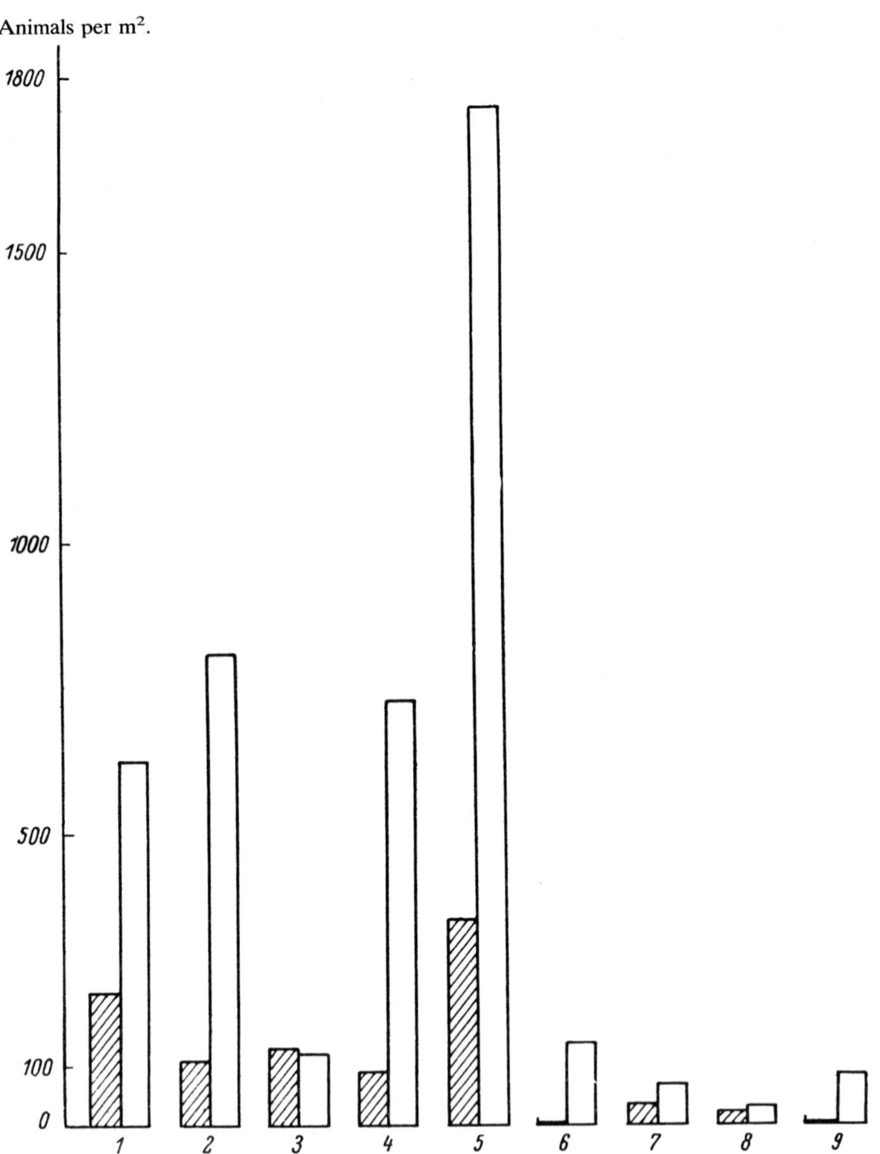

Fig. 85. Numbers of animals for different invertebrate groups per 1 m² in the Aksu River (shaded) and its springs (After L. A. Kustareva.)

quantitative relations between different invertebrate groups in true mountain torrents is obtained when the mean values from Table 51 are compared graphically (Fig. 86).

Counted for different river sections for three, although very close dates (7–18 IV, 23 VI-1 VII, 9–8 VII), the numbers of different invertebrate groups from the Cholpon-Ata River permit comparison of faunal conditions in (i) different sections of the river and (ii) different periods of time (Tables 53 and 54). The difference in the numbers of animals from different invertebrate groups derived for different data is certainly not due to the time period but to the flood in late June which seems to have affected the whole

248

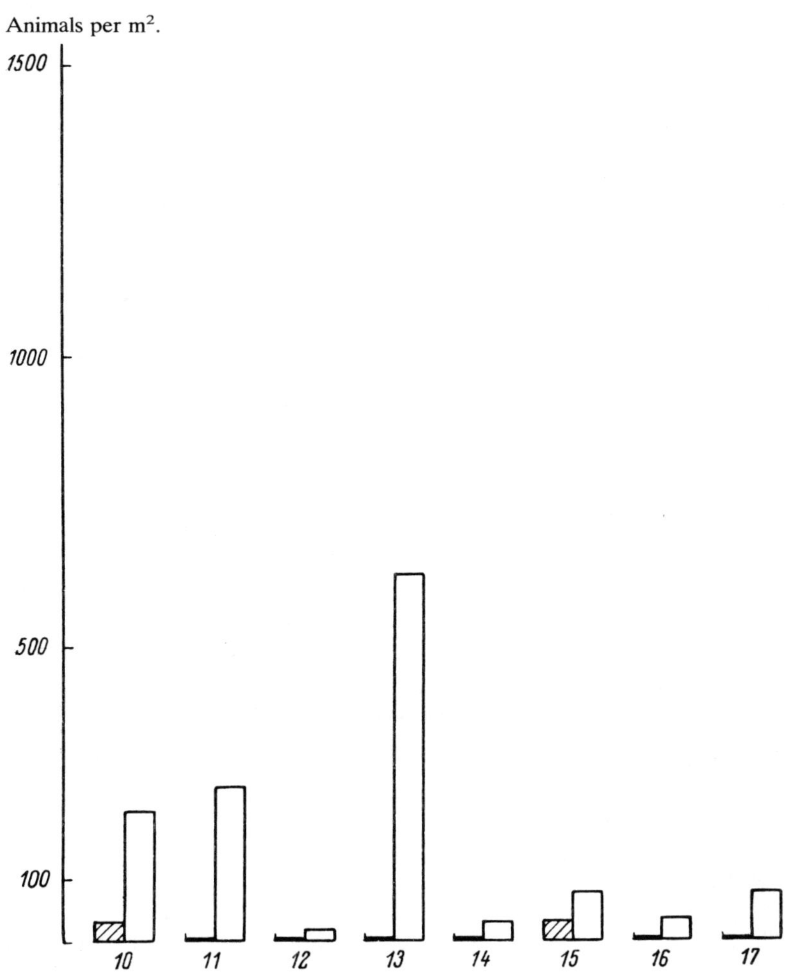

Fig. 85 (continued).
For the serial numbers (1–17) for invertebrate groups, see table 47.

length of the river and caused the decreased numbers of mayflies, caddisflies, stoneflies and other groups (Fig..87) and the increased number of chironomid larvae. Several groups (flatworms, *Oligochaeta* and some others) are not abundant in the Cholpon-Ata River as a typical mountain torrent, and are not shown in Fig. 87 (compare with Fig. 86).

The decreased numbers of animals in the Cholpon-Ata River are hardly because of the flood alone. Perhaps, it is a result of the developmental cycle in the aquatic insects, the emergence and the flight out of imagoes, *etc.* In any event, decreased numbers of the aquatic insects were noticed during the flood in the Akbura and Aksu Rivers (Tables 45 and 46). An important

Table 48. Numbers of animals per 1 m^2 for invertebrate groups in rivers of the Issy-kul Valley, the Terskei-Alatau Range (Kustareva, private communication).

Name of group	Karasu R. 1 XI 1969	Tosor R. 19 VIII 1969	Aksu R. 20 III 1969	Aksai R. 18 III 1969
Ephemeroptera	284	1052	1470	470
Trichoptera	171	154	1840	888
Plecoptera	—	670	80	104
Diptera:				
Simuliidae	464	892	1082	60
Chironomidae	596	4450	1470	1672
Heleidae	—	—	64	4
Blepharoceridae and				
other *Diptera*	60	320	686	564
Rhynchota	—	—	—	4
Coleoptera	8	—	48	—
Hydracarina	28	273	224	52
Amphipoda	96	—	240	4
Copepoda	16	—	16	4
Ostracoda	300	48	64	—
Mollusca	36	—	208	8
Oligochaeta	340	575	112	108
Nematodes	24	—	—	76
Turbellaria	—	—	32	—

Table 49. Numbers of animals per 1 m^2 for the invertebrate fauna in springs at the shores of the Issyk-kul Valley rivers, the Terskei-Alatau Range (Kustareva, private communication).

Name of group	Aksu R. 31 X 1969	Akterek R. 3 XI 1969	Djergalan R. 30 X 1969	Tiup R. 2 XI 1969
Ephemeroptera	2220	2376	2068	564
Trichoptera	2848	3784	596	756
Plecoptera	276	—	4	36
Diptera:				
Simuliidae	4172	7424	6832	112
Chironomidae	72	1428	72	1776
Heleidae	—	20	—	152
Blepharoceridae and				
other *Diptera*	72	292	64	184
Rhynchota	—	4	16	328
Coleoptera	8	96	16	104
Hydracarina	16	564	44	52
Amphipoda	—	24	156	12
Cladocera	—	4	—	32
Copepoda	12	12	44	24
Ostracoda	24	384	16	112
Mollusca	28	20	—	8
Oligochaeta	20	368	8	232
Nematodes	8	12	—	56
Turbellaria	136	12	4	64

250

Table 50. Averaged numbers of animals per square metre in rivers and springs of the Issyk-kul Valley.

Name of group	Rivers	Springs	Note
Ephemeroptera	805	1800	Various species
Trichoptera	614	1990	Various species; caddisflies notably predominate in springs
Plecoptera	189	78	Numerous stoneflies mostly in rivers
Diptera:			
Simuliidae	620	4600	Various species
Chironomidae	2220	834	Various species
Heleidae	17	43	Various species
Blepharoceridae and other *Diptera*	382	153	Various species
Rhynchota	1	87	Various species
Coleoptera	13	53	Various species
Hydracarina	146	169	Various species
Amphipoda	85	48	Various species
Cladocera	—	9	In rivers occasionally
Copepoda	9	23	In rivers occasionally
Ostracoda	103	134	Various species
Mollusca	63	14	Various species
Oligochaeta	250	156	Various species
Nematodes	25	19	Various species
Turbellaria	8	54	Various species
Total	1950	10264	

feature is that despite the considerably decreased numbers of the mayfly nymphs and larvae in a period from 23 June to 1 July, they were more abundant at this time than the caddisfly larvae and pupae. This phenomenon and the abundance of *Blepharoceridae* are typical, as we try to show here, of the true mountain torrent.

Table 53 contains the averaged numbers for eleven invertebrate groups from the upper, the middle and lower Cholpon-Ata River. A general conclusion from this table is the same as for the fauna of the typical mountain torrent: the noticeable predominance of mayflies over caddisflies in all of the river; the abundance of *Blepharoceridae* and *Deuterophlebia*; the abundance of *Chironomidae* and blackflies, although varying with the river site, and, lastly, the extremely minor role of other invertebrate groups.

The quantitative relations of the groups in different river sites are as follows. Mayflies are most numerous in the upper and the lower reaches and much less numerous in the middle reach. In the upper river there predominate larvae and nymphs of genera *Iron* and *Baetis* and in lower reach there appear also those of *Rhithrogena*, then, in the lower reach, of *Ephemerella* (see Table 54). Caddisflies are most numerous in the upper reach, but elsewhere they are less numerous than mayflies. Stoneflies are also numerous in the upper reach, but throughout the river they are much less

Table 51. Numbers of animals per square metre of the lithorheophylous fauna in the Cholpon-Ata River, the southern slope of Kungei-Alatau Range (Kustareva, private communication).

Serial number	Name of group	Number of study point					Averaged numbers of animals per m^2
		5	43	58	65	67	
1	*Ephemeroptera*	1600	1600	1200	1280	728	1280
2	*Trichoptera*	40	40	72	32	8	38
3	*Plecoptera*	80	32	68	—	68	49
	Diptera:						
4	*Blepharoceridae*	—	104	128	448	108	153
5	*Deuterophlebiidae*	—	24	108	40	104	55
6	*Simuliidae*	104	408	68	1360	640	316
7	*Chironomidae*	312	224	120	144	120	184
8	other *Diptera*	40	168	144	40	40	86
9	*Rhynchota*	—	—	8	—	4	2
10	*Coleoptera*	—	—	—	—	4	1
11	*Hydracarina*	—	8	12	16	12	10
12	*Oligochaeta*	32	—	—	—	—	6
13	*Tardigrada*	4	—	—	—	—	1
14	*Nematodes*	16	24	—	—	—	8

Notes: 1) The serial numbers of groups are as in Fig. 86.

2) Point 5, the middle river at the outlet from the gorge, abs. height about 2000 m, 21 VII 1971, air 18°, water 8°5. Point 43, the middle river, 21 VII 1971, air 12°, water 8.4°. Point 58, 3 km downstream of point 43, 17 IX 1971, air 14.5°, water 5.0°. Point 65, the same site as 58, 6 X 1971, air 13.5°, water 4.5°. Point 67, 6 km downstream from point 58 or 65, the lower river.

numerous than mayflies and caddisflies. Such a numerical relation between the three orders in the Cholpon-Ata River is observed in many "true" mountain torrents of the Tien Shan and is likely to be typical of them.

Blepharoceridae are most numerous in the middle reach where in addition to the genus *Asioreas* there appear the genera *Tianschanella* and *Blepharocera*. *Deuterophlebia* is definitely numerous in the upper River, since it is restricted to low water temperatures. Blackflies and *Chironomidae* exhibit a specific distrubution. In complete accordance with their ecology, the former are most abundant in the lower reach and the latter in the upper reach. But conclusions of this kind may be only of a very general nature, since it is difficult to speak about the ecological characteristics of the whole family (even from the swift waters alone): these characteristics vary greatly with the species. Remarkable is the predominance of flatworms in the upper reach, but they and *Oligochaeta* (see Chapter 7) are rare over the whole torrent length.

To illustrate the predominance of mayflies in the torrent consider the field data in the "descending" order of swift Tienshanian watercourses, *i.e.* from the mountain torrent (the Akbura River) to the mountain river (the Chirchik, the Aksai and the Aksu Rivers) and the middle reach of a large river (the Chu) (Table 55). It should be explained that the upper Chirchik River is

252

Table 52. Numbers of animals per $1\,m^2$ (numerator) and biomasses in mg/m^2 (denominator) of the lithorheophylous groups in the Cholpon-Ata River, the southern slope of Kungei-Alatau Range. Sampled in 1971 (after Stankevich).

Name of group	Upper river			Middle river			Lower river		
	7–18 VI	23 VI–1 VII	8–19 VII	7–18 VI	23 VI–1 VII	8–19 VII	7–18 VI	23 VI–1 VII	8–19 VII
Ephemeroptera	$\frac{2428}{7678}$	$\frac{184}{2372}$	$\frac{44}{1197}$	$\frac{280}{968}$	$\frac{184}{1304}$	$\frac{308}{648}$	$\frac{3156}{7476}$	$\frac{108}{1106}$	$\frac{160}{748}$
Trichoptera	$\frac{148}{880}$	$\frac{88}{732}$	$\frac{68}{440}$	$\frac{4}{156}$	$\frac{8}{20}$	—	$\frac{28}{188}$	—	$\frac{16}{48}$
Plecoptera	$\frac{88}{2092}$	$\frac{36}{64}$	$\frac{68}{432}$	—	$\frac{20}{1192}$	—	$\frac{8}{800}$	—	—
Diptera:									
Blepharoceridae	$\frac{8}{40}$	—	$\frac{28}{64}$	$\frac{500}{776}$	$\frac{156}{202}$	$\frac{104}{220}$	$\frac{4}{16}$	$\frac{100}{324}$	$\frac{16}{20}$
Deuterophlebidae	$\frac{256}{68}$	$\frac{36}{22}$	$\frac{20}{60}$	$\frac{36}{20}$	$\frac{4}{8}$	$\frac{16}{8}$	$\frac{4}{12}$	$\frac{24}{16}$	$\frac{4}{8}$
Simuliidae	$\frac{64}{212}$	—	—	$\frac{216}{916}$	$\frac{56}{304}$	$\frac{76}{316}$	$\frac{480}{3062}$	$\frac{640}{1084}$	$\frac{12}{8}$
Chironomidae	$\frac{622}{148}$	$\frac{800}{270}$	$\frac{1280}{200}$	$\frac{8}{12}$	$\frac{12}{4}$	$\frac{96}{24}$	—	$\frac{400}{124}$	$\frac{96}{26}$
Limoniidae	$\frac{20}{88}$	—	$\frac{24}{76}$	—	$\frac{4}{188}$	—	$\frac{4}{76}$	—	$\frac{8}{556}$
Psychodidae	$\frac{28}{32}$	—	—	—	$\frac{16}{106}$	—	—	—	—
Oligochaeta	$\frac{104}{96}$	$\frac{28}{28}$	$\frac{40}{24}$	—	—	—	—	$\frac{28}{16}$	$\frac{12}{6}$
Turbellaria	$\frac{4}{28}$	$\frac{60}{668}$	$\frac{24}{136}$	—	—	$\frac{4}{12}$	$\frac{8}{40}$	—	—

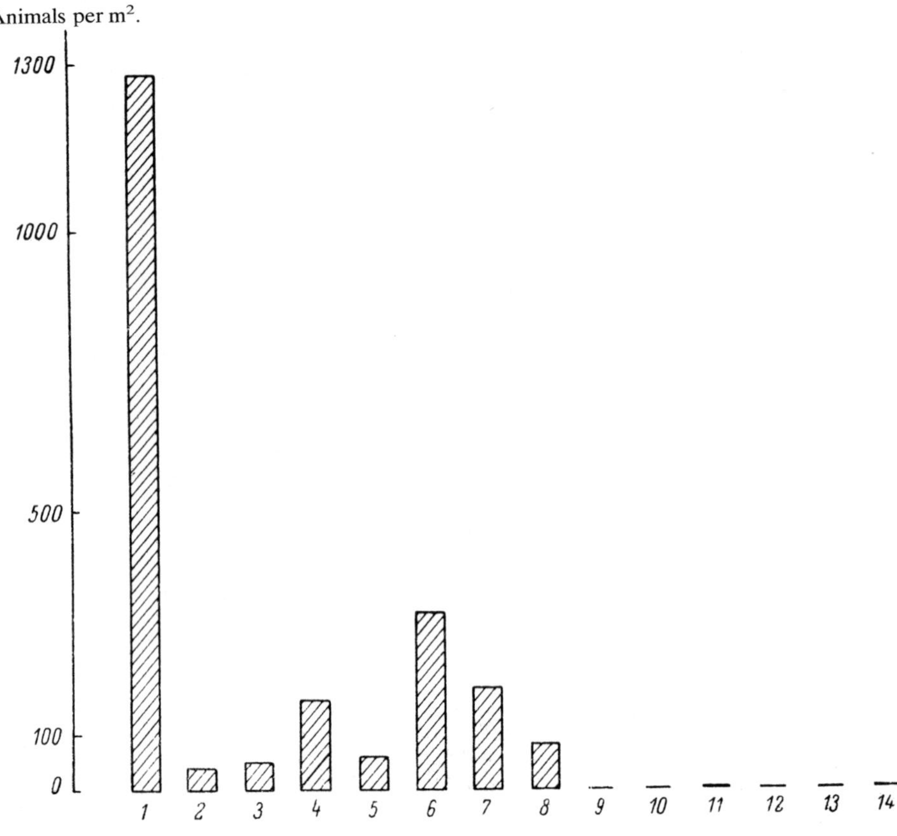

Fig. 86. Numbers of animals of different invertebrate groups per 1 m² in the Cholpon-Ata River. (After L. A. Kustareva.) For the serial numbers (1–14) for invertebrate groups, see Table 51.

Table 53. Invertebrate fauna in the Cholpon-Ata River. Mean numbers of animals per 1 m².

Name of group	Upper river	Middle river	Lower river
Ephemeroptera	892	282	1140
Trichoptera	168	4	15
Plecoptera	64	7	3
Diptera:			
Blepharoceridae	12	253	40
Deuterophlebiidae	140	19	12
Simuliidae	21	116	377
Chironomidae	874	13	50
Limoniidae	16	1	4
Psychodidae	9	5	—
Oligochaeta	57	—	13
Turbellaria	29	1	3

254

Table 54. Numbers of animals per 1 m^2 (numerator) and biomasses mg/m^2 (denominator) of mayfly nymphs in the Cholpon-Ata River, the southern slope of Kungei-Alatau Range. Sampled in 1971 (Stankevich, private communication).

Mayfly genus	Upper river			Middle river			Lower river		
	7–18 VI	23 VI–1 VII	8–19 VII	7–18 VI	23 VI–1 VII	8–19 VII	7–18 VI	23 VI–1 VII	8–19 VII
Iron	92/1214	12/172	140/492	172/1896	84/760	28/708	20/532	8/124	40/304
Rhithrogena	—	—	4/4	12/476	8/900	4/304	16/528	—	—
Ephemerella	—	—	—	—	4/56	—	4/85	4/32	—
Baetis	2244/6184	204/532	2860/6072	—	84/180	64/148	—	296/432	116/404

Averaged numbers of mayflies per 1 m^2 at different reaches along the Cholpon-Ata River.

Mayfly genus	Upper river	Middle river	Lower river
Iron	81	91	23
Rhithrogena	1	8	5
Ephemerella	—	1	4
Baetis	1769	49	137

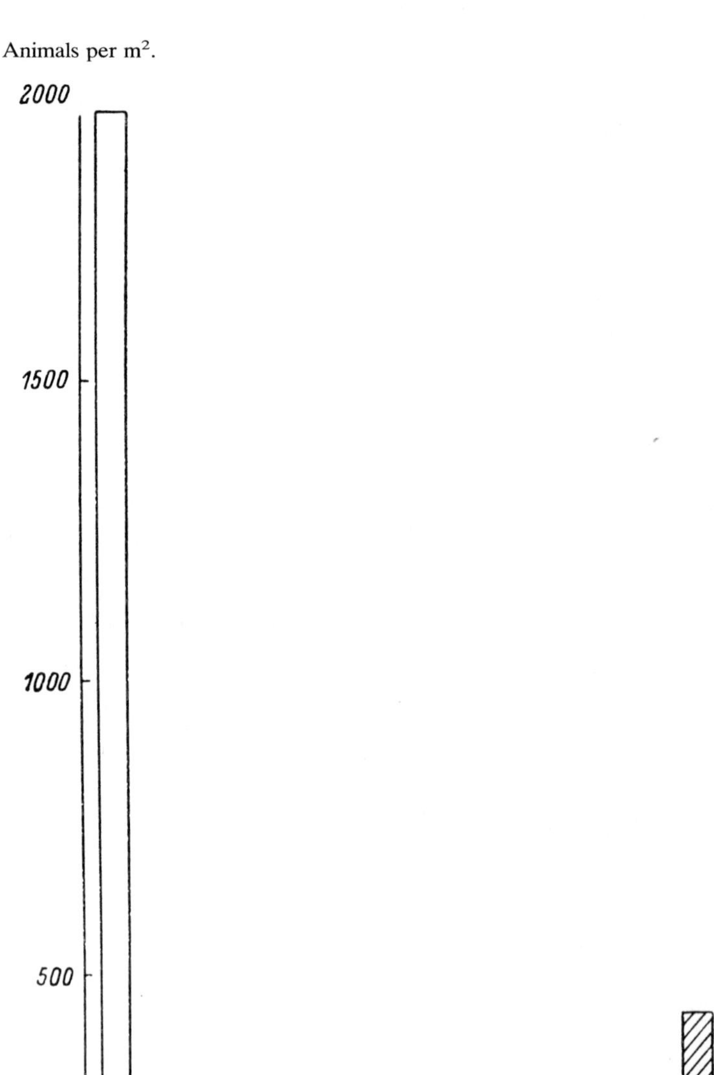

Fig. 87. Numbers of animals of different aquatic insectal groups per m² in the Cholpon-Ata River, 7–18 VI 1971 and 23 VI – 1 VII 1971 (shaded).
1, *Ephemeroptera*, 2, *Trichoptera*, 3, *Plecoptera*, 4, *Blepharoceridae*, 5, *Deuterophlebiidae*, 6, *Simuliidae*, 7, *Chironomidae*.

Table 55. Numbers of animals per 1 m^2 (numerator) and biomasses g/m^2 (denominator) of different insectal orders and families of the lithorheophylous fauna in some Tienshanian river basins. (The Chirchik R. after Sibirtzeva, 1961; the Chu R. after Ovchinnikov, 1936; the Aksu and the Aksai Rivers after Konurbaev and Madjar, 1969; the Akbura orig.).

Name of group	Akbura R. system (Alaiskii Range)					Chirchik R. system (Chatkalskii Range)			Terskei-Alatau Range		Chu River (Zailiiskii Alatau Range)	
	main river	affluents				main river	affluents				main river	affluents
		Kichik-Alai	Chogom	Chalkui-riuk	Laglan		Chatkal	Pskem	Aksai R.	Aksu R.		
Ephemeroptera	620/3.91	745/6.93	531/3.89	894/6.30	261/2.42	151/0.34	313/1.75	379/3.55	148/0.32	97/0.47	—	64/0.40
Trichoptera	46/0.66	61/0.94	90/0.89	14/0.22	72/1.01	446/2.41	350/0.97	395/3.59	252/3.4	478/6.8	96/1.02	389/6.87
Plecoptera	33/0.10	92/0.33	16/0.06	54/0.33	—	—	—	—	24/0.16	3.3/0.12	—	—
Diptera:												
Simuliidae	30/0.12	345/1.00	133/0.37	64/0.89	180/0.44	38/0.06	375/0.28	38/0.08	160/0.12	8/0.01	32/0.48	528/1.00
Chironomidae	—	580/1.50	16/0.06	—	—	139/0.22	63/0.16	75/0.23	51/0.03	223/0.14	32/0.48	540/0.92
other Diptera	61/0.22	73/0.33	174/0.66	82/0.78	44/0.51	—	225/0.52	13/0.08	142/6.83	131/2.48	—	—
Total	820/5.00	1994/11.73	1032/6.85	1325/10.89	657/5.11	837/3.05	1402/3.86	926/7.61	803/7.6	858/10.3	160/1.98	1516/9.19

a true mountain torrent, but the numerical data for it and for the Aksai and Aksu Rivers in Table 55 cover the whole length down the lower reaches, whereas the Akbura, the Cholpon-Ata and the Issyk Rivers maintain the character of a mountain torrent over their entire lengths. That is why the data for the latter rivers in this and the next tables more adequately characterize the structure of the torrential fauna. The data in Table 55 are given separately for the main river and its affluents; but a comparison of averaged data for the whole basins (*i.e.* data summed up for the main channel and affluents: Table 56 and Fig. 88) gives a clear picture of the fast-flowing watercourses in the descending order. Very characteristic are the decreasing numbers of mayflies and the increasing numbers of caddisflies per square metre, as the character of a true mountain torrent becomes less expressed. The numbers of these two orders of insects are inversely proportional, however this rule does not hold for any systematic composition of mayflies but only for just certain genera representative for the mountain torrent and fast-flowing watercourses in general. For example, in the lower reaches there may be abundant mayflies burrowing the banks (*Palingenia, Ephemera* and others) and they may be more numerous than caddisflies, but then each group has other systematic composition. As to the mayflies of swift waters, the data in the table and Figs. 88 and 89 show an increased role of larvae and nymphs of this order in watercourses of the mountain torrent type.

It is important to answer the question whether the mayflies predominate numerically over the caddisflies over the entire length of the torrent. Let us consider the vertical distribution data for the three groups of lithorheophylous fauna – mayflies, caddisflies and stoneflies – in a typical torrent, the Akbura River. The data concern the upper waters (*i.e.* the Kichik-Alai affluent) and the beginning of the middle reach. If arranged consecutively down the stream (Table 57, points 37–26), the numbers and biomasses of the three groups show the predominance of mayflies at every site. In the middle reach they predominate in biomass over caddisflies 30 times and over stoneflies 20 times. The absence of caddisflies and stoneflies from some points does not mean that they are not present here at all, but that they were simply in too small numbers to be found within the measuring area. One can say much against the quantitative fauna sampling method for

Table 56. Averaged numbers of animals per 1 m^2 in some streams of the Tien Shan.

Name of group	Akbura R. system	Chirchik R. system	Aksai R.	Aksu R.	Chu R. system
Ephemeroptera	620	281	148	97	64*
Trichoptera	55	397	252	478	240
Plecoptera	39	4	24	33	—
Diptera:					
Simuliidae	148	150	160	8	280
Chironomidae	119	92	51	223	286
• other *Diptera*	86	119	131	142	—

* Mayflies were only in affluents, not in the main river (?).

258

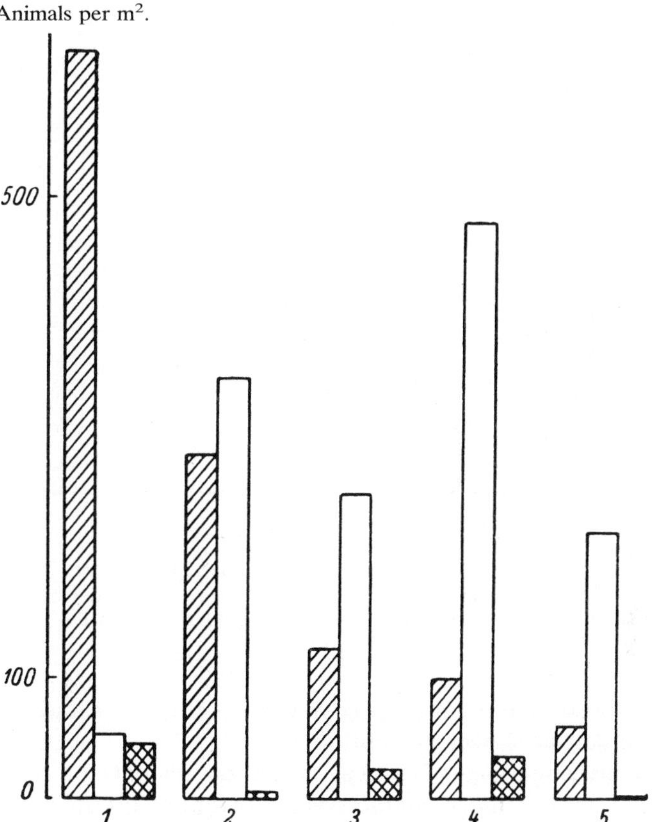

Fig. 88. Numbers of mayflies (shaded columns), caddisflies (white columns) and stoneflies (cross-shaded) per m² in the Akbura (1), the Chirchik (2), the Aksai (3), the Aksu (4) and the Chu Rivers (5).

torrents, but it gives, if anything, numbers which are certainly in favour of mayflies. Let us present also the data for a large affluent of the Akbura River, the Laglan River (Table 57, points 25–12). Despite the general decrease in the numbers of the three groups discussed, the proportions of mayflies, caddisflies and stoneflies remain the same: mayflies predominate over caddisflies and stoneflies both in numbers and biomass.

One more example from Sibirtzeva (1961) can show the predominance of mayflies in the effluents of the Chirchik River which are closer to the mountain torrent type than the main river in its middle reach (Table 58, Fig. 89).

From the data in Table 58 one can see the increased biomass of caddisfly larvae and pupae at slower current (from 1 to 2 m/sec), whereas for mayflies one finds an opposite relation (their biomass tends to increase at faster current, from 2 to 4 m/sec). This is especially noticeable in the Pskem River with very rough and fast current. A possible explanation is the increased size of the mayfly nymphs at faster current because of the more numerous and large nymphs of the genera *Rhithrogena* and *Iron*, as well as of smaller

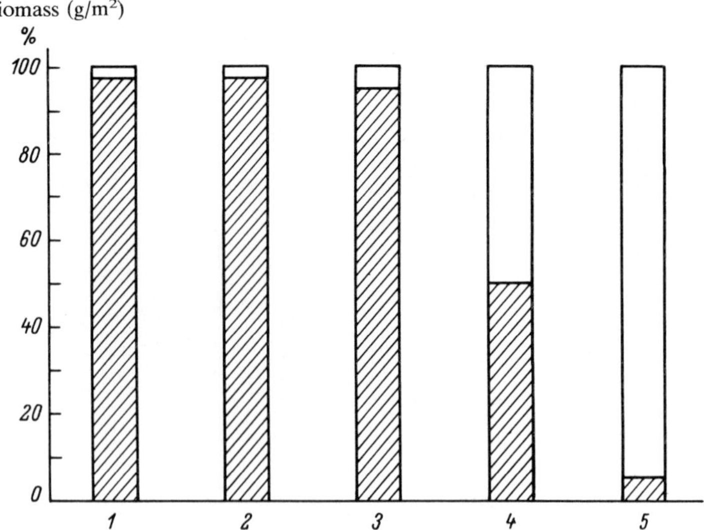

Fig. 89. Biomass ratios for mayflies (shaded) and caddisflies. 1. the upper Akbura River (the Kichik-Alai Range), 2. along the whole length of the Akbura R., 3. the Chatkal River, 4. the Pskem River (an affluent of the Chatkal R.), 5, length of the Chirchik River.

animals like those of the genera *Ecdyonurus, Cinygmula* and *Baetis* which become abundant at slower current.

Thus, the investigations of typical mountain torrents of the Tien Shan have shown that the most characteristic torrential group is the mayflies which are superior to other groups in both numbers and biomass.[2] Mayflies are followed by and sometimes outnumbered by caddisflies. Stoneflies are of very little significance numerically and in biomass (but not as carnivores!). Next to caddisflies, the most abundant groups are the specialized families of the torrential fauna – the *Blepharoceridae, Deuterophlebia* and *Simuliidae* which sometimes outnumber caddisflies or even mayflies. Other invertebrate groups are poorly represented in the Tienshanian mountain torrents; the molluscs and side-swimmers are entirely absent.[3]

The specific features which are inherent to the fauna of the typical Tienshanian mountain torrent enable us to consider the latter as a special type of swift waters different from the brooks, the rivers and the springs. A similar conclusion was made by Hubault from a comparative study of the Vosgesan swift waters in France: there is a torrential fauna, to be more correct, there is no rheophylous fauna ("Il y a une faune torrenticole, il n'y a pas, proprement dit, faune rheophile"). The investigations of the Tienshanian mountain torrents provide firm support for this view.

[2] It should be also taken into account that small mayfly larvae are difficult to sample, but those of nearly microscopic size appear to be extremely numerous in certain periods of the year.
[3] The glacial torrents of the north-western Himalaya have a similar composition of invertebrate groups of the lithorheophylous fauna: mayflies, stoneflies, caddisflies and dipteran families *Chironomidae, Simuliidae, Blepharoceridae* and *Deuterophlebiidae* (Dubey and Kaul, 1971).

260

Table 57. Variation of animal abundance (per 1 m²) and biomasses (mg/m²) with height for mayflies, caddisflies and stoneflies in the upper Akbura River (points 37–26) and its affluent, the Laglan (points 25–12).

Serial number of study points	Abs. height (m)	Mayflies		Caddisflies		Stoneflies	
		numbers	biomass	numbers	biomass	numbers	biomass
37	3550	60	420	—	—	4	360
38	3500	32	32	—	—	—	—
39	3450	680	6.360	—	—	—	—
40	3420	1.280	7.184	—	—	32	144
41	3350	1.328	7.824	16	224	48	160
42	3300	960	9.432	16	96	176	353
43	2860	670	9.120	30	600	20	470
44	2770	400	5.070	—	—	64	224
45	2710	520	3.460	—	—	80	260
46	2700	319	6.000	—	—	44	429
47	2600	1.040	8.288	—	—	96	272
48	2440	528	9.392	208	2.640	48	128
49	2180	1.81	10.062	72	1.134	468	1.424
50	2020	496	8.112	26	494	39	195
28	2010	1.280	11.266	48	816	96	224
27	1890	600	10.280	120	2.300	120	320
26	1800	656	6.608	16	192	16	96
25	2200	390	2.670	10	160	—	—
24	1900	670	5.180	20	670	—	—
23	1850	650	6.840	—	—	—	—
22	1820	570	3.970	30	190	—	—
17	1750	180	3.120	—	—	—	—
16	1600	150	1.326	16	210	—	—
15	1560	179	1.488	66	870	—	—
14	1550	112	720	260	3.340	—	—
13	1530	42	648	78	1.002	6	162
12	1500	76	432	196	1.756	4	48

Note. The points are arranged down the stream, the total distance from 37 to 26 about 97 km and from 25 to 12 about 72 km. The distance between any two points is from 2.5 to 10 km.

Table 58. Variation of animal numbers (per 1 m², numerator) and biomasses (g/m², denominator) with flow velocity for mayflies, caddisflies and stoneflies in the Chirchik River and its affluents (after Sibirtzeva, 1961).

Name of order	Chirchik River		Affluents			
			Pskem River		Chatkal River	
			flow velocity in m/sec			
	1.0–1.9	2.0–4.0	1.0–1.9	2.0–4.0	1.0–1.9	2.0–4.0
Ephemeroptera	$\frac{26}{0.03}$	$\frac{25}{0.03}$	$\frac{100}{0.36}$	$\frac{75}{0.64}$	$\frac{150}{0.75}$	$\frac{125}{0.94}$
Trichoptera	$\frac{400}{2.38}$	$\frac{185}{0.80}$	$\frac{2370}{2.44}$	$\frac{9.0}{0.86}$	$\frac{150}{0.37}$	$\frac{75}{0.12}$
Plecoptera	—	—	—	$\frac{13}{0.02}$	—	—

One of the important geographical conditions of the land where the Tienshanian torrents run, is the climate aridity which makes these torrents and, therefore, their fauna different from those of Europe and North America, but more similar to those of the Western Pamirs, the Hindu Kush, the Karakoram and, partly, the north-western Himalaya. Because of the aridity, many torrents and rivers located on the terminal ranges and in some intermontane depressions and valleys do not reach large aquatic arteries. They represent mostly a closed system. This is stimulated by water expenditure for irrigation. Thus the Issyk-kul Lake is reached at present by only three or four rivers instead of thirty which discharged into the lake earlier. But even without the expenditures the streams, especially those of the terminal ranges, tend to disappear in the semi-desert at the mountain foot. For this reason, the fauna of the Tienshanian torrent may be said to begin at the glacier foots and to end at the mountain foots without creating the intermediate zone peculiar to many rivers of the Altai, Europe and North America.

The last thing to be especially stressed is that we do not consider the mountain torrent as an unchangable, static phenomenon entirely different from other types of the swift waters: mountain rivers, brooks and springs. Watercourses of the intermediate type can be also found. Thus the closer the torrent is to the mountain river or the brook type, the more abundant are the animalistic groups which were rare or absent from the true torrent: beetles, water mites, molluscs, flatworms, aquatic earthworms, *etc.*

The "true" mountain torrent and its typical fauna is a theoretical conception, while in nature we deal with its different modifications.

The significance of the torrential fauna as food for the trout and other fish

In addition to the abiotic factors, of great importance for the formation of the ecological environment of the torrential lithorheophylous fauna is the set of biological relations, in which the carnivores like stonefly nymphs, larvae of some caddisflies, *etc.*, as well as the fishes which are the mass consumers of invertebrates play a great role.

At present even the few data available on the feeding of native fish in the Tienshanian torrents show clearly that the main source of forage for the fish are the mass forms of torrential insects: mayflies, caddisflies and dipteran larvae, including *Chironomidae* and *Blepharoceridae*. This is known from investigations on feeding of fish from the Chirchik and the Zeravshan Rivers (Sibirtzeva, 1961a,b; Sibirtzeva *et al.*, 1961, 1964) and the Akbura River (Omorov, 1973). It applies to marinka *Schizothorax intermedius*, scaleless osman *Diptychus dybowskii*, Tibetan stone loach *Nemachilus stoliczkai*, Amu-dar stone loach *N. oxianus* and Kushakevich's stone loach *N. kushakewitschi*, Turkestan catfish *Glyptosternum reticulatum* and sculpins *Cottus nasalis* and *C. spinulosus*.

The first place in the forage of the marinka is taken by mayflies (50% occurrence) and the second by caddisflies. The same applies to the osman, *i.e.* first come the mayfly nymphs of the *Iron, Ephemerella* and *Baetis*.

262

Blepharoceridae together with *Atherix*, *Bezzia* and *Dixa* make up 37% of occurrence and biomass in marinka stomaches. Stone loaches also feed mainly on mayfly nymphs and larvae. The forage of other fish is mostly mayfly nymphs and larvae and other aquatic insects.

The Amu-dar trout *Salmo trutta oxianus* Kessl. which spends all its life in the river and has no need for lakes to feed and grow, has been acclimatized in the rivers of the Syrdarya basin (previously the trout inhabited only the rivers of the Amudarya basin). Transported from the Kizylsu River affluents (the Amudarya basin) in 1953–54 in numbers over a thousand, the trout has been introduced in the Karaungur and the Djailisu Rivers. In 1957 600 fishes were delivered to the Akbura River. The rainbow trout was acclimatized in the Turgen and the Chilik Rivers (Kazakhstan) which became the first and single reserves of the trout (Kurmangalieva, 1976).

In 1965 a survey was made to find out if the trout had acclimatized to the rivers where it was released. In the Karaungur River the trout was not recovered, but it occurred farther upstream, in an affluent – the Kizylungur River. The trout acclimatized well and became numerous. The recovered individuals weighed from 1 to 3 kg and more (Piskarev, 1968a).

We have no quantitative data on trout abundance in the acclimatization rivers, but in June 1971 we witnessed in the Akbura River six large trout to be caught by ichthiologists in 30 min.

The feeding data on the acclimatized Amu-dar trout from the Syrdarya basin are few, but a general idea can be gained. As for the feeding of the trout in its native waters (*i.e.* in the Amudarya basin) there are two opinions: one is that the aerial insects falling onto the water surface are as important for the trout feeding as the larvae of aquatic insects (Nikolskii, 1938), and the other that the aerial insects are of little importance for it. Shaposhnikova (1950) made a definite conclusion in her work: "The main component of the forage for trout are larvae of *Ephemeroptera* and *Diptera*, among the latter especially important are the *Blepharoceridae*, *Simuliidae* and *Tendipedidae*. The insects falling onto the water play an insignificant role in the trout feeding" (p. 19). She came to this conclusion when examining the feeding of fish from the upper Amudarya area (the Vaksh, Kizylsu, Kafirnigan, Warzob and the Surhandarya Rivers).

K. A. Brodsky and E. O. Omorov examined the stomach content in trouts from the Akbura River (mature fish, Table 59). The primary source of food for these trouts was mayfly nymphs found in the stomach of every analyzed fish, mayflies being the same as in the lithorheophylous fauna of the river: *Iron*, *Rhithrogena* and *Ephemerella*. *Blepharoceridae* were not always recorded but they occurred nearly in every fish.

Some data are also available on the feeding of the Amu-dar trout from the Kara-Kichik River (an affluent of the Kizylungur River on the Ferghana Range; Piskarev, 1968), which are given in Table 59. They show an important place of mayfly larvae and nymphs. *Blepharoceridae* were not recorded.

Few published data are available on the food composition of the Sevan trout acclimatized in the Issyk-kul Lake, chiefly, of the fry which were

Table 59. Food composition (frequency of occurrence, %) of the Amu-dar trout acclimatized in the Akbura River, the Alaiskii Range (after Omorov, 1973), and in the Kara-Kichik River, the Ferghana Range (after Piskarev, 1968a).

Name of group	Akbura River	Kara-Kichik River	
	August	July and August	September
Ephemeroptera	100	38	13
Trichoptera	92	6	13
Plecoptera	32	5	2
Diptera:			
Simuliidae	22	?	?
Chironomidae	48	?	?
Imagoes	From 14	to	68

caught in the river and mostly in springs (Luzhin, 1956; Konurbaev and Madjar, 1969; Kustareva, 1973). Through the courtesy of L. A. Kustareva, we have at our disposal the results of analyses of the fry stomachs (Tables 60 and 61 and Fig. 90). These data indicate a considerable dependence of the trout feeding on the character of the fauna. Thus the presence of side-swimmers in the stomachs shows that the analyzed fishes were not from the

Table 60. Food composition (frequency of occurrence, %) of the Sevan trout ("gegarkuni") acclimatized in the Issyk-kul Lake, of the juveniles from rivers and springs. (The Ton R. after Lukin, 1956; the Aksu R. after Konurbaev and Madjar, 1969; other rivers: Kustareva, private communication).

Name of group	Ton River	Aksu River	Other affluents of the Issyk-kul Lake, and springs				
	VI	VI, VII	9 VI 1969	13 VIII 1969	20 VIII 1969	1 XI 1969	3 VII 1970
*Ephemeroptera**	48	90	100	100	95	67	100
*Trichoptera**	5	46	14	11	56	33	—
Plecoptera	—	—	2	4	—	—	—
Diptera:							
*Simuliidae**	9	18	4	—	26	44	10
*Chironomidae**	13	36	42	8	30	94	—
other *Diptera**	—	—	10	28	9	28	20
Rhynchota	—	—	—	8	21	11	10
Coleoptera	—	—	—	8	4	22	—
Aerial and terrestrial imagoes of insects	?	?	6	8	9	28	—
Hydracarina	—	—	—	4	—	—	—
Amphipoda	48	64	4	—	86	50	—
Nematodes	—	—	2	—	4	11	—
Turbellaria	—	—	2	—	—	—	—
Pisces	—	—	—	—	—	11	10

Note: Here and in Table 61, the asterisk stands for preimaginal phase.

Table 61. Food composition (specimens per trout) of the juvenile trouts ("gegarkuni") from springs of the Djergalan River and from the Karasu River, the Issyk-kul Valley (Kustareva, private communication).

Name of group	Djergalan R.				Karasu R.
	Months				
	VIII	VIII	?	X	XI
	number of analyzed trouts				
	25	16	15	16	18
*Ephemeroptera**	14	17	20	2	4
*Trochoptera**	—	1	2	10	—
Diptera:					
*Simuliidae**	—	4	11	2	1
*Chironomidae**	1	52	—	—	13
other *Diptera**	5	—	—	—	—
Imagoes of aerial insects	—	1	—	31	1
Amphipoda	20	2	16	5	2

main river, but from the subsidiary waterbodies, in particular, from springs which are abundant in the trout fry; it is for them that the food composition is given in the tables.

The intensive investigations being carried out at present at the Issyk-kul Lake Biological Station will hopefully provide evidence about the food composition and its seasonal changes. But even now it can be stated with certainty that the food composition of the native fish and the two, Sevan

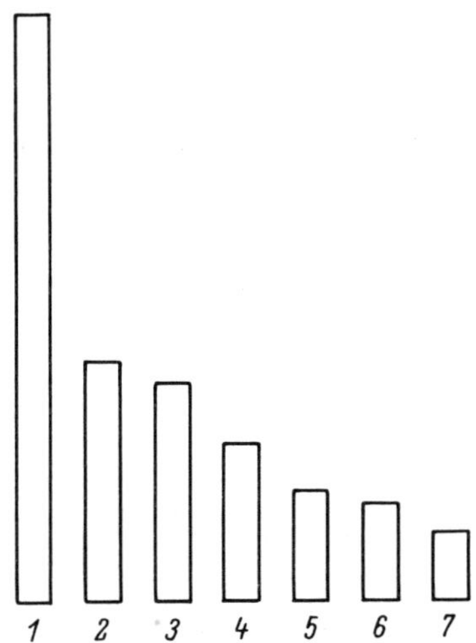

Fig. 90. Frequency of occurrence (%) of invertebrate groups in food of the juvenile Amu-dar trout from the Issyk-kul Valley rivers and springs (not shown if less than 10% occurrence).
1. *Ephemeroptera* (86%), 2. *Chironomidae* (35%), 3. *Amphipoda* (32%), 4. *Trichoptera* (23%),
5. *Simuliidae* (16%), 6. *Blepharoceridae* and other *Diptera* (14%), 7. aerial imagoes of insects (10%).

265

and Amu-dar, acclimatized trouts is affected by the composition of the lithorheophylous fauna which is typical of the mountain torrent; and that mayflies and *Diptera* (both in the preimaginal stage) are most important, but all other groups of the fauna, including caddisflies, are of minor importance as food. The only exception to the rule are the juveniles from springs where side-swimmers represent an important component of their food.

It is quite evident that the evaluation of the native fish resources and, in particular, the acclimatization measures cannot go on effectively without the qualitative and quantitative, but chiefly the ecological investigations of invertebrates (mostly aquatic insects; Sadovskii, 1946; Zelinka, 1971) from mountain torrents and rivers with their subsidiary waterbodies. Indeed, the potentialities of fish acclimatization, above all the trout, in the Tienshanian mountain torrents are very great. Despite the relatively small water mass in an individual torrent, their numerosity provides in total a large water basin which can be used for acclimatization of valuable fishes. Taking into account the large biomass of lithorheophylous fauna in mountain rivers and torrents, we may expect a large yield of fish, primarily of such a valuable breed as the trout.

Faunistic-ecological vertical zonality of the Tienshanian torrents

Mountain streams of the Tien Shan and Middle Asia in general are very peculiar and can not be characterized with the schemes proposed for Western Europe (Steimann, 1907; Kühtreiber, 1934; Thienemann, 1936; Ohle, 1937; Dittmar, 1955; Kownacka and Kownacki, 1972). More appropriate is the classification by Botosaneanu (1959) for the Carpathians. Some points of similarity can also be found in classifications for North American swift waters (Dodds and Hisaw, 1925; Muttkowsky, 1929, *etc.*). However, the Middle Asian streams are so specific that they are worth being placed in a special type. Their specificity consists in that they run down from the high mountains with a large area of glaciers and perennial snows, that their middle reaches flow through the arid zone and that most of them are expended for irrigation at the debris cone. Thus they are relatively short, have only few vertical zones typical of the streams on the Alps and the Cordillera, and represent an integral system of few vertical zones, the most lengthy of them being the mountain torrent proper. It is most appropriate to confer the mountain torrents of Middle Asia with the steep stretches of mountain rivers of the Alps and the Cordillera. We do not mention the mountain torrents of the Himalaya, the Hindu Kush and the Karakoram, since the vertical zonality of their fauna is still unknown.

Among the Soviet classifications of mountain watercourses mention should be made of the schemes by Muzafarov (1958, 1965) and Shadin (1950d). Following V. L. Shultz, E. M. Timofeev and A. M. Nadezhdin (1936), V. I. Shadin divides watercourses vertically into sections (drainage zones). Using the Amudarya as an example of the Middle Asian rivers, he distinguishes between five drainage zones: (i) high-mountain plateau, (ii) humid area, (iii) subarid zone, (iv) arid zone and (v) delta (1950d, p. 246; 1951, p. 266). The zones are distinguished by their height above sea level, hydrological features and fish population (1950d, p. 12). But we noted previously that different types of watercourses (springs, brooks, rivers) having different faunas occur in the same zone and are not differentiatied further (1950d, p. 7). The same is true of Muzafarov's scheme. From the study of the *Simuliidae* fauna E. O. Konurbaev proposed his own scheme of vertical zonality for the Middle Asian swift waters (Konurbaev, 1974, 1977). He groups the watercourses into three ecological categories and subdivides these, in turn, into types. Thus he distinguishes:

 (i) spring brooks of four types: those of the high-altitude plains, of the middle-altitude mountains, of the low-altitude mountains and, lastly, those of the piedmonts and lowlands;

 (ii) small and medium-size affluents of large rivers. These are subdivided

into affluents of rivers of the high-altitude valley with glacial alimentation, affluents of rivers of the middle-altitude mountains with glacial and snowy alimentation, and affluents of large rivers on the lowlands;
(iii) large rivers of the high-altitude mountains with glacial alimentation, of the middle-altitude mountains with glacial and snowy alimentation, and of the piedmonts and the lowlands with mixed alimentation.

This scheme is certainly a step forward compared to that of Shadin. Konurbaev not only subdivides the watercourses vertically, but he distinguishes between different types in every vertical zone, the basis for the subdivision into the types being the strength (the discharge) of the waterbody. In our opinion, one disadvantage of the scheme is the absence of the genetic relation between the types. For example, "large rivers" of the high-altitude mountains are rather vague objects. Perhaps, the transition of one type into another should also be taken into account, *e.g.* brooks into torrents, torrents into rivers, *etc.* Unfortunately, despite certain merits, this scheme does not distinguish the type of the mountain torrent as an integral system. As a result, one cannot use any scheme mentioned for the subdivision of one type of swift waters – the mountain torrent.

Without considering all the ways of vertical zoning of torrents, which were proposed for the European rivers by Illies and Botosaneanu (1963), one can find some similarity (only some!) in the schemes based on the fish distribution and suggested by Thienemann (1925; the brook trout zone), Macan (1961b, 1963) and Macan and Worthington (1951; the upper and the middle salmon zones) and Banarescu (1956; the trout zone).

The criteria used for the vertical zonation are very diverse. Most frequently, it is the distribution of fish (the above cited authors), then it may be the interpretation of the same "ichthyological" zones in terms of the bed slope and the channel width (Huet, 1946). The zonation based on the fish distribution has been subjected to some criticism owing to the high mobility of fish (Schmitz, 1957), and the need has arisen for different criteria of zoning. Lastly, Illies (1961) has developed a theoretical scheme of zonality, every zone being named specifically: epirhithron, metarhithron, hyporhithron, *etc.* For distinguishing between these zones, he proposed a statistical method to estimate the number of invertebrate species, which allows to represent graphically the fauna alternation in the longitudinal profile of the river. The graphs consists of lines which show varying numbers of species at successive points in the river. The method was mathematically developed and a number of formulae was proposed.

Without going into too much detail, we think that this method does not take into account the ecological features of the species, so we offer our considerations concerning the criteria for the vertical zonation of the torrent. First, it must be asserted what kind of zones are being discussed. As is seen from the title of the Chapter, these are the ecologo-faunistic zones, and this fact determines the zonation criteria which will be the fauna and the ecological evaluation of the fauna environment: aquatic vegetation, topo-

268

graphic and hydrological zones and all those ecological conditions which are accessible for investigation in the field or semi-stationary conditions. In other words, a torrent as a system embraces the whole abiotic and biotic situation. In literature there is an opinion that the zonation must consider the total fauna (Illies and Botosaneanu, 1963, Fig. 4). However it is so inadequately known that even the fauna lists for the European mountain torrents contain many unidentified species. Still worse is the situation with the Middle Asiatic torrents, and we have tried to use as indicators the few groups which are relatively better known than others.

For vertical zoning, we have at our disposal rather detailed data from two streams of the Tien Shan: the Issyk River in the Zailiiskii Alatau Range, the Northern Tien Shan, and the Akbura River in the Alaiskii Range, the Southern Tien Shan. On the basis of less detailed data from other torrents of the Tien Shan, we suggest that the zonality given below is typical of most torrents running down the Tienshanian mountains. The intermontane torrents will possible have a different extension of some zones, in particular, a more stretched middle reach.

The schemes we suggest use the data only for the main river, while those for the affluents are excluded from consideration. Judging by the available data however, the zonality of the affluents basically repeats the zonality of the main river.

What factors help distinguish the boundaries between the zones? Account was taken of the physical and chemical conditions, the distribution of animals, mostly those accepted as indicative ones, the number of species and the abundance of every species. Generally, the question of the boundary identification appears to be of great complexity, and Illies seems to be right in marking the boundary at the end of the large affluents where, as he notes, there occurs an abrupt change of the hydrological conditions (Illies, 1961). The boundaries of the vertical zones in the Issyk and the Akbura Rivers, in their upper and middle reaches, are also restricted to the sites of confluence.

We do not give special names for the vertical zones of the Tienshanian mountain torrent; we use the commonly accepted terminology which describes the position of the zones in respect to the headwaters. But it should be borne in mind that the "upper" and the "middle" reaches are categories of the ecologo-faunistic classification rather than of the hydrological or any other scheme.

As far as we know, the vertical zonality scheme for the Issyk and Akbura Rivers is a first attempt of such a kind with respect of the Middle Asian torrents, and we have grounds to expect much similarity with the vertical zonality of torrents in the Hindu Kush, the Karakoram and the Himalaya.[1] Whether this scheme is universal will be seen from further studies, but the scheme and the invertebrates representative for the different zones were found to be valid for many mountain torrents of the Tien Shan, although the numerical data are available only for the Issyk and the Akbura Rivers.

[1] Mani (pers. comm., 1973) informed me that the vertical zonality recognized by us for the glacial torrents in the Tien Shan was also found in the north-western Himalaya (between absolute heights of 1800 and 4000 m).

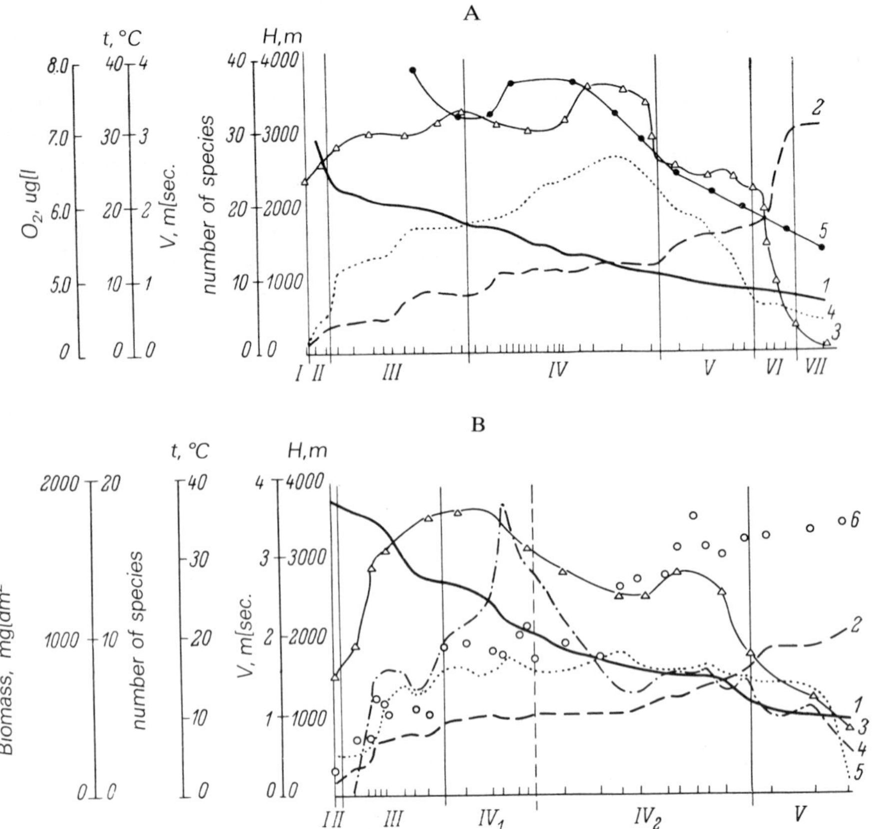

Fig. 91. Change of some physical, chemical and biological parameters along the longitudinal profile of the Tienshanian torrents. The Roman figures stand for the zones and their subdivisions. On the abscissa are the points of study, successively from headwaters (from left to right) to lower reach. A. the Issyk River: 1, absolute height H in m; 2. water temperature t in °C, August; 3. mean surface flow velocity, V in m/sec; 4. the number of species of the lithorheophylous fauna (mostly aquatic insects); 5. dissolved oxygen O_2 in mg/1. B, the Akbura River: 1–3, as above; 4, the number of species of mayflies, stoneflies and caddisflies; 5, their total biomass in mg/dm²; 6, air temperature in °C, August.

Figure 91 shows the change of some physical, chemical, and biological parameters. The graphs give an idea about the longitudinal profiles of the two torrents, the water temperature, the mean and the surface flow velocities, the oxygen content and the air temperature in the Akbura region. All the data are for a short period in August, *i.e.* they are nearly similtaneous. The data characterize the entire length of the rivers, from the glacial headwaters to the site of disappearance in the desert (the Issyk River) or the site of inflowing into the arterial canal (the Akbura River). The position of the study sites is shown in Figs. 31 and 32 and is marked by verticals on the abscissa in Fig. 91.

Perhaps there is no need to comment on these graphs in detail, as they speak for themselves. Both rivers have a similar pattern of water temperature

270

Table 62. Vertical zonality of the Issyk River.

Name of zone and its approximate extent in km	Geographical description	Ecological conditions	Representative insects
I Glacial, 0.2	Glaciers, the perennial snow limit	Abs. height 4000–3000 m, water temperature about 0° (0.3°). Almost distilled water. Ice, debris	No animals in the water. Only near the glacier foot solitary *Phoenocladius* sp.
II Glacial headwaters, brooks, 0.4	Moraine turning into alpine zone	Abs. height 3000–2900 m, water temp. 2.0–3.0°. Relatively slow velocity, about 2 m/sec. Small discharge, under 1 m³/sec. Fragments, sand, coarse pebbles, very mobile substrate. Turbid water because of glacial silt	DIPTERA *Chironomidae* * *Phoenocladius* (2 species) EPHEMEROPTERA * *Ameletus alexandrae* Brodsky TRICHOPTERA *Rhyacophila* sp. PLECOPTERA *Leuctra* sp.
III Upper reach of mountain torrent, 8.0	Coniferous forest zone (*Picea schrenkiana*)	Abs. height 2900–1800 m, water temp. 3.0–8.0°. Very large flow velocity, over 3–4 m/sec. Larger water discharge than in zone II, over 1–2 m³/sec. Rocks, fragments, boulders, pebbles. At places aquatic moss on rocks. Less mobile substrate than in zone II	DIPTERA *Deuterophlebiidae* * *Deuterophlebia mirabilis* Edwards *Blepharoceridae* * *Asioreas tianschanica* (Brodsky) EPHEMEROPTERA * *Iron montanus* Brodsky, *Rhithrogena tianschancia* Brodsky, *Ecdyonurus* sp. TRICHOPTERA * *Hymalopsyche gigantea* Martynov PLECOPTERA *Nemoura* sp.

Table 62. (Continued).

Name of zone and its approximate extent in km	Geographical description	Ecological conditions	Representative insects
IV Middle reach of mountain torrent, 12.0	Lower limit of coniferous forest, below it the zone of hardwood forest and shrubs. In the lower part of the zone the torrent goes out onto the debris cone	Abs. height 1800–1000 m, water temp. 8.0–12.0°. Large velocity, over 3–4 m/sec. Largest water discharge, up to 10 m³/sec. Rocks, fragments, boulders, rounded pebbles. Aquatic moss. Mobile substrate in some places. Solitary flowering plants	DIPTERA *Deuterophlebiidae* *Deuterophlebia mirabilis* Edwards *Blepharoceridae* * *Tianschanella monstruosa* Brodsky, * *Blepharocera asiatica* Brodsky EPHEMEROPTERA *Iron montanus* Brodsky, * *Ir. rheophilus* Brodsky, *Rhitrogena tianschanica* Brodsky, * *Ephemerella submontana* Brodsky TRICHOPTERA *Rhyacophila extensa* Martynov, *Dinarthrum reductum* Martynov, *Apatelia copiosa* McLachlan, *Agapetus tridens* McLachlan, *Ag. kirgisorum* Martynov PLECOPTERA *Neoperlina capnoptera* McLachlan, *Amphinemura* sp.
V Lower reach of mountain torrent, 5.4	Lower limit of shrubs, the beginning of the steppe zone. Piedmont debris cone of torrent	Abs. height 1000–800 m, water temp. 12.0–17.0°. Flow velocity drops from 3–4 to 2 m/sec. Still large water discharge decreases because of water expenditure through irrigative system. Large and small pebbles, sometimes silted. Flowering plants	DIPTERA *Blepharoceridae* *Blepharocera asiatica* Brodsky EPHEMEROPTERA *Ephemerella submontana* Brodsky, *Ephemerella* sp., * *Iron nigromaculatus* Brodsky, * *Ecdyonurus rubrofasciatus* Brodsky, * *Rhithrogena* sp., *Caenis* sp., *Baetis transiliensis* Brodsky

| VI Reduction into brook, 1.7 | Lower limit of debris cone. Steppe zone | Abs. height 800–700 m, water temp. 17–29° (rapidly rising downstream). Flow velocity and water discharge tend to decrease from 2.0 to 0.41 m/sec. and from 5.0 to 0.09 m³/sec. Pebbles, gravel, sand, silt in pools. Wandering channel, flat and broad | TRICHOPTERA
Hydropsyche sp.
No representative aquatic insects. At the beginning of the zone:
DIPTERA
Blepharoceridae
Blepharocera asiatica Brodsky
EPHEMEROPTERA
* *Baetis issykuvensis* Brodsky
Others are eurybionts, in particular, from *Chironomidae* and *Simuliidae* |
| VII End of the brook, 1.4 | Plain, semi-desert zone | Abs. height from 700 m and less. Water temp. about 30°. Very slow current, 0.06 m/sec; small water discharge, 0.014 m³/sec. Sand, silt, rare flat pebbles. Reeds flowering plants | No members of the torrential fauna. Dwellers of standing waters (*Nepa cinerea* L.) |

Note: The asterisk shows most representative and abundant species in a zone.

Table 63 Vertical zonality of the Akbura River.

Name of zone and its approximate extent in km	Geographical description	Ecological conditions	Representative insects
I Glacial		Not examined	
II Glacial headwaters, brooks, 2.5	Moraine	Abs. height 4000–3600 m, water temp. 0.5–1.5°. Flow velocity 1.5 m/sec. Small water discharge, under 1 m³/sec. Fragments, coarse pebbles	DIPTERA *Blepharoceridae* * *Asioreas nivia* (Brodsky) EPHEMEROPTERA *Iron montanus* Brodsky
III Upper reach of mountain torrent, 41.0	Moraine turns into the alpine zone	Abs. height 3600–2700 m, mean water temp. about 6.2°. Flow velocity, in average, 2 m/sec. Water discharge over 1 m³/sec. Fragments, pebbles, boulders	DIPTERA *Blepharoceridae* * *Asioreas nivia* (Brodsky) *Simuliidae* *Cnephia kirjanovae* Rubzov, *Cn. jankowskae* Rubzov EPHEMEROPTERA * *Iron montanus* Brodsky, *Rhithrogena tianschanica* Brodsky TRICHOPTERA * *Hymalopsyche gigantea* Martynov PLECOPTERA *Capnia elongata* Zhiltzova
IV Middle reach of mountain torrent – the upper division IV₁, 37.0	Juniper zone (*Juniperus turkes-tanika*)	Abs. height 2700–1900 m, mean water temp. about 9.9°. Flow velocity 3.2 m/sec. in average. Large water discharge, over 3–5 m³/sec. Fragments, rounded pebbles, rocks, boulders	DIPTERA *Deuterophlebiidae* * *Deuterophlebia mirabilis* Edwards *Blepharoceridae* *Blepharocera asiatica* Brodsky (upper limit), *Tianschanella monstruosa* Brodsky

			EPHEMEROPTERA * *Iron montanus* Brodsky, * *Rhithrogena tianschancia* Brodsky
the lower division IV$_2$, 87.0	Hardwood forest, lower limit of Juniper zone at 1700 m	Abs. height 1900–1100 m, water temperature about 11.6°. Flow velocity 2.5 m/sec. in average. Large water discharge, about 30 m^3/sec. Rounded pebbles, not silted	DIPTERA *Deuterophlebiidae* *Deuterophlebia mirabilis* Edwards *Blepharoceridae* * *Blepharocera asiatica* Brodsky, * *Tianschanella monstruosa* Brodsky EPHEMEROPTERA *Iron montanus* Brodsky, *Ir. rheophilus* Brodsky, *Rhithrogena tianschanica* Brodsky, * *Ephemerella submontana* Brodsky, *Ecdyonurus rubrofasciatus* Brodsky, *Ameletus alexandrae* Brodsky (lower limit at 1560 m)
V Lower reach of mountain torrent (a part of it, 41 km)	Debris cone, hardwood plants	Abs. height 1100–900 m, water temp. about 19°. Mean flow velocity 1.47 m/sec. Large water discharge, about 15–20 m^3/sec, becomes decreased because of irrigation. Silted rounded pebbles.	DIPTERA *Blepharoceridae* * *Blepharocera asiatica* Brodsky (numerous), *Tianschanella monstruosa* Brodsky (rare) *Simuliidae* *Wilhelmia mediterranea* Puri EPHEMEROPTERA * *Iron nigromaculatus* Brodsky, *Heptagenia perflava* Brodsky, * *Ephemerella* sp. TRICHOPTERA *Agapetus tridens* McLachlan, *Ag. kirgisorum* Martynov, *Hydropsyche gracilis* Martynov, *Glosossoma dentatum* McLachlan

Note: The asterisk shows most representative and abundant species in a zone.

variation: the curve abruptly goes up near the glacial headwaters, then tends to flatten in the upper reach and be relatively horizontal in the middle reach and, lastly, shows a sharp increase in the lower reach. Similar also are the curves for the mean surface velocities: the ascent, the maximum and the descent. The curves for the number species and the biomass of the lithorheophylous fauna are also very characteristic. From negligible values (near the headwaters) these approach a maximum in the middle river and then drop to a minimum in the lower reach. In our opinion, all this indicates that the torrential fauna, as well the Tienshanian torrent itself, represents a self-consistent system. This is true even of the torrent (the Akbura) which has turned into the arterial canal: the torrential fauna disappears entirely, having been replaced by another kind of fauna.

The results of comparison of changes in the physical and chemical parameters of the torrent and its lithorheophylous fauna, mostly of the faunal groups studied by the author (mayflies, *Blepharoceridae* and *Deuterophlebia*), are represented in Tables 62 and 63 as a scheme of vertical zonality of the Issyk and the Akbura Rivers. From the consideration of the graphs and the scheme, one concludes that the most significant physical factors affecting the torrential fauna seem to be the water temperature and the flow velocity. However their effects are not direct, but through the change of the whole set of conditions (the formation of the substrate, the dissolved oxygen content, the character of aquatic vegetation, *etc.*) and, of course, through the formation of adaptations and biocentic relations.

The zones characterized in Tables 62 and 63 can be found in most of the Tienshanian torrents, but their lengths may vary greatly with the water volume (the discharge). The larger the volume, the more stretched will be the longitudinal zones. For example, the Talgar River (the Zailiiskii Alatau) with a much larger volume of water than in the Issyk, at an elevation of 1500 m, had the water temperature 5.5°, whereas the Issyk River with smaller water volume at the same elevation had much higher water temperature, 11.0°. The vertical zones in the Talgar River are shifted downstream, if compared with those in the Issyk River. The same thing is observed when the Issyk and the Akbura Rivers are compared. Thus, the general feature of the Tienshanian mountain torrents, although the same scheme of the vertical zonality remains valid, will be different extension and altitudinal position of the zones, depending on the water discharge. The smaller the discharge, the more shortened and upwardly displaced are the zones. Naturally, the same applies not only to the ecological conditions in the torrent, but to its fauna as well.

Conclusion

Practical activity has refuted older, very limited ideas that the invertebrate fauna of mountain torrents is only of theoretical value in the study of adaptations of organisms to life in fast and rough current. It is no coincidence that the lotic, especially swift waters have lately received much attention from hydrobiologists in Western Europe and the United States.

The rapid development of rhithrobiology is characteristic of all countries. In the Soviet Union, however, the torrential invertebrates have for a long time been outside the interests of investigators who have directed their efforts primarily to the study of standing water fauna.

Large-scale construction of reservoirs, acclimatization of fish in rivers, streams and irrigative canals, the problem of clean water and, lastly, the use of the swift water dwellers as indicators of hydrological conditions – all this necessitates the study of the composition of the invertebrate fauna and fish populations, especially, the ecological characteristics of mountain torrents.

At present, detailed investigations of the fauna in rivers of the Issyk-kul Valley are being performed at the Biological Station, Academy of Sciences, Kirghiz SSR (Cholpon-Ata City at the Issyk-kul Lake). Its activity concerns the feeding habits of the juvenile trout (the "gegarkuni" which has been acclimatized in the Issyk-kul Lake) and its habitats in rivers and springs. Studies are made on the statics and seasonal dynamics of the forage biomass and on other questions concerning the river life cycle of the trout.

The faunistic investigations in the Akbura River are under way at the Pedagogical Institute, Osh City. They will be hopefully continued at the Ferghana Pedagogical Institute, and at the Academy of Sciences, Uzbek SSR, and at the State University of Tashkent. The Institute of Zoology, the USSR Academy of Sciences, investigates the species composition, although of only few invertebrate groups from the Middle Asian mountain torrents (stoneflies and *Blepharoceridae*).

Of course, this work is still insufficient for a comprehensive ecological study of the torrential fauna, which is hampered by inadequate knowledge of the invertebrate systematics. The study of the systematics of all faunistic groups, especially, of the aquatic insects is as important as the quantitative and ecological investigations, since the latter are of little effectiveness without the exact knowledge of the species composition.

Investigations of the invertebrate fauna are of value also for the reservoir designing. The value is of two kinds. First, the fauna of a torrent or a river is a source of the reservoir fauna. A good example is the faunistic analysis of the Chirchik River which is a source of the fauna for the Charvak reservoir (Sibirtzeva 1964; Sibirtzeva *et al.*, 1972). The data are available for the Papan reservoir which will be constructed on the Akbura River (Omorov and Ibragimov, 1972). Studies on the composition and the ecology of the benthic forms from mountain torrents are of great value also for the use of specific forms as biological indicators. The contents of this book show that the indicator species may help reveal such important aspects of the dynamics and statics of a torrent or a river as the constant or temporary character of the channel, the substrate mobility, the salt composition of the water, the oxygen content and the mudflow danger. For prediction of the torrent dynamics on the basis of fauna analysis, it is insufficient to study only the mass forms: single ones can tell much about the torrent. It should be kept in mind that larvae of mayflies, caddisflies, *Blepharoceridae*, *Deuterophlebia* and others sometime live in a torrent for more than one year and, therefore, their species composition, numbers of animals and their distribution pattern can give an idea of the year-round dynamics of the torrent or river. These considerations are essential to reservoir designing (Maltzev, 1964).

The theoretical value of the study of torrential fauna, we hope, is clear from the content of this essay. It should be noted only that bionics may profit greatly from the analysis of adaptations of torrential animals, in particular, of such a perfect device as the suckers in *Blepharoceridae* larvae.

Torrential ecology may become instructive and useful for the development of general biological problems. Earlier we drew the reader's attention to the fact that the torrential conditions are extreme conditions, which also permits investigation of ethological and physiological problems which are poorly understood, if at all. These include the problem of the imago hatching from pupa under conditions of rough fast current when the delicate winged insects are put to the test of the hydraulic impact; also that of the imago flight over a limited place of all the length of torrent; the behavior of larvae in conditions of highly variable water level, *etc.*

Of paramount importance is the study of biocenotic conditions which contribute to the formation of "monotonic" groups and biocenoses which differ so much from the faunistically rich biocenoses of the slow and standing waters.

For the solution of these problems the field studies alone will be insufficient; they must be supplemented by stationary studies which can be done only on the base of a specialized biological station concerned exclusively with the torrential fauna and flora and their ecology. Such biological stations have been working fruitfully for many years at one, sometimes small, lake; there is an urgent need in setting up such biological stations on typical mountain torrents.

This country has vast areas occupied by mountain massives (the Tien Shan, the Carpathians, the Caucasus, the Eastern Siberia, the Far East) and the establishment of biological stations on mountain torrents, primarily, in Middle Asia[2] would make an important contribution to the knowledge of torrential life and to the development of general biological problems.

[2] On the Northern Caucasus, near Ordjonikidze City, a biological station, organized by Prof. D. A. Tarnogradskii, has been active for many years. Its activity includes the study of the fauna of mountain torrents. At present, studies of torrents are under way in Georgia, Adjaria and other regions of the Caucasus (Zosidze, 1972, 1973).

278

When evaluating the significance of mountain torrents as a habit of specific biota, one must keep in mind such aspects as their incessant amenity and as an object of nature conservation.

From the aesthetic point of view, the mountain streams are a unique phenomenon of nature, without which one cannot imagine a reservation, a national park, nor the mountain locality in general. If rivers, streams, cascades and waterfalls were eliminated from a mountain landscape it would become a desert and lose all its aesthetic and emotional value. Such a prospect is quite feasible, since in addition to the regulated flow of mountain rivers, some planners raise the question of the use of fast-flowing watercourses for pasture irrigation at the very heads of the streams. As an object of nature conservation, the mountain torrents, especially, the typical ones, must be preserved for detailed study and as a standard for comparison with the regulated streams.

In conclusion, we should like to express our hope that the study of the fauna and ecology of mountain streams of this country, in particular, of Middle Asia, will receive adequate attention from the hydrobiological and ecological sciences, which will correspond to the great extensions of mountain massives in the Soviet Union and to the problems confronted by the ecology of specific environments so vividly represented in mountain torrents. I hope that this essay will stimulate studies of the mountain torrent and its fauna and flora.

References

In Russian

Agharkar, M. A. 1914. On a new species of Blepharocerid fly from Kashmir, together with a description of some larvae from the same locality. Rec. Indian Mus. 10: 159–164.

Albrecht, M.-L. 1961. Ein Vergleich quantitativer Methodes zur Untersuchung der Makrofauna fliessender Gewässer. Verh. Internat. Verein. Limnol. 14: 486–490.

—— 1966. Beitrag zur quantitativen Erfassung der makroskopischen Bodenfauna fliessender Gewässer. Limnologie (Berlin) B.4, 42: 351–358.

Alexander, Ch. P. 1958. Geographical distribution of the netwinged midges (*Blepharoceridae, Diptera*). Proceed. 10th Congr. Ent. Monreal, 1: 813–828.

*Alimov, V. F., Vasilieva, L. V., Kachalova, O. L., Pankratova, V. Ya., and Finogenova, N. P. 1977. A hydrobiological study of the Tiup River and the Tiup Bay of Issyk-kul Lake. Leningrad: Zool. Inst., Ac. Sci. USSR. 145p.

Allan, J. D. 1975. The distributional ecology and diversity of benthic insects in Cement creek, Colorado. Ecology, 56: 1040–1053.

Allen, K. 1971. New Asian *Ephemerella* with notes (*Ephemeroptera, Ephemerellidae*). The Canadian Entomologist, v.103, N4, 1911: 512–528.

Ambühl, H. 1959. Die Bedeutung der Strömung als ökologischer Faktor. Schweizer. Zeitschr. Hydrologie, v.31, 2: 133–264.

—— 1962. Die Besonderheiten der Wasserströmung in physikalischer, chemischer und biologische Hinsicht. Schweizer. Zeitschr. Hydrologie, v.24, 2: 367–382.

Anderson, N. H., and Lehmkuhl, D. M. 1968. Catastrophic drift of insects in a woodland stream. Ecology 49: 198–206.

Annandale, N. 1919. Fauna of certain small streams in the Bombay Residency. Calcutta Museum, Rec. of the Indian Museum, 16: 109–161.

——, and Hora, S. L. 1922. Parallel evolution in the fish and tadpoles on mountain torrents. Rec. Indian Mus. 24, 4: 505–509.

Anthon, H., and Lyneborg, L. 1968. The cuticular morphology of the larval head capsule in *Blepharoceridae* (*Diptera*). Spolia Zool. Mus. haun., Copenhagen, 27: 1–56.

Baker, F. C. 1928. Influence of a changed environment in the formation of new species and varieties. Ecology, 9, 3: 271–283.

Banarescu, P. 1956. Importance des espéces de goujons (genre *Gobio*) comme indicateurs des zones biologique des rivières. Bull. Institului de Cercetari Piscicole 15(3): 53–56.

Behning, A. 1928. Das Leben der Wolga. Binnengewässer, Theinemann, 5: 1–162.

*Beklemishev, V. N. 1949. Turbellarian worms (*Turbellaria*). In: Freshwater life of the Soviet Union, vol. 2: 10–34.

*Berg, L. S. 1949. Freshwater fishes of the Soviet Union and adjacent countries. Moscow-Leningrad: Ac. Sci. USSR (1948–1949). Parts 1–3, 1381 p.

Berner, L. 1950. The mayflies of Florida. Univ. Fla. Publ. biol. Sci. Ser. 4, 4: 267.

Bertrand, H. P. 1972. Larves et nymphes des Coleopteres aquatique du globe. Paris: 804 p.

Bishop, J. E., and Hynes, H. B. 1969a. Downstream drift of the invertebrate fauna in a stream ecosystem. Arc. Hydrol. 66, 1: 56–90.

——, ——. 1969b. Upstream movements of the benthic invertebrates in the Speed River, Ontario. J. Fish. Res. Can. 26: 271–298.

Bjarnov, N., and Thorup Jens. 1970. A simple method for rearing running-water insects, with some preliminary results. Arch. Hydrob. 67, H. 2: 201–209.

280

Bogoescu si I. Tabacaru, 1957. Contribut la studiul sistematic al nimfelor de Ephemeroptere Din R. P. R. Ac. Rep. Pop. Romane, Bull. St. 3, IX: 1–279.

Bornhauser, K. 1912. Die Tierwelt der Quellen in der Umgebung Basels. Int. Revue der Hydrob. u. Hydrogr. Biol. suppl., ser. V: 1–90.

Botosaneanu, L. 1959. Recherches sur les trichoptères du massif de Retezat et des Monts du Banat. Biol. Biol. Animala (Bucarest) I: 1–165.

Bournaud, M. 1963. Le courant, facteur écologique et étologique de la vie aquatique, Hydrobiologia, 21: 125–165.

—— 1972. Influence de la vitesse du courant sur l'activité de locomotion des larves de *Micropterna testacea* (Gmel.) (*Trichoptera, Limnophilidae*). Ann. Limnologie, t.8, f. 2: 141–216.

*Brodsky, A. L., and Brodsky, K. 1926. *Deuterophlebia* in Middle Asian mountains. Bull. Gosud. Sredneasiat. Univ., 13: 23–26.

Brodsky K. A. 1930a. Zur Kenntniss der mittelasiatischen Ephemeropteren I (Imagines). Zool. Jahrb. 59: 681–720.

——. 1930b. Zur Kenntniss der Wirbellosen fauna der Bergströme Mittelasiens. II. *Deuterophlebia mirabilis* Edw. Ztschr. Morph. Ökol. Tiere 18, 1/2: 290–321.

——. 1930c. Zur Kenntniss der Wirbellosenfauna der Bergströme Mittelasasiens III. *Blepharoceridae* I (Imagines). Zool. Anz. 90, 5/6: 129–146.

*—— 1930. A contribution to the knowledge of *Ephemeroptera* of South Siberia. Russk. Entomol. Obozr., 24(1–2): 31–40.

*—— 1935. A contribution to the knowledge of the invertebrate fauna of Middle Asiatic mountain streams. I. The Issyk River. Trudy Gosud. Sredneasiat. Univ., ser. 8-a, Zoologia, 15: 1–112.

*—— 1936. A contribution to the knowledge of the invertebrate fauna of Middle Asiatic mountain streams. IV. *Blepharoceridae* II. Trudy Zool. Inst., 4: 72–105.

*—— 1954. *Blepharoceridae* (*Diptera*) of the Altai and South Primorie. Trudy Zool. Inst., 15: 229–257.

*—— 1957. Calanoid fauna and the zoogeographical regionalization of the North Pacific and adjacent seas. Moscow-Leningrad: Ac. Sci. USSR. 222 p.

*—— 1965. Systematics of marine planktonic organisms and oceanography. Oceanologia, 5(4): 577–591.

*—— 1972a. *Asioreas* (gen. n.) *altaicus* (Brodsky)-*Blepharoceridae* (*Diptera*) from Mongolia. Meshdun. Soviet. Mongol. Exped., 1, Nasekomie. Leningrad: Nauka. 741–750.

*—— 1972b. New species and refined status of previously described Middle Asiatic species of dipterans from the family *Blepharoceridae* (*Diptera*). Entomol. Obozr., 51(3): 637–645.

*—— 1972c. Philogeny of the family *Calanidae* (*Copepoda*), based on comparative morphological analysis. Issled. fauni morei, 12(20): 5–110.

*—— 1975. On the protection and study of aquatic insects as a major group of the mountain lithorheophylous fauna. Dokl. vtor. sov. ob ohrane nasekomyh. Min. sel. hoz. Arm. SSR, Erevan: 26–30.

*—— 1976. Mountain torrent of the Tien Shan. An ecologo-faunistic essay. Leningrad: Nauka. 243 p.

*——, and Omorov, E. O. 1972a. Ecologo-faunistic vertical zonality of mountain torrents of the Tien Shan. Dokl. Acad. Nauk SSSR, 206(3): 759–762.

*——, —— 1972b. Distribution of net-winged and mountain midge larvae of families *Blepharoceridae* and *Deuterophlebiidae* (*Diptera*) in the Akbura River (the Alai Mts.). Entomol. Obozr., 51(1): 66–73.

*——, —— 1973. Ecologo-faunistic vertical zonality in mountain torrents of the Tien Shan with reference to the distribution of representative aquatic insects. Gidrobiol. Jurn., 2: 40–51.

Brundin, L. 1966. Transantarctic relationships and their significance, as evidenced by Chironomid midges. Kungl. Svenska vetenskapsakad Handlingar, 4 ser. B.II. 1: 1–472, 30 pl.

*Burtzev, P. N., and Baryshnikova, M. M. 1969. On possibility of using the current meters for measuring turbulent flow. Trudy Gosud. Inst. Hydrol., 172: 51–69.

*Bykov, V. D., and Vasiliev, A. V. 1972. Hydrometry. 3nd ed. Leningrad: Gidrometeoizdat. 448 p.

Carpenter, K. 1927. Faunistic ecology of some Cardiganshire streams. Journ. Ecology, XV, N 1: 33–54.

Carpenter, K. E. 1928. On the distribution of freshwater *Turbellaria* in the Aberyswith district, with especial reference to the ice age relicts. Jour. Ecol., 16, 1: 105–122.

Černosvitov 1930. Oligochaetaen aus Turkestan. Zool. Anz., Bd. 91, H. 1/4: 7–15.

Comstock, I. H. 1895. Manual for the study of insects. Ithaca, N 4.

Creaser, Ch. W. 1930. Relative importance of hydrogen-ion concentration, temperature, dissolved oxygen, and carbondioxide tension on habitat selection by brook-trout. Ecology, 11, 2: 246–262.

Cummins, K. W. 1962. An evaluation of some techniques for the collection and analysis of benthic samples with special emphasis on lotic waters. Am. Midl. Nat. 67: 477–504.

——. 1966. Trophic relationships in a small woodland stream. Verh. int. Ver. Limnol. 16: 627–638.

Cushing, C. E. 1964. An apparatus for sampling drifting organisms in streams, Journ. Wildlife Management, 28: 592–594.

*Davidov, L. K. 1927. Materials on glaciation of Middle Asian mountains. Trudy Gidromet. otdela, 1(1): 67–100.

*—— 1929. Fluctuation of streamflow rates in Middle Asian rivers. Trudy Gidromet. otdela, 1(2): 5–48.

*—— 1933. Classification of Middle Asian rivers by alimentation. Zap. Gosud. Hydrol. Inst., 10: 143–51.

*—— 1953. Hydrography of the Soviet Union (Terrestrial waters). Part 1. General description of the waters. Leningrad: Gosud. Univ. 184 p.

*—— 1955. Ibid. Part 2, Hydrography of regions. 600 p.

Davis, H. S. 1938. Instructions for conducting stream and lake surveys. Fishery Circ. Bur. Fish. 26, pp. 4, 247.

Demoulin, G. 1964. Mission H. G. Amsel in Afghanistan (1956): *Ephemeroptera*. Bull. Ann. Soc. Roy. Ent. Belg., 100(28): 351–63.

Dittmar, H. 1955. Ein Sauerlandbach, Untersuchungen an einem Wissen-Mittelgebirgsbach Arch. Hydrob. 50: 307–544.

Dodds, G. S., and Hisaw, F. S. 1924a. Ecological studies of aquatic insects I. Adaptation of mayfly nymph to swift streams. Ecology, 5, 2: 137–148.

——, ——. 1924b. Ecological studies of aquatic insects II. Size of respiratory organs in relations to environmental condition. Ecology 5, 3: 262–271.

——, ——. 1925. Ecological studies on aquatic insects IV. Altitudinal range and zonation of mayflies, stoneflies and caddisflies. Ecology 6: 380–390.

Dorier, A. 1937. La faune des eaux courantes alpines Verh. int. Verein. theor. angew. Limnol. 8: 33–41.

——, et F. Vaillant. 1955. Sur le facteur vitesse du courant. Verh. int. Verein. theor. angew. Limnol., 12: 593–597.

Douglas, B. 1958. The ecology of the attached diatoms and other algae in a stony stream. J. Ecol. 46: 295–322.

Dubey, O. P. 1971. Torrenticole insects of the Himalaya VI. Description of nine new species of Ephemerida from the Northwest Himalaya. Oriental Insects, 5: 521–548.

——, and Kaul, B. K. 1971. Torrenticole insects of the Himalaya IV. Some observations on the ecology and the character insect communities of the r. Alhni. Oriental insects, Univ. of Delhi, v. 5(1): 47–71.

Edington, I. M., and Hildrew, A. G. 1971. Experimental observations relating to the distribution of *Trichoptera* in streams. Limnologorum Conventus XVIII, Leningrad: 19.

Edington, I. M., and Molyneux, L. 1960. Portable water velocity meter. J. scient. Instrum, 35: 455–7.

Edwards, F. W. 1922. *Deuterophlebia mirabilis* gen. et sp. n. a remarkable dipterous insect from Kashmir. Ann. a. Mag. Nat. Hist. 9, 52: 379–387.

282

Enderlein, G. 1936. Notizen zur Klassifikation der Blepharoceriden (Dipt.) Mitt. Deutsch. Ent. Ges. 7(3): 42–43.

*Essays on the hydrography of rivers of the Soviet Union. 1953. Ed. Lvovich, M. I. Moscow. 324 p.

*Freshwater Life of the Soviet Union. Eds. Shadin, V. I., and Pavlovskii, E. N. Ac. Sci. USSR. Vol. 1, 1940, 460 p.; vol. 2, 1949, 537 p.; vol. 3, 1950, 910 p.; vol. 4, part 1, 1956, 470 p.; vol. 4, part 2, 320 p.

*Geller, S. Yu., and Ranzman, E. E. 1968. Topographic description. In: Middle Asia. Moscow: Nauka. p. 39–74.

*Gerasimov, I. P., Zimina, R. P., and Rodin, L. E. 1964. Landscape subdivision of Middle Asian mountains and geographical problems of management. In: Problemy landshaft. gorn. stran. Alma-Ata. p. 19–31.

Gessner, F. 1955. Hydrobotanik I. Energiechaushalt. Veb. Deutsch. Ver. Wissensch., Berlin: 1–517.

Gibbies, M. T. 1949. Notes on *Ephemeroptera Baetidae* from India and South. East Asia. Trans. R. ent. Soc. London, 100, 6: 161–177.

Grenier, P. 1949. Contribution à l'étude biologique des **Simuliides** de France. Physiologia comp. Oecol. 1: 165–330.

*Grib, A. V. 1950. *Oligochaeta* of Middle Asia. Trudy Zool. Inst., 9(1): 199–254.

*Gritzai, T. Ya., and Zhiltzova, L. A. 1973. Contribution to the knowledge of stoneflies (*Plecoptera*) of Tadjikistan. In: Fauna i ecol. chlenistonogih Tadjikistana. Dushanbe: Tadj. Univ. p. 17–39.

Grünberg, K. 1910. *Diptera*. Zweiflüger, Brauer. Süssw. Deutschl. Hf 2A: 1–312.

*Gurvitsch, V. F. 1966. On the Pamir-Tibetan biolimnological region. In: Biol. resursi vod., puti ih reconstr. i ispolz. Vses. Gidrobiol. Obshch., Akad. Nauk SSSR. p. 93–99.

——. 1971. La caracteristique zoogeographique de la faune des lacs de hautes montagnes de l'Asie moyenne sovietique. Limnologorum conventus XVIII, Leningrad: 27.

Harris, E. K. 1967. Further results in the statistical analysis of stream sampling. Ecology, 38: 463–468.

Harrison, A. D., and Agnew, I. D. 1961. The distribution of invertebrates endemic to acid streams, in the Western and Southern Cape Province. Nat. Inst. water Res. South African Counc. f. Sci. Industr. Res. Ms.: 1–11.

Hora, S. 1923. Observations on the fauna of certain torrential streams in the Khasi Hills. Rec. Indian Mus. 25: 579–600.

—— 1930. Ecology, bionomics and evolution of the torrential fauna, with special reference to the organs of attachment. Phil. Trans. Roy. Soc. London 218: 171–282.

Hubault, E. 1927. Contribution à l'étude des invertébrés torrenticoles. Bull. biol. France et Belg. Suppl. 9: 388 p.

Huet, M. 1946. Note préliminaire sur les relations entre le pente et les populations piscicoles des eaux courantes. Règle des pentes. Dodonaea 13: 232–243.

*Hydrobiological papers on Middle Asia. 1950. Trudy Zool. Inst., 9(1): 1–355.

Hynes, H. B. 1967. A key to the adults and nymphs of British stoneflies (*Plecoptera*). Freshw. Biol. Ass., Sci Publ. N 17: 1–90.

——. 1970. The ecology of running waters. Liverpool Univ.: 555 p.

——. 1976. Biology of *Plecoptera*. Ann. Review Entom., 21: 135–153.

*Ibragimov, I. I., and Omorov, E. O. (In print.) Hydrobiology of the middle Akbura River and adjacent waterbodies.

*Iliasov, A. T. 1959. Formation of the river runoff in the mountainous Middle Asia and the role of snows, and glacial and underground waters in this process. Trudy 3 vses. hydrol. siezda, 2: 692–697.

Illies, I. 1961. Versuch einer allgemeinen biozönotischen Gliederung der Fließgewässer. Internat. Rev. ges. Hydrobiol. 46: 206–213.

——. 1962. Die Bedeutung der Strömung für die Biozönose in Rhithron und Potamon. Schweizer, Zeitschr. v. 24, 2: 433–435.

——, et Botosaneanu, L. 1963. Problemes et méthodes de la classification et de la zonation écologique des eaux courantes, considerées surtout du point de vue faunistique. Intern. Verein. theor. und angew. Limnologie, Mitt. 12: 1–57.

*Ivanova, L. M. 1973. Hydrobiological description of the Aksu River (the Issyk-kul Lake basin). In: Ihtiol. issled. v Kirghizii. Frunze. p. 64–76.

Jones, I. R. E. 1951. An ecological study of the River Towy. I. Anim. Ecol. 20: 68–86.

Kalman, L. 1966. Beitrag zur Messung von Fliessgeschwindigkeiten in Klärbecken mit Thermosonden. Schweiz. Hydrol., 28: 69–87.

Kapur, A. P., and Kripalani, M. B. 1961. The mayflies (*Ephemeroptera*) from the North-Western Himalaya. Rec. Indian Mus. 59, 1, 2: 183–221.

*Karimova Burul. 1972. Algal flora of waterbodies of the Alai Valley and the Kurshab River basin. Avtoref. kandid. dissert. Akad. Nauk Uzbek SSR. Tashkent. 27 p.

Kaul, B. K. 1971. Torrenticole insects of the Himalaya V. Description of some new *Diptera*: *Psychodidae* and *Blepharoceridae*. Orientental insects, Univ. of Delhi, v. 5(3): 401–434.

Kawai, Teizi, 1963. Stoneflies (*Plecoptera*) from Afghanistan, Karakoram and Punjab Himalaya. Res. Kyoto Univ. Sci. exp. Karakoram and Hindukush, 1955, 4 (Insect fauna of Afghanistan and Hindukush): 53–86.

——. 1966. *Plecoptera* from the Hindukush. Res. Kyoto Univ. sci. exp. Karakoram and Hindukush, 1955. 8 Addit. Rep.: 203–216.

*Kazakhstan. 1970. In the series: The Soviet Union. Moscow: Misl. 408 p.

Kellogg, V. I. 1903. The net-winged midges (*Blepharoceridae*) of North America. Proc. Calif. Ac. Sci. 3 ser. Zool. 3: 187–223.

*Kemmerikh, A. O. 1958. The surface waters. In: Middle Asia. Moskow: Akad. Nauk SSSR. p. 167–200.

Kennedy, H. D. 1958. Biology and life history of a new species of mountain midge *Deuterophlebia nielsoni*, from Eastern California (*Diptera, Deuterophlebiidae*). Trans. Am. microsc. Soc. 77: 201–228.

——. 1960. *Deuterophlebia inyoensis*, a new species of mountain midge from the alpine zone of the Sierra Nevada range, California (*Diptera: Deuterophlebiidae*). Trans. Am. microsc. Soc. 79: 191–210.

Kimmins, D. E. 1942. Keys to the British species of *Ephemeroptera*, with Keys to the genera of the nymphs. Sci. Publ. N 7, Freshwater Biol. Assoc. Brit. Emp.: 1–64.

——. 1950. The 3rd Danish Expedition to Central Asia. Zoological results 4. *Odonata, Ephemeroptera* and *Neuroptera* (Insecta) from Afghanistan. Vidensk. Medd. dansk naturh. Foren. Kbh, Copenhagen, 112: 235–241.

——. 1972. A revised Key to the adults of the British species of *Ephemeroptera* with notes on their ecology. Freshwater Biol. Ass. Sci. Publ., 15: 1–75.

*Kiselev, I. A., and Vozjennikova, T. F. 1950. Materials on the algal flora of waterbodies in the Amu-darya basin. Trudy Zool. Inst., 9(1): 281–343.

Komarek, J. 1914a. Die Morphologie und Physiologie der Haftscheiben der Blepharoceriden Larven. Sitz. Ber. Keis. böhm. Gesel. d. wiss. II Klasse XXV: 1–28.

——. 1914b. Ueber die Blepharoceriden aus dem Kaukasus und Armenien. Sitzber. Kais. böhm. Gesellschaft d. wiss. N 9: 1–319.

*Konurbaev, A. O., and Madjar, L. A. 1969. The forage resources in some spawning rivers and their utilization by gegarkuni juveniles, the Issyk-kul trout. In: Ihtiol. i hydrobiol. issled. v Kirghizii. Frunze: Ilim. p. 9–27.

*Konurbaev, E. O. 1965. Blackflies (*Diptera, Simuliidae*) of the Issyk-kul Valley. In: Entomol. issled.v Kirghizii. Akad. Nauk Kirghiz SSR. p. 75–81.

*——. 1970. Materials on the fauna of the dipteran blood-suckers from some regions of Kirghizia. Sbornik trudov Oshskogo pedinstituta. 8: 112–19.

*——. 1973. Variability of some quantitative characters in blackfly larvae (*Diptera, Simuliidae*) from the mountainous Middle Asia. Entomol. obozr., 52(4): 915–923.

*——. 1976a. Morphological and functional analysis of diagnostic structures in the preimaginal blackflies (the *Simuliidae* family) from Middle Asia. Trudy Prshevalskogo Ped. Inst., seria estestv. nauk, 17: 14–34.

*——. 1976b. Comparative zoogeographical description of the blackfly fauna of Middle Asia. Trudy Prshevalskogo Ped. Inst., seria estestv. nauk, 17: 14–34.

*——. 1977. Ecological classification of Middle Asian running waters and the blackfly distribution (*Diptera, Simuliidae*) in different streams. Entomol. obozr., 56(4): 734–780.

*——, Omorov, E. O., and Tadjibaev, D. T. 1972. The fauna, biology, and vertical distribution of blackflies (*Diptera, Simuliidae*) in the Akbura River, Kirghizia. Entomol. Obozr., 51(1): 59–65.

*——, and Tadjibaev, D. T. 1970. Materials on the blackflies (*Diptera, Simuliidae*) of the Osh region. Sbornik trudov Oshskogo Ped. Inst., 8: 119–23.

*——, ——. The landscape-related and vertical faunistic complexes of blackflies (*Simuliidae, Diptera*) of Middle Asia. Sbornik trudov Oshskogo Ped. Inst., (In print.)

*Kovalev, V. 1973. The most daring insect. Yunii naturalist, 3: 8–10.

Kownacka Marta, Kownacki Andrzei. 1972. Vertical distribution of zoocenoses in the streams of the Tatra, Caucasus and Balkans Mts. Verh. Int. Ver. theoret. und angew Linnol., 18, P. 2: 742–750.

Kühtreiber, I. 1934. Die Plecopterenfauna Nordtirols. Ber. Naturwiss. Med. Ver. Innsbruck 43/44: 1–219.

*Kurmangalieva, Sh. G. 1974. On the chironomid fauna (*Diptera, Chironomidae, Orthocladiinae*) of the Turgen River, the Zailiiskii Alatau Range. In: Biol. nauki, 6. Kazakh. Gosud. Univ.

*——. 1976a. The quantitative characterization of lithorheophylous fauna of waterbodies in the Turgen River basin, the Zailiiskii Alatau Range. In: Biol. osnovi rib. hoz. Sred. Azii i Kazakhstana, Dushanbe. p. 103, 104.

*——. 1976b. The benthic invertebrates (*Ephemeroptera, Trichoptera, Diptera* and others) of waterbodies of the Zailiiskii and the Kungei Alatau Ranges. Avtoref. kandid. dissert., Kazakh. Univ., Alma-Ata. 24 p.

*Kustareva, L. A. 1973. The invertebrate fauna of the Karasu River (the Issyk-kul Lake basin) and its use by the Issykkul trout juveniles. In: Ihtiol. hydrobiol. issled. Kirghizii. Frunze. p. 51–63.

*—— 1976a. The forage resources of the Akterek River and their use by juveniles of the Issyk-kul trout. In: Biol. osnovi rib. hoz. Sred. Azii i Kazakhstana, Dushanbe. p. 102, 103.

*—— 1976b. Mayflies (*Ephemeroptera, Ephemerellidae, Heptageniidae*) from rivers of the Issyk-kul Valley. Entomol. Obozr., 55(1): 58–68.

*—— 1978. Mayflies (*Ephemeroptera, Heptageniidae*) in rivers of the Issyk-kul Valley. II. Entomol. obozr., 57(1): 92–96.

*——, and Ivanova, L. M. 1970. A contribution to the knowledge of the invertebrate fauna of rivers of the Issyk-kul Valley. In: Biol. processi v mor. i continent. vodoemah. Akad. Nauk SSSR, Vses. Hydrobiol. Obshch., Kishenev. p. 208.

*Kuzin, P. S. 1960. Classification of rivers and hydrological regionalization of the Soviet Union. Leningrad: Gidrometeoizdat. 455 p.

*Lastochkin, D. A. 1949. *Chaetopoda*. In: Freshwater Life of the Soviet Union, vol. 2: 111–30.

Lehmkuhl, D. M. and N. H. Anderson, 1972. Microdistribution and density as factors affecting the downstream drift of mayflies. Ecology, 53, N 4: 661–667.

*Lepneva, S. G. 1927. Insectaries for studing metamorphosis in aquatic insects. Russk. Hydrobiol. Jurn., 6: 236–237.

*—— 1945. Remarkable larvae of genus Rhyacophila Pict. in streams of Middle Asia. Entomol. Obozr., 28(3–4): 64–74.

*—— 1949. The benthic fauna of the Teletskoe Lake. Trudy Zool. Inst., 4: 7–118.

*—— 1951. Caddisflies of the Kondara Gorge. In: The Kondara Gorge. p. 152–157.

*——. 1956. A new caddisfly larva (*Trichoptera, Rhyacophilidae*) from the high altitudes of the Southern Caucasus. Entomol. Obozr., 35(4): 899–911.

*——. 1960. The species of *Himalopsyche* (*Trichoptera, Rhyacophilidae*) in Tadjikistan and southern Kirghizia. Izv. Otd. selskohoz. i biol. nauk, Akad. Nauk Tadjik SSR, 2(3): 103–109.

*——. 1964. Caddisflies. In: The Fauna of the Soviet Union, vol. 2(1), Larvae and pupae of Suborder *Annulipalpia*. Moscow-Leningrad: Inst. Zool., Akad. Nauk SSSR. 562 p.

*——. 1966. Caddisflies. In: The fauna of the Soviet Union, vol. 2(2), Larvae and pupae of Suborder *Integripalpia*. Ibid. 562 p.

*Levanidova, P. M. 1968. Benthos of the Amur River affluents (ecological and faunistic description). Izv. TINRO, 64: 181–289.

Linduska, X. P. 1942. Bottom type as a factor influencing the local distribution of mayfly nymphs. Can. Ent. 74: 26–30.

*Lipskii, V. N. 1902, 1905. The mountainous Buchara. Results of three expedition to Middle Asia in 1896, 1897 and 1899. Parts 1–3. Izd. Russk. geograph. Obshch. 735 p.

*Logacheva, L. C. 1959. The blackfly fauna of northern Kirghizia. Sbornik trudov nauch.-issled. inst. epidem., microbiol. i gigieni, 4: 273–76.

Longhurst, A. R. 1959. The sampling problem in benthic ecology. Proc. N. Z. ecol. Soc. 6: 8–12.

*Luzhin, B. P. 1956. Gegarkuni, the Issyk-kul trout. Akad. Nauk Kirghiz SSR, Frunze. p. 3–135.

*Lvovich, M. I. 1938. An attempt at classification of rivers of the Soviet Union. Trudy Gosud. Hydrol. Inst., 5: 58–108.

Lyman, F. E. 1955. Seasonal distribution and life cycles of *Ephemeroptera*. Ann. ent. Soc. Am. 38: 234–236.

Macan, T. T. 1958. Methods of sampling the bottom fauna in stony streams. Mitt. int. Verein. theoret. angew. Limnologie, 8: 21.

——. 1961a. The *Ephemeroptera* of a stony stream. Journ. Anim. Ecol., 26: 317–342.

——. 1961b. A review of running water studies. Verh. Intern. Verein. Limnol. XIV: 587–602.

——. 1961c. A key to the nymphs of the British species of *Ephemeroptera*. N 20: 1–64.

——. 1962. Biotic factors in running water. Schweiz. Ztschr. 24, 2: 386–407.

——. 1963. Freshwater ecology. Longmans, London: 1–338.

——, and Worthington, E. B. 1951. Life in lakes and rivers. London: 1–272.

Madsen, B. L., Bengtson, J. B. I., Butz, I. 1973. Observations on upstream migration by imagines of some *Plecoptera* and *Ephemeroptera*. Limnol. and Oceanogr. 18, 4: 678–681.

*Maltzev, A. E. 1964. Study of environmental conditions for designing reservoirs on mountain rivers. Frunze: Kirghizstan. 112 p.

Mani, M. S. 1968. Ecology and biogeography of high altitude insects. Series entomologica, 4, Hague: 1–527.

Mannheims, B. 1954. Die Blepharoceriden Griechenlands und Mitteleuropes (Dipt.). Bonn. zool. Beitr. Sonderband, I: 87–110.

*Martynov, A. 1927a. Contribution to the entomofauna of Turkestan. I. *Trichoptera annulipalpia*. Eshegod. Zool. Muz. Akad. Nauk SSSR, 23: 162–93.

*——. 1927b. A contribution to the aquatic entomofauna of Turkestan. II. *Trichoptera integripalpia*, with a note on a new species of *Rhyacophila* Pict. Ibid. 28: 457–95.

*——. 1929. Ecological prerequisites for zoogeography of freshwater benthos. Russk. zool. jurn., 9(3): 3–38.

——. 1930a. On the Trichopterous fauna of China and Eastern Tibet. Proc. Zool. Soc., London, 1(5): 65–112.

*——. 1930b. Caddisflies. Trudy Pamir. exp. 1928, 2, Zoology: 79–81.

*——. 1935. A contribution to the knowledge of *Amphipoda* from the running waters of Turkestan. Trudy Zool. Inst., 2: 411–508.

McConnell, W. I., and Sigler, W. F. 1959. Chlorophyll and productivity in a mountain river. Limnol. Oceanogr. 4: 335–351.

Mecom, John O. 1972. Feeding habits of *Trichoptera* in a mountain stream. Oicos, v. 23, N 3: 401–407.

Miall, L. C. 1922. The natural history of aquatic insects. London: 396 p.

*Middle Asia. 1968. Moscow: Nauka. 484 p.

Minckley, W. L. 1963. The ecology of a spring stream Doe Run, Meade County, Kentucky. Wildl. Monogr. Chestertown, 11: 1–124.

Moffett, I. W. 1936. A quantitative study of the bottom fauna in some Utah streams variously affected by erosion. Bull. Univ. Utah biol. ser. 26, 9: 1–33.

Müller, K. 1954. Faunistisch-ökologische Untersuchungen in nordschwedischen Waldbächen. Oikos, 5: 77–93.

———. 1962/63. Eine Methode zur Gewinnung von Naturfutter für Forellenzuchten. Z. Fischerei NF. 11: 743–745.

Mundie, I. H. 1964. A sampler for catching emerging insects and drifting materials in streams. Limnol. Oceanogr. 9: 459–469.

*Murzaev, E. M. 1968. Factors in differentiation of the environmental conditions and as a basis for regionalization. In: Middle Asia. Moskow: Nauka. p. 319–392.

*Muzafarov, A. M. 1958. Algal flora in mountain waterbodies of Middle Asia. Tashkent: Inst. Botan., Akad. Nauk, Uzbek SSR. 378 p.

*———. 1965. Algal flora in waterbodies of Middle Asia. Tashkent: Inst. Botan., Akad. Nauk, Uzbek SSR. 569 p.

Muttkowsky, R. A. 1929. The ecology of trout streams in Yellowstone National Park. Roosev. Wild Life Ann. 2, 2: 155–240.

Needham, I. C., and Lloyd, I. T. 1916. The life of inland water. Ithaca, N 4: 438 p.

Nielsen, A. 1950. The torrential invertebrate fauna. Oikos, 2: 176–196.

———. 1951. Is dorsoventral flattening of the body an adaptation to torrential life? Verh. int. Verein. theor. angew. Limnol. 11: 264–267.

*Nikolskii, G. V. 1948. The Amur River and its freshwater fishes. 95 p.

Ohle, W. 1937. Kalksystematik unserer Bienengewasser und der Kalkgehalt Rügener Bäche. Geol. Meere u. Binnengew. 1: 291–316.

*Oldekop, E. M. 1917. On prediction of water discharge in rivers of Turkestan. Bull. Gidrom., Tashkent. Parts 1–3.

*Omorov, E. 1973. Invertebrate fauna of the Akbura River basin. Its composition, vertical distribution, seasonal changes and biomass of the mass groups. Avtoref. kandid. dissert., Gosud. Univ., Tashkent. 29 p.

*———. 1975. On the ecology of caddisflies (*Trichoptera*) from mountain streams of the Southern Tien Shan (the Akbura River). Entomol. Obozr., 54(4): 773–79.

*——— (In print.) Vertical distribution of mayflies (*Ephemeroptera*) in the Akbura River and its affluent, the Kichik-Alai. Entomol. Obozr.

*———, and Ibragimov, I. I. 1972. Zooplankton and macrozoobenthos in waterbodies of the Papan reservoir flood zone. Tez. dokl. konf. biol. osnovi rib. hoz. resp. Sred. Azii i Kazakhstana. Tashkent-Ferghana. p. 124, 125.

*Ovchinnikov, I. F. 1936. Hydrological and hydrobiological review of major waterbodies in the Chu River valley. Trudy Kirgh. kompl. exp., 3(1): 115–144.

*Palgov, N. N. 1954. Glacial-snowy runoff in rivers of the Zailiiskii Alatau Range. Geogr. sbornik, Akad. Nauk SSSR, 4: 20–26.

*Pankratova, V. Ya. 1950. Larval fauna of the family *Tendipedidae* in the Amu-darya basin. Trudy Zool. Inst., 9(1): 116–98.

*———. 1970. Larvae and pupae of midges of Subfamily *Orthocladiinae* (*Diptera, Chironomidae-Tendipedidae*) of the Soviet Union. Keys to the Fauna of the Soviet Union issued by the Institute of Zoology, Academy of Sciences, USSR, no. 102: 343 p.

*Pavlov, V. N. 1972. The botanical and geographical regionalization of the Western Tien Shan. Bull. Moskow. obshch. ispit. prirodi, otd. biologii, 76(6): 99–110.

Pennak, R. W. 1945. Notes on mountain midges (*Deuterophlebiidae*) with a description of the immature stages of a new species from Colorado. Am. mus. novitates, N 1276: 1–10.

———. 1953. Fresh-water invertebrates of the United States. Ronald Press, N 4: 769 p.

Percival, E. and Whitehead, H. 1929. A quantitative study of the fauna of types of stream-bed. J. Ecol. 17: 382–314.

———, ———. 1930. A biological survey of the River Wharfe. II. Report on the invertebrate fauna. J. Ecol., 18: 286–302.

Phillipson, J. 1956. A study of factors determining the distribution of the blackfly *Simulium ornatum* Mg. Bull. ent. Res. 47: 227–238.

*Piskarev, K. V. 1968a. On acclimatization of the Amu-darya trout in rivers of the Osh region, Kirghiz SSR. Uchen. zap. biol. facult., Oshskii Ped. Inst., 7: 7–12.

*———. 1968b. Results of acclimatization of trout in rivers of the Osh region. In: Tezisi dokl. konf. rib. hoz. resp. Sred. Azii i Kazakhstana, 1968, Frunze. p. 117–119.

Pleskot, G. 1962. Strömung, Bodenstruktur und Besiedlungsdichte. Schweizer. Zeitschr. Hydrologie v. 24, 2: 383–385.

Pomeisl, E. 1953a. Der Mauerbach. In Beiträge zur Limnologie der Wienerwaldbache Wetter. u. Leben, Sonderheft 2: 103–121.

——. 1953b. Über die Plecopteren des Mauerbaches. In: Beiträge zur Limnologie der Wienerwaldbache. Wetter a. Leben, Sonderheft, 2: 177–178.

Powers, E. B. 1929. Fresh water studies. The relative temp., oxyg. cont. alk. res. carb. diox. tens. and pH of the water of certain mountain streams. Smoky mountain, National park. Ecology, 10, 1: 97–111.

Pulikowsky, N. 1924. Metamorphosis of *Deuterophlebia* sp. (*Diptera, Deuterophlebiidae* Edw.). Trans. Ent. Soc. 30. London: 45–62.

*Raspopov, I. M. 1963. On the basic concepts and trends in hydrobotany in the Soviet Union. Uspehi sovr. biol., 55(3): 453–464.

*Reihardt, A. N., and Ogloblin, D. A. 1940. Beetles (*Coleoptera*). In: Freshwater Life of the Soviet Union, vol. 1, p. 158–186.

Reisen, W. K. and Prins, R. 1972. Some ecological relationships of the invertebrate drift in Praters creek, Pickens county, South Carolina. Ecology, 53, N 5: 876–884.

Robins,C. R. and Crawford, R. W. 1954. A short accurate method for estimating the volume of stream flow. J. Wildl. Mgmt, 18, 366–194.

*Rodendorf, V. V. 1959. Phylogenetic relicts. Trudy Inst. morphol. zivotnih, 27: 41–51.

*Rubzov, I. A. 1951a. Blackflies. In: The Kondara River Gorge. p. 149–151.

*——. 1951b. A contribution to blackfly systematics (*Simuliidae, Diptera*) of Middle Asia. Trudy Zool. Inst., 9: 743–860.

*——. 1951c. Ecological and biological characteristics of the Middle Asiatic blackflies. Parazit. sbornik Inst. Zool. Akad. Nauk SSSR, 13: 328–342.

*——. 1956. Blackflies (family *Simuliidae*). The Fauna of the Soviet Union. Dipteran insects. 2nd ed., 6(6). Moscow-Leningrad, Ac. Sci, USSR. 859 p.

*——. 1970. The adaptive character of variability of taxonomic features. Zool. Jurn., 49(4): 635–646.

*——. 1972. Phoresia in blackflies (*Diptera, Simuliidae*) and new species discovered with mayfly larvae. Entomol. Obozr., 51(12): 403–410.

*——. 1972, 1974. Aquatic *Mermithoidea*. Part 1, 1972, 254 p. Part 2, 1974, 222 p. Inst. Zool., Acad. Sci., USSR.

*——. 1974. Variability of taxonomic features in blackflies (*Diptera, Simuliidae*). Problems and methods. Entomol. Obozr., 53(1): 24–37.

*——, 1976. New blackfly species (*Diptera, Simuliidae*) from Kazakhstan. Trudy Inst. Zool., Akad. Nauk Kazakh SSR, 36: 5–17.

*——, and Shakirzianova, M. S. 1976. On the blackfly fauna (*Diptera, Simuliidae*) of Kazakhstan. Trudy Inst. Zool., Akad. Nauk, Kazakh SSR, 36: 18–35.

Ruttner, F. 1953. Fundamentals of Limnology. Transl. D. G. Frey and F. E. J. Fry, Univers. of Toronto Press: 242 p.

*Sadovskii, L. A. 1946. Data on the forage base for commercial fishes in the upper and middle Kura River. Trudy Zool. Inst., Georgian SSR, 6: 119–164.

*——. 1948. The benthometer, a new device for quantitative sampling zoobenthos in mountain rivers. Soobshch. Akad. Nauk Georgian SSR, 9(6): 365–368.

Santokh Singh, 1961. Ent. survey of Himalaya, p. 32. A note on the larva of an apparently undescribed sp. of *Deuterophebia* Edw. (*Deuterophelebiidae, Diptera-Nematocera*) from the North-West (Punjab) Himalaya. Agra Univers. Journ. of Res. 10, 1: 109–113.

*Sapozhnikov, V. V. 1907. Essays on the Semirechie. 2. The Djungarskii Alatau and an expedition to the Zailiiskii Alatau. Tomsk. 107 p.

*Savchenko, E. N. 1974. Another species of midges (*Diptera, Tanyderidae*) from Soviet Middle Asia. Zool. Jurn., 53(12): 1892–1894.

Schmitz, W. 1957. Die Bergbach-Zoozönosen und ihre Abgrenzung, dargestellt am Beispiel der oberen Fulda. Arch. Hydrobiol. 53: 465–619.

——. 1961. Fliesswässerforschung-Hydrographie und Botanik Verh. int. Verein, theor. angew. Limnol. 14, 541–586.

Schoenemund, E. 1925. Beiträge zur Biologie der Plecopteren-Larven insbesonders Berüksichtigung der Atmung. Arch. Hydrob. 15: 339–369.

———. 1930. Eintagsfliegen oder *Ephemeroptera*, in Dahl. Die Tierwelt Deutschlands, 11: 1–106.

*Schtakelberg, A. A. 1951. *Diptera*. In: The Kondara River Gorge. p. 129–148.

Schwoerbel, J. 1961. Die Bedeutung der Wassermilben für die biozönotische Gliederung. Verh. int. Verein theor. angew. Limnol. 14: 355–361.

Scott, D. 1958. Ecological studies on the *Trichoptera* of the River Dean, Cheshire. Arch. Hydrobiol. 54: 340–392.

*Shcheglova, O. P. 1960. Alimentation of Middle Asiatic rivers. Samarkand. Gosud. Univ., Tashkent. 244 p.

*Sergeeva, E. N. 1925. Essay on the geology of the Djetysu. In: The Djetysu. Ed. Shnitnikov, V. N. Tashkent. p. 93–121.

*Shadin, V. I. 1950a. Freshwater molluscs from the Amu-darya Basin. Trudy Zool. Inst., 9(1): 55–74.

*———. 1950b. Theoretical and practical importance of hydrobiological study of the Amu-darya basin. Trudy Zool. Inst., 9(1): 7–15.

*———. 1950c. Life in Rivers. In: Freshwater Life of the Soviet Union, vol. 3: 113–256.

*———. 1950d. An experience of hydrobiological classification of rivers of the Soviet Union and their subdivision. In: Freshwater Life of the Soviet Union, vol. 3: 244–256.

*———. 1956. Methods of study of aquatic bottom fauna and ecology of benthic invertebrates. In: Freshwater Life of the Soviet Union, vol. 4, part 1: 279–382.

*———, and Gerd, S. V. 1961. Rivers, lakes and reservoirs of the Soviet Union, their fauna and flora. Moscow: Uchpedgiz. 599 p.

*———, Kiselev, I. A., Rodina, A. G., Pankratova, V. Ya., and Grib, A. V. 1951. Life in the waters of the Kondara. In: The Kondara River Gorge, p. 245–264.

*Shaposhnikova, G. H. 1950. Fishes of the Amu-darya River. Trudy Zool. Inst., 9(1): 16–54.

Shelford, V. E., and Eddy, S. 1929. Methods for the study of stream communities. Ecology 10: 382–391.

*Shultz, V. L. 1944. On classification of Middle Asiatic rivers by alimentation. Bull. Akad. Nauk Uzbek SSR, 5–6: 3–6.

*———. 1965. Rivers of Middle Asia. Leningrad: Gidrometeoizdat. 692 p.

*———, Timofeev, E. M., and Nadezhdin, A. M. 1936. Major features of hydrology of Middle Asia. Tashkent. 104 p.

*Sibirtzeva, L. K. 1958. Materials on the fauna of the Chirchik River. Caddisflies. In: Sbornik rabot aspirantov, otd. biol. nauk, Akad. Nauk Uzbek SSR, Tashkent. p. 169–179.

*———, 1961. The bottom fauna of the Chirchik River and its use by fishes. Vopr. biol. kraev. meditzini, Taskent, 2: 417–428.

*———. 1964. The invertebrate fauna of the Chirchik basin. Avtoref. kand. biol. nauk Akad. Nauk Uzbek SSR, Tashkent: 3–14.

*———, Kiseleva, E. V., Abdullaev, M. A. 1961. Hydrobiology of the upper Zeravshan River. Trudy Samarkand. Univ., Novaia seria, 10, Zoologia, p. 97–117.

*———, Nikolaenko, V. A., and Afanasieva, L. A. 1972. Hydrochemical and Hydrobiological regimes of the Charvak reservoir in the first year of its existence. In: Tezisi dokl. konf. biol. osnovi rib. hoz. resp. Sred. Azii i Kazakhstana, Tashkent-Ferghana: 135, 136.

*Sinichenkova, N. D. 1973a. A contribution to the knowledge of the genus *Rhithrogena* Eaton (*Ephemeroptera, Heptageniidae*). Vest. Mosk. Univ. biol. pochv., 3: 16–22.

*———. 1973b. Larvae of palaearctic mayflies of the genus *Rhithrogena* Eaton (*Ephemeroptera, Heptageniidae*). Vest. Mosk. Univ. biol. pochv., 5: 9–17.

*Sribnii, M. F. 1960. A formula for mean flow velocity and hydraulic classifications of rivers by resistance. In: Study and multi-purpose use of water resources. Moscow: Acad. Sci., USSR. p. 204–290.

Steine, I. 1972. The number and size of drifting nymphs of *Ephemeroptera, Chironomidae* and *Simuliidae* by day and night in the river Strande, West. Norway. Norsk. Entom. Tidskr. 19: 127–131.

Steinmann, P. 1907. Die Tierwelt der Gebirgäche. eine faunistisch-biologische Studie. Ann. Biol. Lacustre, 2: 30–164.

——. 1915. Praktikum der Susswasserbiologie: I Teil. Die Organismen des flissenden Wassers. Berlin: 1–184.

*Stepanova, and Ledyaeva. 1957. Hydrobiology of the Ferghana Valley irrigative system. Tashkent, Acad. Sci., Uzbek SSR. 116 p.

Strenger, A. 1953. Zur Kopfmorphologie der Ephemeridenlarven. Erster Teil. *Ecdyonurus* und *Rhithrogena*. Öst. zool. 4: 191–228.

*Taubaev, T. 1970. Flora of Middle Asiatic waterbodies and its value for the national economy. Inst. Botan., Akad. Nauk Uzbek SSR. Tashkent: FAN. 491 p.

*The Kondara River Gorge. 1951. Eds. Shadin, V. I., and Pavlovskii, E. N. Zool. Inst., Acad. Sci., USSR. 421 p.

Thienemann, A. 1912. Der Bergbach des Sauerlandes. Faunistisch-biologische Untersuchungen. Intern. Rev. Biol. Suppl. 4: 1–125.

——. 1925. Die Binnegewässer Mitteleuropas. Die Binnengewäss. 1, Stuttgart: 1–255

——. 1926. Hydrobiologische Untersuchungen an den Kalten Quellen und Bächen der Halbinsel Jasmund auf Rügen. Arch. Hydrobiol. 30: 167–262.

——. 1936. Alpine Chironomiden. Arch. Hydrobiol. 30: 167–262.

Tonnoir, A. L., 1923a. Australian *Blepharoceriidae*. Larvae and Puppae. Austr. Zool. III: 47–59.

——. 1923b. Australian *Blepharoceridae*. Corr. and addit. p. I, 5; II, Austr. zool. III: 135–142.

——. 1924. La *Blepharoceridae* de la Tasmanie. Ann. Biol. lac. XIII: 5–67.

——. 1930. Notes on Indian Blepharocerid larvae and pupae, with remarks on the morphology of Blepharocerid larvae and pupae in general. Rec. Indian Mus., 32(2): 161–214.

——. 1931. Notes on some types of Indian *Blepharoceridae*. Rec. Indian Mus. 33: 283–289.

——. 1932. Notes on Indian *Blepharoceridae*, III. Rec. Indian Mus. 34: 269–275.

Traver, I. R. 1939. Himalayan mayflies (*Ephemeroptera*). Ann. Mag. Nat. Hist. London, 11, 4: 49–56.

*Tshernova, O. A. 1972. Some new Asiatic mayfly species (*Ephemeroptera, Heptageniidae, Ephemerellidae*). Entomol. Obozr., 51(3): 604–614.

*—— 1972. The generic composition of mayflies of the family *Heptageniidae* (*Ephemeroptera*) in the Holarctic and Oriental regions. Entomol. Obozr., 53(4): 801–14.

*—— 1976. A key to the larvae of the genera of *Heptageniidae* (*Ephemeroptera*) from the Holarctic and Oriental regions. Entomol. Obozr., 55(2): 332–346.

*Tshupakhin, V. M. 1964. Physical geography of the Tien Shan. Alma-Ata: Acad. Sci. Kazakh SSR. 373 p.

Uéno, M. 1955. Fauna and flora of Nepal Hymalayas. Mayfly nymphs. Sci. Res. Japanese Exped. to Nepal Himalaya 1 Kyoto: 301–316.

—— 1966. Mayflies (*Ephemeroptera*) collected by the Kyoto University Pamir-Hindukush Expedition 1960. Res. Kyoto Univers. Sci Exp. to the Karakoram and Hindukush, 1955. 8, Addit. Rep.: 299–326.

*Uklonskii, A. S. 1925. Materials for hydrochemical characterization of the waters of Turkestan. Tashkent. 72 p.

Ulmer, G. 1935. "*Ephemeroptera*"-Visser, Karakorum, 1: 232.

Van Emden, F. I. 1957. The taxonomic significance of the characters of immature insects. Ann. Rev. Ent. 2: 91–106.

*Vyhodzev, I. V. 1956. Vertical zonality of vegetation in Kirghizia (the Tien Shan and the Alai Mts.). Moscow: Acad. Sci. USSR. 84 p.

Watanabe, N. P., and Haroda Saburo. 1976. Some problems in sampling stream fauna, with particular reference to the variability of samples. Jap. J. Limnol., 37(2): 47–58.

Waters, T. F. 1961. Standing crop and drift of bottom organisms. Ecology 42: 532–537.

——. 1962a. A method to estimate the production rate of a stream bottom invertebrate. Trans. Amer. Fish. Soc. 91: 243–250.

——. 1962b. Diurnal periodicity in the drift of stream invertebrate. Ecology 43: 316–320.

——. 1964. Recolonization of denuded stream bottom areas by drift. Trans. Amer. Fisheries, Soc. 93: 311–315.

Wu, Y. F. 1931. A contribution to the biology of *Simulium*. Pap. Mich. Acad. Sci., 13: 543–599.

Wurts, C. B. 1960. Quantitative sampling. Nautilus, 73: 131–135.

*Yankovskaia, A. I. 1948. Materials on the hydrobiology of the Karadarya basin. Izv. Akad. Nauk Uzbek SSR, 1: 60–69.

*——. 1950. The water bodies of the Pamirs. Priroda, 2: 46–49.

*Zabusova-Shdanova, Z. I. 1970. The flatworm distribution pattern in the Soviet Union. In: Vopr. ecol. morph. biocen. posv . . . , Kazakhskii Univ. p. 164–172.

Zelinka, M. 1971. Die Eintagsfliegen (*Ephemeroptera*) in Forellenbächen der Beskiden. II, Produktion. Limnologorum conventus XVIII, Leningrad: 24.

*Zhiltzova. L. A. 1969. New and rare species of stoneflies of the family *Capniidae* (*Plecoptera*) from Middle Asia. Entomol. Obozr., 48(3): 593–611.

*——. 1970. A revision of Middle Asiatic species of stoneflies from the genus *Mesoperlina* Klap. (*Plecoptera, Perlodidae*). Entomol. Obozr., 49(3): 578–591.

*——. 1971a. A contribution to the knowledge of Middle Asiatic stoneflies (*Plecoptera*): new and little known species of the family *Nemouridae*. Entomol. Obozr., 50(2): 347–365.

*——. 1971b. Genus *Filchneria* Klap. and its taxonomic position in the family *Perlodidae* (*Plecoptera*). Zool. Jurn., 50(7): 1034–1040.

*——. 1972a. A stonefly family (*Plecoptera, Leuctridae*) new to Middle Asia. Zool. Jurn., 51(12): 1815–1822.

*——. 1972b. A stonefly family *Taeniopterygidae* (*Plecoptera*) new to the fauna of Middle Asia. Zool. Jurn., 51(12): 1815–22.

*——. 1974. New and little known stonefly species of the family *Capnidae* (*Plecoptera*) from Middle Asia. Entomol. Obozr., 53(1): 137–149.

*——. 1976. A contribution to the stonefly fauna of the family *Nemouridae* (*Insecta, Plecoptera*) in Middle Asia. I. Zool. Jurn., 55(10): 1476–1481.

——, and Zwick, P. 1971. Notes on Asiatic *Chloroperlidae* (*Plecoptera*), with description of new species. Entomologist Tidskrift, h 3–4, Lund: 183–197.

Zimmerman, P. 1961. Experimentelle Undersuchungen über die Ökologische Wirkung der Strömungsgeschwindigkeit auf die Lebensgemeinschaften des fliessenden Wassers. Schweizer. Zeitschr. Hydrologie, 23: 1–81.

*Zosidze, R. Sh. 1972. On vertical distribution of zoobenthos in mountain rivers of the Adjar ASSR. Soobshchenie Akad. Nauk Georgian SSR, 68(2): 461–463.

*——. 1973. The benthic fauna of rivers of the Adjar ASSR. Avtoref. kandid. dissert. Tbilis. Gosud. Univ., Tbilisi. 22 p.

Zwick, Peter. 1968. Zur Kenntnis der Gattung Dioptopsis (*Dipt. Glepharoceridae*) in Europa. Mitt. Schweiz. Ent. Ges. XLI: 253–265.

——. 1970. *Blepharocera fasciata* ssp. *gynops* nov. ssp. aus Sardinien (*Diptera*). Mitt. Schweiz. Ent. Ges. XLIII, 1: 56–58.

Taxonomic Index

296

Subject Index

Index of Proper Names

IV. *Organizations*

Index of Authors

310